Building a Successful Board-Test Strategy

Building a Successful Board-Test Strategy

Second Edition

Stephen F. Scheiber

Boston   Oxford   Johannesburg   Melbourne   New Delhi

Library of Congress Cataloging-in-Publication Data

Scheiber, Stephen F.
    Building a successful board-test strategy / Stephen F. Scheiber.
        p.   cm.
    Includes bibliographical references and index.
    ISBN 0-7506-7280-3 (pbk. : alk. paper)
    1. Printed circuits–Testing.   I. Title.
    TK7868.P7 S34   2001
    621.3815'310287–dc21

                                    2001032680

British Library Cataloguing-in-Publication Data
A catalogue record for this book is available from the British Library.
The publisher offers special discounts on bulk orders of this book.
For information, please contact:
Manager of Special Sales
Butterworth-Heinemann
225 Wildwood Avenue
Woburn, MA 01801-2041
Tel: 781-904-2500
Fax: 781-904-2620

For information on all Newnes publications available, contact our World Wide Web home page at: http://www.newnespress.com

10 9 8 7 6 5 4 3 2 1

Printed in the United States of America

Contents

Preface to the Second Edition

When I wrote the first edition of *Building a Successful Board-Test Strategy*, my intent was to avoid (as much as possible) the malady that plagues many books in our industry: like the products they deal with, they become obsolete before release to the public. To accomplish this goal, the book discussed tools, alternatives, and ways to evaluate and select test strategies, rather than dictating what those strategies should be.

In many respects, I succeeded. Most of the comments in the original edition are as true today as when they were written. Nevertheless, the industry refuses to stand still. Test has undergone something of a transformation in the past few years. The migration of production capacity away from traditional manufacturers toward contractors continues to accelerate. Today's army of contractors ranges from "garage shops" catering to complex very-low-volume products to multi-billion-dollar megaliths handling board volumes in the millions. This continuing evolution brings with it new challenges, the most significant of which is how to select a contract manufacturer. Such vendors are not like commodity products. As with pieces of test equipment, contractors offer a wide range of strengths and areas of expertise. Choosing one requires finding a combination of skills and capabilities that best matches your needs. Discussions throughout the new edition take this trend into account.

One development that I missed completely in the first edition was the plague that open circuits bring to our surface-mount world. Hidden nodes, board coplanarity (flatness), and other characteristics of today's boards require another look at test methods. A new section of Chapter 2 explores these issues.

The concept of what constitutes a "test" strategy is evolving as well. Various forms of inspection, once mere adjuncts to the quality process, have become intimately linked with more traditional forms of conventional test. Then, too, inspection is not a single technique, but in fact a menu of approaches, each of which has advantages and drawbacks. The new Chapter 3, "Inspection as Test," examines this solution in considerable detail.

This new edition also updates information in many places, adding examples and figures to prior discussions. Much of the additional material comes from seminars that I have given in the past few years—both my own work and material from attendees. Some of those contributions are attributed to their sources. Other examples must remain by their nature anonymous. Nevertheless, I appreciate all of the assistance I have received.

At the risk of leaving out some important names, I would like to thank certain people explicitly for their help. Bob Stasonis at GenRad, Jim Hutchinson at Agilent Technologies, Charla Gabert at Teradyne, and Robin Reid at CyberOptics provided considerable assistance for the chapter on inspection. Jon Titus, Editorial Director at *Test & Measurement World,* has provided constant encouragement along with a stream of contact suggestions and source recommendations over many years. And, of course, my family has once again had to endure my particular brand of craziness as I rushed to complete this project.

Stephen F. Scheiber
December 18, 2000

CHAPTER 1

What Is a Test Strategy?

This book examines various board-test techniques, relating how they fit into an overall product design/manufacturing/test strategy. It discusses economic, management, and technical issues, and attempts to weave them into a coherent fabric. Looking at that fabric as a whole is much more rewarding than paying too close attention to any individual thread. Although some of the specific issues have changed in the past few years, the basic principles remain relatively constant.

Printed-circuit boards do not exist in a vacuum. They consist of components and electrical connections and represent the heart of electronic systems. Components, boards, and systems, in turn, do not spring to life full-blown. Designers conceive them, manufacturing engineers construct them, and test engineers make sure that they work. Each group has a set of tools, criteria, and goals. To be successful, any test strategy must take all of these steps into account.

Test managers coined the briefly popular buzzword "concurrent engineering" to describe this shared relationship. More recently, enthusiasm for concurrent engineering has waned. Yet the ideas behind it are the same ones that the test industry has been touting for as long as anyone can remember. The term represents merely a compendium of techniques for "design-for-marketability," "design-for-manufacturability," "design-for-testability," "design-for-repairability," and so on. The fact that the term "concurrent engineering" caught on for awhile was great. A company's overall performance depends heavily on everyone working together. Regardless of what you call it, many manufacturers continue to follow "design-for-whatever" principles. For those who do not understand this "we are all in it together" philosophy, a new term for it will not help.

Concurrent engineering boils down to simple common sense. Unfortunately, as one basic law of human nature so succinctly puts it, "Common sense isn't." In many organizations, for example, each department is responsible only for its own costs. Yet, minimizing each department's costs does not necessarily minimize costs across an entire project. Reducing the costs in one department may simply push them off to someone else. Achieving highest efficiency at the lowest cost requires that all of a project's participants consider their activities' impact on other departments as well as their own.

The test-engineering industry is already feeling the effects of this more global approach to test problems. Trade shows geared exclusively toward testing electronics—aside from the annual International Test Conference sponsored by the IEEE—have largely passed into the pages of history. Instead, test has become an integral part of trade shows geared to printed-circuit-board *manufacturing*. There are two basic reasons for this phenomenon. When test shows first appeared, test operations enjoyed little visibility within most organizations. The shows helped focus attention on testing and disseminated information on how to make it work. In addition, most companies regarded testing as an isolated activity, adopting the "over-the-wall" approach to product design. That is, "I designed it, now you figure out how to test it."

Today, neither of those situations exists. Everyone is aware of the challenges of product test, even as they strive to eliminate its huge costs and its impact on time to market. Managers in particular dislike its constant reminders that the manufacturing process is not perfect. They feel that if engineering and manufacturing personnel had done their jobs right the first time, testing would not be necessary.

Also, in the past few years, product-manufacturing philosophy has migrated away from the vertically integrated approach that served the industry for so long. Companies still design and market their creations, but someone else often produces them and makes sure that they work. Even within large companies that technically perform this task themselves, production flows through one or a few dedicated facilities. These facilities may differ legally from contract manufacturers, but from a practical standpoint they serve the same purpose, possessing both the same advantages and the same drawbacks.

Because of the popularity of at least the concept of concurrent engineering, considering test activities as distinct from the rest of a manufacturing process is no longer fashionable. Design engineers must deliver a clean product to either in-house or contract manufacturing to facilitate assembly, testing, and prompt shipment to customers. Depot repair and field-service engineers may need to cope with that product's failure years later. With the constant rapid evolution of electronic products, by the time a product returns for repair, the factory may no longer make it at all.

Therefore, although this book is specifically about building board-test strategies, its principles and recommendations stray far afield from that relatively narrow venue. The most successful board-test strategy must include all steps necessary to ship a quality product, whether or not those steps relate directly to the test process itself.

The aim of this book is not to provide the ultimate test strategy for any specific situation. No general discussion can do that. Nobody understands a particular manufacturing situation better than the individuals involved. This book will describe technical and management tools and fit them into the sociology and politics of an organization. You must decide for yourself how to adapt these tools to your needs.

1.1   Why Are You Here?

What drives you to the rather daunting task of reading a textbook on board-test strategy? Although reasons can vary as much as the manufacturing techniques themselves, they usually break down into some version of the following:

- The manufacturing process is getting away from you,
- Test represents your primary bottleneck, and
- Test has become part of the problem rather than part of the solution.

The design-and-test process must treat "test" as an ongoing activity. Its goal is to furnish a clean product to manufacturing by designing for manufacturability and testability, while encouraging the highest possible product quality and reliability. (Product *quality* means that it functions when it leaves the factory. *Reliability* refers to its resistance to failure in the field.)

The purpose of manufacturing is to provide:

- The most products
- At the lowest possible cost
- In the shortest time
- At the highest possible quality

Debate has raged for years over the relative importance of these goals. Certainly, test people often maintain that quality should be paramount, while management prefers to look first at costs. Nevertheless, a company that cannot provide enough products to satisfy its customers will not stay in business for very long.

Suppose, for example, that you have contracted to provide 100,000 personal-computer (PC) motherboards over some period of time, and in that time you can deliver 50,000 perfect motherboards. Despite superior product quality, if you cannot meet the contract's volume requirements, the customer will fly into the arms of one or more of your competitors who can.

Using similar reasoning, the purpose of "test" is to maximize product throughput, reduce warranty failures, and enhance your company's reputation—thereby generating additional business and keeping jobs secure. We get there by designing the best, most efficient test strategy for each specific situation.

1.2   It Isn't Just Testing Anymore

Therein lies part of the problem. What is "test"? Unless we broaden the concept to include more quality-assurance activities, verifying product quality through "test" will soon approach impossible.

Inspection, for example, is usually considered part of *manufacturing*, rather than test. Simple human nature suggests that this perception tends to make test engineers less likely to include it in a comprehensive strategy. Yet inspection can identify faults—such as missing components without bed-of-nails access or insufficient solder that makes proper contact only intermittently—that conventional test will miss.

Similarly, design for testability reduces the incidence of some faults and permits finding others more easily. Feeding failure information back into the process allows adjustments to improve future yields. These steps also belong as part of the larger concept of "test."

Embracing those steps in addition to the conventional definition of "test" allows test people to determine more easily the best point in the process to identify a particular fault or fault class. Pushing detection of certain faults further upstream reduces the cost of finding and repairing them. In addition, not looking for those same faults again downstream simplifies fixture and test-program generation, shortens manufacturing cycles, and reduces costs.

Test strategies have traditionally attempted to find every fault possible at each step. Adopting that approach ensures that several steps will try to identify at least some of the same faults. A more cost-effective alternative would push detection of all faults as far up in the process as possible, then avoid looking for any fault covered in an earlier step later on.

Self-test, too, forms part of this strategic approach. Many products include self-test, usually some kind of power-on test to assure the user that the system is functioning normally. Such tests often detect more than a third of possible fault mechanisms, sometimes much more. Which suggests the following "rules" for test-strategy development:

- Inspect everything you can, test only what you must.
- Avoid looking for any problem more than once.
- Gather and analyze data from the product to give you useful information that allows you to improve the process.

1.3 Strategies and Tactics

Test strategies differ significantly from test tactics. In-circuit test, for example, is a tactic. Removing manufacturing defects represents the corresponding strategy. Other tactics for that strategy include manual and automated inspection, manufacturing-defects analysis (a subset of in-circuit test—see Chapter 2), and process improvement.

A strategy outlines the types of quality problems you will likely experience, then describes which of those problems you choose to fight through the design process during design verification, which you assign to test, and which you leave for "Let's wait until the product is in the field and the customer finds it."

The difference between strategies and tactics boils down to issues of term and focus. A test strategy lasts from a product's conception until the last unit in the field dies. During that time, the manufacturer may resort to many tactics. Also, a tactic addresses a particular place and time in the overall product life cycle. A strategy generally focuses on the whole picture.

In building a test strategy, we are always looking for "digital" answers to "analog" problems. That is, we must decide whether the product is good or bad. But how good is good? How bad does "bad" have to be before the circuit will not function?

Suppose, for example, that the manufacturer of a digital device specifies a "0" as less than 0.8 volts. Based on that specification, a parametric measurement of a logic low at 0.81 volts would fail. Yet will the system actually not perceive a voltage of 0.81 as a clean "0"? How about 0.815 volts? The answer, of course, is a firm "It depends." The situation resembles the century-old conundrum: How many raindrops does it take before a baseball field is wet enough to delay the game? In that context, the question seems absurd. You can't count raindrops! Yet at some point, someone must make a value judgment. Most baseball people accept the fact that delaying the start of a game usually requires less rain than does stopping play once the game has begun. Similarly, the question of how closely a circuit must conform to published specifications may depend on surrounding circumstances. The *real* question remains: Does the product work? As product complexity continues to skyrocket, the necessity to accept the compromises implicit in this approach become glaringly apparent.

Compounding the challenge, issues of power consumption, heat dissipation, and portable-product battery life have required drastically reducing operating voltages for most digital systems. The 5V transistor-transistor logic (TTL) parts of the past have yielded to devices operating at less then 3V, with more to come. As a result, the gap between a logic "1" and a logic "0" narrows every day. Devices must perform more precisely; boards and systems cannot tolerate electromagnetic interference (EMI) and other noise that were commonplace only a few years ago. New generations of test equipment must cope with these developments, and test strategies must take them into account.

1.3.1 *The First Step*

Consider a (loaded) question: What is the single most important consideration in developing a test strategy? The answer may seem obvious. Yet in board-test-strategy seminars from New York City to Singapore, responses range from budgets to design-for-testability to "Do we need to test?" to "Do we choose in-circuit or functional test?" Before facing any of these issues, however, designing a successful test strategy requires determining the nature of the product. That is, what are you trying to test? What is the product? What does it look like? How does it work? What design technologies does it contain? Who is designing it? Who is manufacturing it? Who is testing it? In many organizations, one obstacle to arriving at an effective test strategy is that the people involved decide on test-strategy components and tactics before answering these simple questions.

Test engineers do not design products. If nobody tells them what the product is, how it is designed, and what it is supposed to do, their decisions may make no sense. They might arrive at a correct strategy, but only by accident, and it would rarely represent both the most successful and the most economical approach.

Test components or test strategies that work for one company or product line may not be appropriate in another situation. If you do not know what you are trying to test, you cannot systematically determine the best strategy. Even if you find a strategy that works, thoroughly knowing the product will likely help you suggest a better one. Which brings us to the following definition: A successful test strategy represents the optimum blend of test methods and manufacturing processes to produce the best-quality boards and systems in sufficient number at the lowest possible cost.

Selling-price erosion among electronic products and electronic components of larger products exerts ever-increasing pressure on manufacturing and test operations to keep costs down. As they have for more than two decades, personal computers (PCs) provide an excellent case in point. The price of a particular level of PC technology declines by more than two-thirds every four years.

Looking at it another way, today's PCs are about 40 times as powerful as machines of only five years ago, at about the same price. Bill Machrone, one of the industry's leading analysts, describes this trend as what he calls "Machrone's Law": The computer you want will always cost $5000. One could argue the magnitude of the number, which depends partly on the choice of printers and other peripherals, but the idea that a stable computer-equipment budget will yield increasingly capable machines is indisputable. Machrone's reputation as an industry prognosticator remains intact—especially when you consider that he coined the law in 1981!

Reasonably equipped PCs priced at under $1000 have become increasingly common. Peripherals have reached commodity-pricing status. Even the microprocessors themselves are experiencing price pressure. PCs based on new microprocessors and other technologies rarely command a premium price for more than a few months before competition forces prices into line. Meanwhile, product-generation half-lives have fallen to less than a year, less than 6 months for some critical subsystems such as hard disk drives and CD-ROM readers and burners. Even flat-panel liquid-crystal displays (LCDs), once exorbitantly expensive, are beginning to replace conventional monitors—the last remaining tubes in common use.

Customers are forcing companies to cut manufacturing costs, while test costs remain stable at best and rise dramatically at worst. Test costs today often occupy one-third to one-half of the total manufacturing cost. Every dollar saved in testing (assuming an equivalent quality level) translates directly to a company's bottom line. For example, if manufacturing costs represent 40 percent of a product's selling price (a reasonable number) and test costs represent one-third of that 40 percent, then a strategy that reduces test costs by 25 percent reduces overall manufacturing costs to 36.67 percent, a difference of 3.33 percent. If the company was making a 10 percent profit, its profit increases to 13.33 percent, a difference of one-third. No wonder managers want to reduce test costs as much as possible!

1.3.2 Life Cycles

A successful test strategy is a by-product of overall life-cycle management. It requires considering:

- Product development
- Manufacturing
- Test

- Service
- Field returns
- The company's "image of quality"

Note that only test, field returns, and service involve testing at all, and field returns do so only indirectly. Reducing the number of failures that get to the test process or the number of products that fail after shipment to customers also simplifies test activities, thereby minimizing costs.

Test-strategy selection goes far beyond merely choosing test techniques. Design issues, for example, include bare-board construction. An engineer once described a 50-layer board that was designed in such a way that it could not be easily repaired. To avoid the very expensive scrapping of bad boards, his colleagues borrowed a technique from designers of random access memory (RAM) components and large liquid-crystal-display (LCD) panels—they included redundant traces for most of the board's internal logic paths. Paths were chosen by soft switches driven by on-board components individually programmed for each board.

Although this solution was expensive, the board's $100,000 price tag made such an expensive choice viable, especially because it was the only approach that would work. Without the redundancy, board yields would have been unacceptably low, and repair was impossible. Unfortunately, the solution created another problem. The board's components contained specific instructions to select known-good paths. The bare board defied testing without component-level logic. Therefore, the engineers created a test fixture that meshed with the sockets on the board and mimicked its components. In addition to pass or fail information, the test would identify a successful path, then generate the program with which to burn the "traffic-cop" devices as part of its output. Including the redundancy as a design choice mandated a particular extremely complicated test strategy. Sometimes test-strategy choices reduce to "poor" and "none."

The acceptability of particular test steps depends on whether the strategy is for a new or existing facility, product, product line, or technology. In an existing facility, is there adequate floor space for expansion? Is the facility already running three work shifts, or can a change in strategy involve merely adding a shift?

Test managers must also decide whether to design their own test equipment or buy it from commercial vendors, whether they should try to "make do" with existing equipment, and whether new equipment must be the same type or from the same manufacturer as the installed base.

A test strategy's success also depends on aspects of the overall manufacturing operation. For instance, how does a product move from test station to repair station or from one test station to the next? Are there conveyors or other automated handlers, or do people transfer material manually? Concurrent-engineering principles encourage placing portions of the manufacturing process physically close to one another, thereby minimizing bottlenecks and in-transit product damage. This arrangement also encourages employees who perform different parts of the job to communicate with one another, which tends to increase manufacturing efficiencies and lower costs.

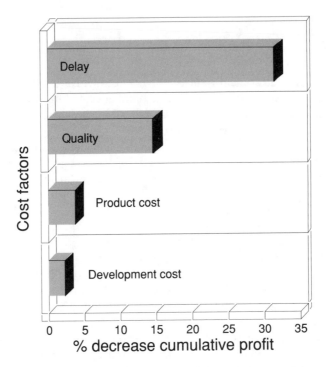

Figure 1-1   The percent decrease in a product's overall profit potential resulting from a six-month delay in product introduction, a 10% product price discount to accommodate quality problems, a total product cost 10% higher than expected, and a 50% higher-than-expected development cost. Note that delaying the product has by far the most pronounced effect. Individual bar values are (from top to bottom): 31.5%; 14.9%; 3.8%; 2.3%. (Prang, Joe. 1992. "Controlling Life-Cycle Costs Through Concurrent Engineering," ATE & Instrumentation Conference, Miller-Freeman Trade-Show Division, Dallas, Texas.)

It seems fairly clear that company managers will not accept even the best test strategy if it exceeds allowable budgets. Yet, evidence suggests that increasing manufacturing costs is less damaging to a product's long-term profitability than is bringing the product to market late. Figure 1-1 shows such an example.

This figure assumes that the product has competitors. If a product is first in its market, delays shorten or eliminate the time during which it is unique and can therefore command a premium price. If a competitor gets in ahead of you because of delays, you lose all of your premium-price advantage. If you are addressing a market where someone else's product got there first, delays will reduce your new product's impact and may mean having to fight one or more additional competitors when it finally arrives.

Therefore, performing the absolute last test or getting out the very last fault may not be worthwhile. The commonly quoted Pareto rule states that the last 20 percent of a job requires 80 percent of the effort. In testing, the last 10 percent taking 90 percent of the effort might be more accurate.

This analysis is not meant to advocate shoddy product quality. Every company must establish a minimum acceptable level of quality below which it will not ship product. Again, that level depends on the nature of the product and its target customers.

Companies whose operations span the globe must also consider the enormous distances between design and manufacturing facilities. Language, time zones, and cultural differences become barriers to communication. In these situations, manufacturing must be fairly independent of product development. Design-to-manufacture, design-to-test, and similar practices become even more critical than for more centralized organizations.

Therefore, selecting an efficient, cost-effective test strategy is a mix of engineering, management, and economic principles, sprinkled with a modicum of common sense. This book tries to create a successful salad from those ingredients.

1.4 The Design and Test Process

There are only three ways in which a board can fail.

Poor-quality raw materials result from inadequacies in the vendor's process or design. This category includes bad components; that is, components (including bare boards) that are nonfunctional or out of tolerance when they arrive from the vendor, rather than, for example, delicate CMOS components that blow up from static discharge during handling for board assembly.

If the board design is incorrect, it will not function properly in its target application, even if it passes quality-control or test procedures. An example would be a bare board containing traces that are too close together. Very fast signals may generate crosstalk or other kinds of noise. Impedance mismatches could cause reflections and ringing, producing errors in edge-sensitive devices ranging from microprocessors to simple flip-flops and counters. In addition, if the traces are too close together, loading components onto the board reliably while avoiding solder shorts and other problems may be impossible, so that even if the bare board technically contains no faults, the loaded board will not function.

A board can also fail through *process variation*. In this case, the board design is correct but may not be built correctly. Faults can result from production variability, which can include the compounding of tolerances from components that individually lie within the nominal design specifications or from inconsistent accuracy in board assembly. Sometimes substituting one vendor's component for an allegedly equivalent component from another vendor will cause an otherwise functioning board to fail. Also in this group are design specifications, such as requiring components on both board sides, that increase the likelihood that the process will produce faulty boards. In addition, even a correct and efficient process can get out of control. Bent or broken device legs and off-pad solder or surface-mount parts fall into this category. The culprit might be an incorrectly adjusted pick-and-place machine, paste printer, or chip shooter. In these cases, test operations may identify the problem, but minimizing or preventing its recurrence requires tracing it back to its source, then performing equipment calibrations or other process-wide changes.

The relative occurrence of each of these failure mechanisms depends on manufacturing-process characteristics. Board-to-board process variation, for example, tends to be most common when assembly is primarily manual. You can minimize the occurrence of these essentially random failures by tightening component specifications or automating more of the assembly process.

More-automated manufacturing operations generally maintain very high consistency from board to board. Therefore, either almost all of a given board lot will work or almost all of it will fail. Examining and correcting process parameters may virtually eliminate future failures, a potent argument for feeding quality information back into the process. Under these conditions, quality assurance may not require sophisticated test procedures. For example, a few years ago, a pick-and-place machine in an automated through-hole line was miscalibrated, so that all of the device legs missed the holes. Obviously, the crimper failed to fasten the legs in place, and when the board handler picked the board up, all of the components from that machine slid off. Even a casual human visual inspection revealed that the board was bad, and the pattern of failures identified the correct piece of equipment as culprit.

It is important to recognize that even a process that remains strictly in control still produces some bad boards. However diligently we chase process problems as they occur, perfection remains a myth. Test professionals can rest assured that we will not be eliminating our jobs or our fiefdoms within the foreseeable future.

You must determine your own failure levels and whether failures will likely occur in design, purchased parts, or assemblies. Fairly low first-pass yields, for example—perhaps less than 80 percent—often indicate assembly-process problems. Very high board yields suggest few such problems. Those failures that do occur likely relate to board design or to parts interactions. If board yields are very high but the system regularly fails, possible causes include board-to-board interactions, interactions of a board with the backplane, or the backplane itself. Understanding likely failure mechanisms narrows test-strategy choices considerably.

1.4.1 Breaking Down the Walls

Test activities are no longer confined to the "test department" in a manufacturing organization. Design verification should occur even before prototyping. It represents one of the imperatives of the simulation portion of the design process, when changing and manipulating the logic is still relatively painless. In addition, inspection, once considered a manufacturing rather than a test step, can now reduce burdens on traditional test. In creating a test strategy, you must therefore take into account the nature and extent of inspection activities.

In fact, test is an integral part of a product's life from its inception. Consider the sample design-and-test process in Figure 1-2. It begins with computer-aided research and development (CARD). A subset of computer-aided design and computer-aided engineering, CARD determines the product's function and begins to formulate its physical realization.

The output from CARD proceeds to schematic capture, producing logical information for design verification and analysis. Next comes logic simulation, which requires both a schematic (along with supporting data) and stimulus-and-response vectors. Either human designers or computer-aided engineering (CAE) equipment can generate the vectors. The logic-simulation step must verify that the theoretical circuit will produce the correct output signals for any legitimate input.

The question remains, however, how many stimulus vectors are enough? The only answer is that, unless the input-stimulus set includes every conceivable input combination, it is possible that the verification process will miss a design flaw.

If logic simulation fails, designers must return to schematic capture to ensure correct translation from design concept to schematic representation. If no errors are evident from that step, another pass through CARD may become necessary.

If logic simulation passes, indicating that the theoretical circuit correctly expresses the designers' intentions, the next step is a design-for-testability (DFT) analysis. Notice that this analysis occurs long before a physical product is available for examination. At this point, there is not even a board layout. Design-for-testability attempts to confirm that if a logic fault exists, there is a place in the circuit to detect it.

If DFT analysis fails, engineers must return to schematic capture. Although logic simulation has shown the schematic to perform as designers intended, the circuit's logical structure prevents manufacturing operations from discovering if a particular copy of the circuit is good or bad.

This tight loop of DFT analysis, schematic capture, and logic simulation continues until the DFT analysis passes. The product then proceeds to a fault simulation. The analysis has determined that testing is possible. Fault simulation must determine if it is practical. For example, consider the test sequence necessary to verify on-board memory. The test must proceed from a known state. If there is no reset function, however, initializing the circuit before beginning the test may be difficult or time-consuming. The test may require cycling the memory until it reaches some known state before the test itself can begin. Similarly, if the memory array is very large, the test may take too long to warrant its use in high-volume production.

Similarly, fault simulation must determine the minimum number of functional test vectors required for confidence that the circuit works. Each fault may be testable, but achieving an acceptable fault coverage in a reasonable time during production may not be practical.

Like logic simulation, fault simulation requires a set of vector inputs and their expected responses. At the same time, however, these two techniques are fundamentally different. Logic simulation attempts to verify that the design works,

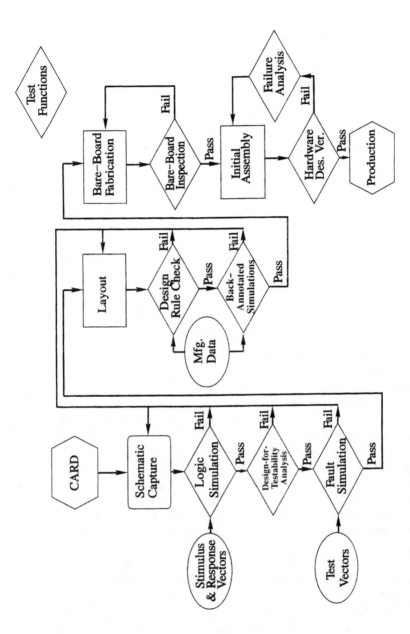

Figure 1-2   Sample design and test process.

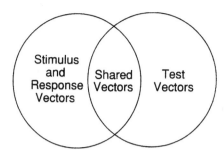

Figure 1-3 Because the assumptions are different, stimulus-and-response vector sets for logic simulation and test vector sets for fault simulation resemble the overlapping circles of a Venn diagram. Some vectors are unique to logic simulation, others are unique to fault simulation. Certain vectors will be common to both techniques.

assuming that the product is built correctly. It therefore seeks only to ensure that an input stimulus that the board will experience in actual operation will produce the output pattern that the designer has specified.

Fault simulation works the other way around. It assumes that the design is correct and attempts to determine whether the assembly process has built it properly. If the board is built correctly, a particular set of inputs will produce a specific set of outputs. If it contains faults, some input vectors must produce incorrect signatures. Those incorrect signatures should pinpoint the exact cause of the problem.

Because the assumptions are different, input vector sets are different as well. They resemble the overlapping circles of a Venn diagram, as shown in Figure 1-3. Some vectors are unique to logic simulation, others are unique to fault simulation. Certain vectors will be common to both techniques.

When simulating a circuit for design verification, an engineer can apply vectors of unlimited flexibility. That is, issues such as resolution, access, and test times border on irrelevant. The object is to verify that, if the manufacturer produces the board according to specification, that it will work the first time.

When simulating for production test, the selected test equipment must be able to execute whatever vectors the engineer has proposed. Circuit access must be available either directly through a bed-of-nails or from the edge connector or indirectly using a technique such as boundary scan. (See Chapter 5.)

Perhaps the biggest difficulty with fault simulation is determining when it has covered all possible faults. Generally, there are very few ways to build a circuit correctly and a plethora of faulty alternatives. For most complex circuits, the number of input vectors required to cover every conceivable failure mechanism is far greater than a simulator can apply in a reasonable amount of time.

To estimate the fault coverage in a fault simulation, an engineer routinely injects faults into the simulated circuit, then ensures that the simulation process will uncover them. At some point, someone must decide that the fault-coverage percentage is the best it can be and move on.

Consider a board on which U14 pin 12 is tied to ground. Injecting that fault into the simulation produces one of three results. If the input vector never causes pin 12 to go high, the output vector will be correct and the simulation does not detect that fault. If the vector tries to drive the pin high, the output will be incorrect. If the faulty output pattern is unique, the simulation has located the specific failure. If, however, the faulty output signature is identical to the output when the board contains a different fault, then that vector cannot be used to identify either fault. Another vector or vectors must be selected, adding to or replacing that vector in the set.

No matter how carefully an engineer performs fault simulation, some faults will always go undetected. The philosopher Voltaire once postulated that God is a comedian playing to an audience that is afraid to laugh. Every now and then, God plays a little joke to remind us that we are not as smart as we think we are. Very high estimated fault coverages often fall into that category.

At the device level, in its day the 80386 microprocessor was one of the most carefully designed components in the industry's history. Intel laid out design-for-testability rules, made sure that many of the most critical functions were verified by internal self-test, and otherwise took elaborate steps to guarantee the product's quality. They estimated a better than 97 percent fault coverage. Yet, despite all of their efforts, some of the earliest devices exhibited a very peculiar bug. If software was running in protected mode and the device resided in a particular temperature range, multiplying two numbers together from a specific numerical range might produce the wrong answer. The only saving grace was that there was no software at that time that required the protected mode. Everything had been designed for the 80286-class environment, which lacked that capability.

The now-classic problem with the original Pentium processors also illustrates this point, while graphically demonstrating how customer perceptions of such problems have changed. After the experience with the 80386, Intel redoubled its efforts in design and test to ensure that subsequent devices (the Pentium contained a then-staggering 3.2 million transistors on a piece of silicon the size of a dime) would work properly. To maximize processor calculation speeds, the Pentium employs a multiplication lookup table, rather than performing each calculation individually. This table, however, contained a small error in one of the entries. After thoroughly researching the problem, Intel contended at first that the error was so tiny that aside from astronomers and others making extremely precise empirical calculations, almost no computer users would ever notice the discrepancy.

Unfortunately for Intel, this adventure occurred after the proliferation of the Internet. Electronic bulletin boards everywhere carried news of the problem. Letters of complaint began to appear from people claiming to have reproduced the bug. Further investigation showed that these people had indeed performed a multiplication that had yielded the small discrepancy. This result was unavoidable. The exact location of the error had been posted on the Internet, and engineers had downloaded the information to check it out for themselves. Since the error resided in a lookup table, *any* calculation that included those specific locations in the table would exhibit the problem. The resulting public-relations nightmare forced Intel to provide a replacement for any such Pentium at no cost to the computer customer—but an enormous expense for the company. The entire episode vividly illustrated that although electronic systems had become vastly more complex than they were only a few years ago, buyers of those systems have no tolerance at all for known bugs.

(Despite the fact that I understood how small the error really was, I must admit that I replaced the Pentium on my own computer. Like many engineers, I have a philosophical aversion to enduring a known bug when the solution is offered *gratis*. On the other hand, I was not so naïve as to believe that by doing so I had necessarily eliminated the only possible bug in that device, merely the only one that I knew about.)

At the board level, if devices start out good, the problem seems simpler. Yet, densely populated boards with hundreds of tiny surface-mounted components on both sides can exhibit problems that elude simulation steps to reveal them because they represent that last 2 or 3 percent of uncovered faults. At any stage, a new problem may emerge, which is why the process diagram in Figure 1-2 contains so many feedback loops. In addition, suppose instead of multitudes of individual components, the board contains highly integrated application-specific integrated circuits (ASICs) or similar alternatives. In that case, someone has to construct a test for the ASICs themselves, which in many respects resembles a functional board test, so test development becomes a series of separate but interrelated steps.

Again, if fault simulation fails, the design process loops back to schematic capture. Complex designs often require several iterations before passing all of these early steps. Once past fault simulation, the product moves into layout.

1.4.2 Making the Product

The layout step creates a physical board description, including placement of traces, solder pads, vias (through-holes), and components. A design-rule check verifies that the layout conforms to the company's testability and manufacturability requirements for a reliable, high-quality product. Problems can include logic nodes under ball-grid arrays (BGAs) and other components, prohibiting either bed-of-nails fixturing of the loaded board or guided probing for functional-failure analysis. Perhaps test pads are too small or node spacing is 25 mils (.025") when the design specifies no less than 50, again to facilitate probing. Design rules may demand that all through-hole components reside on one side of the board or that the circuit's logic partitioning permit in-circuit cluster testing.

Whereas DFT analysis depends on the board's logic, a design-rule check examines its physical implementation. (One industry expert once observed that the term "design-rule checking" itself might be the greatest piece of marketing that test engineers in the electronics industry have ever done. Design rules are constraints that facilitate manufacturing and testing, two activities that many design engineers would rather leave to their manufacturing-engineering counterparts.

Calling them design rules makes them a bit more palatable.) Failure at this step requires returning to layout, if not all the way to schematic capture.

Once the layout passes the design-rule check, the engineer must back-annotate the circuit simulations to ensure that they accurately reflect any design changes to this point and that the circuit still works. These two steps require data on the manufacturing operation, such as paste-printer and chip-shooter accuracies.

The design then proceeds to initial bare-board fabrication and inspection. Third-party suppliers usually manufacture the boards, but most companies (or their contract manufacturers) perform their own inspection of these early boards to guarantee that they conform to the final design. Even after production ramp-up, some manufacturers inspect the bare boards themselves, rather than relying on vendors to provide good products.

If the preceding steps have been done correctly, a hardware design-verification step after adding components to these early bare boards should not reveal any surprises. Manufacturers should also perform a functional test on the boards and analyze any failures. Theoretically, the only faults uncovered at this stage result from the assembly process. Once design and manufacturing engineers are confident that the board functions as intended and that the manufacturing process will produce good boards, full production ramp-up can begin.

Notice the number of test activities that occur even before the first production-type test after bare-board fabrication. Test never really was an isolated activity.

1.4.3 New Challenges

Today's components present greater challenges than ever before. As the gap between a digital device's logic "0" and logic "1" continues to narrow, manufacturers must reduce noise at both the device and the board level. Therefore, new designs tolerate much less noise than older ones did.

For example, with Gunning Transceiver Logic (GTL) and GTL+ (found in many Pentium-class devices), open-drain signals require termination to a V_{TT} supply that provides the high signal level. Receivers use a threshold voltage (V_{REF}) to distinguish between a "1" and a "0". Unlike prior-generation PC systems, these designs include a tolerance requirement for V_{TT} of $1.5\,V \pm 3\%$ while the system bus is idle. A reliable system requires proper design techniques and a consistent, reliable termination. In this case, termination resistors must be $56\Omega \pm 5\%$.

Ever-lower noise margins also increase the circuit's vulnerability to "ground bounce." As speeds increase, transient-switching currents increase as well, creating a brief ground-level bounce as the current flows through the inductance of the ground pins. Through modeling, designers can predict this phenomenon. Nevertheless, you should permit no open ground pins during manufacturing.

Another issue complicating test-strategy decisions relates to the much stricter standards on EMI that international regulators have imposed on electronics manufacturers. In the U.S., the Federal Communications Commission demands no EMI between consumer products. Companies selling to customers in Europe must comply with even stricter requirements, as the European Community tries to ensure that products designated for sale in one country do not interfere with broadcast and cellular-phone frequencies in another. Products must undergo strict compliance testing and receive compliance engineering (CE) certification before they can be sold in the target country.

Other trends that will complicate test-strategy decisions include a tendency to pack ever more components onto a particular sized board, or (conversely) to shrink board size without comparably reducing component count. In either case, the resulting layouts offer much less space for test points and nodes for bed-of-nails probing than before. Where companies are integrating more functions onto ASICs and other complex devices, real estate is not an issue, and the *lower* parts count reduces the complexity of board-level tests. At the same time, however, this development pushes more test responsibility back to the device level, burdening device designers and manufacturers with the need to ensure that only good devices reach customers' hands. As in the past, manufacturers should not assemble boards from other than known-good components. That rule brings its own challenges.

In addition, the classic design-for-testability guideline that attempts to mandate components on only one board side is often impractical. Complex handheld products such as cellular phones must not exceed a certain size. Designers must therefore often get somewhat creative to cram sufficient functionality into the small permissible space.

Not all products suffer from ever-shrinking footprint and declining nodal access, however. As Figure 1-4 shows, telephone company switches, aerospace electronics, and other products continue to enjoy through-hole access. At the other end of the spectrum, wireless cell-phone handsets, many notebook PCs, and personal digital assistants (PDAs—so-called "hand-held PCs") increasingly defy conventional access, and consequently confound traditional test methods.

1.5 Concurrent Engineering Is Not Going Away

An old industry adage states that there is never enough time to do a job right but always enough time to do it again. In its purest form, concurrent engineering is an attempt to put paid to that attitude by setting forth a set of principles that maximize the likelihood that a job will be done right the first time. Sammy Shina, in his *Concurrent Engineering and Design for Manufacture of Electronic Components* (1991), defines the term as: "Integrating a company's knowledge, resources, and experience into development, manufacturing, marketing, and sales activities as early in the cycle as possible."

Companies that have adopted this philosophy have reaped substantial benefits. In the early 1990s, for example, AT&T managed to reduce its development time for new telephone products from two years to one. Hewlett-Packard printers that once took $4\frac{1}{2}$ years from drawing board to production now get there in less than half that time.

In many cases, the time from customer order to product shipment has fallen even more dramatically. Concurrent engineering helped Motorola pagers to get out

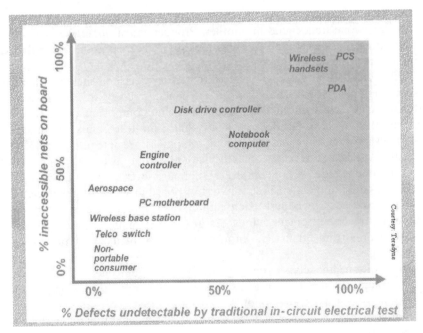

Figure 1-4 Not all products experience "incredible shrinking electronics." (Courtesy Teradyne.)

the door in 2 hours instead of 3 weeks, and what is now Agilent Technologies reduced the time to ship test equipment from 4 weeks to 5 days. Accompanying these improvements were drastic reductions in inventory and associated holding costs, which translate directly to the company's bottom line.

Despite the fact that the term "concurrent engineering" has somewhat fallen out of favor, the imperative to embrace its concepts has not changed. Successfully introducing these concepts (whatever you choose to call them) into a manufacturing operation requires that designers learn to be proactive rather than simply reactive to the impact of their decisions on downstream activities. As wide a range as possible of design and manufacturing technologies deserves consideration if the product is to achieve its performance and economic goals. For example, incorporating boundary-scan circuitry into ASICs tends to minimize test-development times for devices, boards, and systems. Early adoption of these techniques minimizes their inconvenience without sacrificing their benefits.

Designers must solicit input from people in all departments involved in a product's life cycle. Marketers have to sell concept changes to customers, while warranty and field-service people contend with quality problems that escape the factory. Early involvement from these people drastically reduces the number of costly design changes late in the development process. Products destined for automobiles or other hostile environments, for example, must withstand very unfavorable conditions without breaking down. Service people's experiences can help designers make new products more reliable.

For companies whose manufacturing operations are half a world away from design teams, concurrent engineering makes the transfer of responsibility much smoother. Disk-drive manufacturers often develop their boxes and perform early production in their own country, then send the entire package to Asia (for example) for high-volume production. The single biggest impediment to this arrangement is a lack of communication across many time zones and thousands of miles. If manufacturing and test engineers participate in the design, the Asian side of the operation can become relatively self-sufficient.

Picture a Texas company with a manufacturing facility in south Asia running 30,000 boards a day. The Asian arm must be capable of dealing with most problems, from the trivial to the fairly serious, independently. Phone calls, faxes, and e-mails could take days to achieve results, during which time the production line is idle or showing unacceptable quality levels. If a stateside engineer has to hop an airplane to Asia to address the problem, delays get much worse and costs skyrocket. Unfortunately, a problem with the design itself may necessitate such drastic steps.

Perhaps the least disruptive time to introduce concurrent engineering into an existing facility is during new-product development. Extending the practice to other product lines can occur gradually later. The first step must document the capabilities and constraints of the existing manufacturing process. Eliminating redundant or superfluous operations and optimizing what remains represent a good beginning. Unfortunately, this step can be very difficult. Some organizations do not fully understand their manufacturing processes because the processes evolved over long periods and follow no master plan.

As stated earlier, one way to reduce the impact of increasing product complexity on test is to take advantage at each step of information that previous steps have generated and to avoid looking for faults that previous steps have already found. The principal drawback to this approach, however, is that test-program-generation tools often cannot automatically know what faults or fault mechanisms to ignore. Therefore, engineers must allow the tools to generate complete test programs and eliminate redundant sections later or generate the programs manually—neither alternative being particularly attractive. Fortunately, new tools are emerging that keep track of which test-and-inspection steps identify which fault types. These tools, properly implemented, can drastically reduce test-generation efforts as well as test times without sacrificing product quality.

Designers must be sure that CAD/CAE information transfers easily into manufacturing and test terms, reducing data-translation times and minimizing transcription errors. In addition, design requirements must not include specifications that manufacturing or test activities cannot achieve.

Part of the concurrent-engineering effort must include statistical process control (SPC), an online technique for determining whether the process is likely to create high-quality products and whether quality problems result from consistent process failures or random variations. Statistical process control is discussed further in Section 1.9.

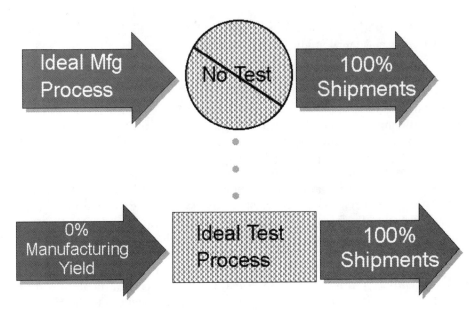

Figure 1-5 100% test effectiveness is possible even with no test or 0% manufacturing yield.

The effectiveness of any test strategy depends on the corresponding effectiveness of the manufacturing process. Consider, for example, a perfect manufacturing process that produces only good products. In that case, any test strategy—even no test at all—will produce 100 percent good products, as in Figure 1-5. Assuming that the process remains in control, any expenditure on test would prove—at best—superfluous.

In contrast, a perfect *test strategy* would ship 100 percent good products, even if every board and system off the manufacturing line contained defects. Of course, the economics of finding and repairing those defects might present a considerable challenge. Nevertheless, from the customer's standpoint, the result would not change. The *test effectiveness* in both cases (a combination of manufacturing and test processes) is 100 percent, although the strategies will likely be very different.

Real manufacturing operations lie somewhere in between. They never achieve the perfection of the first case, so some test strategy will be necessary. Nor do they experience the horrors of zero yield, so a less-than-perfect test strategy will probably suffice. The goal is always to strive for 100 percent test effectiveness, regardless of results from the individual segments.

One drawback to eliminating redundancy between test steps is concern that *over*elimination of test overlap will result in missed faults and, therefore, reduced product quality. To some extent, this concern is justified—as always, depending on the nature of the product under test. As with most engineering decisions, the line between thorough fault coverage and test overlap represents a tradeoff.

Each test engineer must consider that potential danger in designing the overall strategy, striking the most effective compromise for that particular combination of product, customer, and application, within the scope of the available test budget and test tools.

The complexity of today's electronic products makes facing these issues more difficult than ever before. Unfortunately, creating a qualitative list of the requirements is far easier than quantifying them.

1.6 The Newspaper Model

A board-test strategy consists of three basic elements: test techniques, diagnostic techniques to determine failure mechanisms or for statistical process control, and product flow from one process step to the next. Constructing a successful strategy requires asking the same questions one asks when writing a good newspaper story—the so-called "five Ws and an H": Who? What? When? Where? Why? How?—although not necessarily in that order.

The first important question is, Why test? Too many companies consider a test strategy with an artificial goal in mind, such as 97 percent yields or 99.3 percent yields, attaching a number that has no real-world basis. Ultimately, a company tests its products to provide the highest possible quality to customers.

Unfortunately, quantifying the "highest possible quality" depends as much on economics and the nature of both the products and their customers as it does on altruistic goals. Quality targets for a $3 hand calculator would be different from those for a sophisticated PC or one of the million-dollar computer systems aboard the space shuttle. Rewards for changing the process or repairing the product vary in each case, as do the consequences of product failure after receipt by customers.

1.6.1 Error Functions

Electronic-product failures generally follow the classical Gaussian function for the normal distribution of a continuous variable depicted in Figure 1-6. For example, for a $10\,\mathrm{k\Omega}$ resistor with a process standard deviation σ of 1 percent, 68 percent of the device resistances would fall between $9.9\,\mathrm{k\Omega}$ and $10.1\,\mathrm{k\Omega}$. Similarly, 95.4 percent would fall between 9.8 and $10.2\,\mathrm{k\Omega}$, and 99.7 percent would fall between $9.7\,\mathrm{k\Omega}$ and $10.3\,\mathrm{k\Omega}$. Consider a product requiring $10\,\mathrm{k\Omega}$ 3 percent resistors. In a lot of 15,000 boards, each of which contains 20 such resistors, 900 resistors would fail. If each of the resistors resides on a different board, the lot yield from this one failure would be only 94.0 percent.

In practice, statistical analysis predicts actual yields closer to 94.2 percent because some boards will contain more than one bad resistor, according to the following equation:

$$Y = P_1, P_2, \ldots, P_n$$

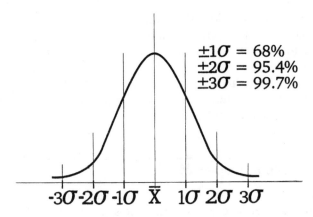

Figure 1-6   A Gaussian normal distribution of a continuous variable with standard deviation σ.

where Y is the expected board yield, and P_n is the probability that component or process step n is good.

If all probabilities are equal,

$$Y = P^n$$

In this example,

$$Y = (.997)^{20} = .9417 = 94.17\%$$

Therefore, 875 boards will fail because of a faulty resistor.

However, if each board contained 300 components and each component had a probability of failure of .003, board yield would total only 40.6 percent. Suppose the process includes 5000 steps (place component 1, place component 2, . . . solder component 1, solder component 2, etc.) and that when the process functions perfectly each step has the same .003 chance of error. The probable overall board yield would be only .0000121 percent. This is referred to as a *three-sigma process.*

When a comparable *four-sigma process* functions perfectly, each component and process step has a failure rate of only 63 parts per million (ppm). Total yield for 5000 process steps and 300 components is 71.6 percent, somewhat better, but still too low to avoid testing altogether. On the other hand, a six-sigma process produces only .002 ppm failures. For the same product, barely one product in 10,000 would fail. In this case, routine testing—especially when calculating costs on the basis of failures found—would prove exorbitantly expensive.

A decrease in vendor-component quality or an increase in process problems (such as inadequate temperature in the reflow oven) will significantly degrade board yields. Therefore, even in a six-sigma environment, manufacturers must constantly monitor process variables and test lot samples, analyzing data and watching for unexplained deviations.

If the process mean does not coincide with the nominal value of the test, the number of board-level and system-level failures will increase. To allow for that situation, Davis [1994] suggests applying a variance of $\pm 1.5\sigma$ to the 6σ nominal. As a result, the process will produce 3.4 ppm failures. With 5000 parts and process steps, approximately 2 systems out of every 100 will fail, too high to avoid testing in most cases.

It is impossible to overemphasize the necessity for tight process control. Achieving a six-sigma process and maintaining it are not the same. In an automated-assembly operation, if someone loads a reel of incorrect components onto a pick-and-place machine and no one catches it, a large number of consecutive boards will fail, and the process will no longer qualify for six-sigma status.

In addition, no one can achieve a six-sigma process overnight. A manager cannot legislate .002 ppm failures per process step. People will react by throwing up their hands and saying, "We can't get there." The factory floor is no place for lofty theoretical goals. Rather, opt for incremental improvements.

Process analysis will indicate steps and components most likely to fail. Acting on those elements first reduces overall board and system failure rates. Over a long time, the total process will begin to approach six-sigma levels.

1.6.2 What Do You Test?

As stated earlier, before creating a test strategy, you must clearly examine what you will be testing, both initially and during the life of the product, the product line, and the facility. Will the boards be heavily analog, as with electro-mechanical systems such as automobile engine controllers, for example? Will boards be primarily digital, destined for PCs and related products? Will there be a lot of mixed-signal technology, as with communications boards? Will product requirements confine the board to a maximum size, as with boards for cellular phones and other hand-held products?

How complex are the electronic designs, and what portion of the overall product manufacturing cost do the electronics represent? How high are total production volumes?

An elevator, for example, is a fairly expensive piece of low-volume hardware that must be extremely reliable. Its electronic technology is both primitive and inexpensive compared to, say, a high-end PC. The system motor responds to the pressing of a button either in the elevator car or on the building floor. It needs to know whether to go up or down, where to stop, when to open the door, and how long to leave it open. A sensor on the door's rubber bumper has to know to open the door if it encounters an object (such as a person) while it is trying to close.

The main controller may rely on a many-generations-old microprocessor that is fairly easy to test either in-circuit or functionally (even with relatively inexpensive and unsophisticated equipment) and has a very low failure rate. Circuit logic is fairly shallow, facilitating test-program development.

In addition, although an elevator system may cost hundreds of thousands of dollars, costs for the electronics and for electronics test constitute a very small percentage of that total. Therefore, changes will not affect the system's overall manufacturing cost very much. An appropriate test strategy for this product tends to be fairly inexpensive but must be reasonably exhaustive. A successful strategy must ensure that the product will work, but many strategic choices can satisfy that criterion. Test engineers should opt for the easiest implementation, rather than selecting the ultimate in state-of-the-art test equipment.

In contrast, a piece of magnetic resonance imaging (MRI) equipment, another high-priced, low-volume product, requires edge-of-technology computer engines, extremely fast data-transmission rates, digital-signal processing, lots of memory, and other high-end capabilities. Testing such products is both more difficult and more expensive than testing elevators. Also, electronics make up the lion's share of the product's technology, so that test represents a substantial portion of overall manufacturing costs. Therefore, test-strategy decisions will directly affect the company's bottom line.

As with the elevator, test strategies must stress quality, but this time test equipment will likely be both state of the art and complex. Fortunately, low production volumes permit long test cycles with little impact on schedules.

Sometimes the end use dictates the nature of the test strategy. Automotive electronics are of the same level of complexity as, say, a PC. But an automobile—which contributes heat, vibration, voltage spikes, and EMI—represents the most hostile common environment in which electronics must function. Therefore, the test strategy must stress ruggedness and durability. Also, the electronic content of today's cars far exceeds the level of only a few years ago. So testing has become more complex, more expensive, and represents an ever-larger percentage of the overall manufacturing cost.

High-volume products, such as PCs, disk drives, and printers, also require sophisticated test strategies. In these cases, however, fast test times are imperative to prevent production bottlenecks. Also, because product generations change very quickly, test-strategy flexibility and fast program development are prime concerns.

Corporate philosophy can also influence the design of a test strategy. Most automobile companies, for example, will go to extraordinary lengths to avoid increasing the cost of each board produced by even a fraction of a cent. They prefer spending the money on test steps and test generation up front to avoid higher material costs later. Advocating extra components or test nodes in that environment will generally fall on deaf ears. Instead, test engineers must find ways to invest during the development phase to reduce material costs during production as much as possible.

Other companies, especially in high-mix situations, take the opposite approach. They would prefer to increase material costs slightly to reduce the cost of test development and setup.

What types of components will the boards contain? Will all components come off the shelf, or will you use ASICs? Will all digital components require similar voltages? Will there be ECL parts? Will microprocessors be relatively slow and uncomplicated, or will they be fast, highly integrated, and state of the art? Will the board include mixed-signal devices?

What types of memory devices does the system contain? Read-only memory chips (ROMs), non-volatile RAM chips (NVRAMs)? How large are the single inline memory modules (SIMMs)? How many different SIMM configurations must the board accommodate?

Including ASICs in a design opens up a host of additional questions. Do you test them at all, or do you depend on device suppliers to provide working parts? Theoretically, board and system manufacturers should be able to rely on suppliers and forego incoming inspection of ASICs because of its prohibitive cost. On the other hand, you do not want to build a board with parts that do not work. The most efficient approach usually involves cooperation between parts makers and customers. Vendors can generally test ASICs less expensively than their customers can, and they have more experience and greater expertise in quality assurance and failure analysis.

In addition, just-in-time manufacturing techniques include small production lots and tight inventory control. Holding up operations to accommodate component testing rapidly becomes unacceptable.

Regardless of who tests the parts, where do test sets come from? System designers who are responsible for designing first-pass ASICs should propose initial test approaches, but again final versions should represent a cooperative effort between system designers and ASIC vendors. When presented with this recommendation, one company cited an example where an ASIC vendor's final production test provided insufficient fault coverage. In that case, the ASIC customer had to either work with the vendor to improve the test, or, as a last resort, find another vendor. Quality of the incoming ASIC remains the vendor's responsibility.

If manufacturing engineers do not have confidence that the incoming ASICs work correctly, they must test the devices before board assembly. Aside from the conventional wisdom that says isolating a fault is more expensive at the board level than at the device level, given the complexity of today's devices, adequately testing them on boards is, at best, very difficult. A large company with a high-volume product may elect to set up a full incoming-inspection operation. For most smaller manufacturers or lower-volume products, finding an independent test house to screen the devices offers a more cost-effective alternative.

Pushing ASIC testing back to the device level and assuming that the ASICs worked prior to board assembly simplifies board-level test considerably. It is unlikely that a working device will develop a subtle failure. Therefore, only a limited number of failure mechanisms need testing: Is the device still alive? That is, did some step during handling or board assembly develop sufficient electrostatic charge to blow it up? Can it execute some basic functions? Do all the pins work? Did something get crunched during part placement? Do all the solder joints work reliably? Looking for a failure in bit 12 on the fourth add register in the arithmetic logic unit (ALU) is unnecessary.

Test vectors for this sort of gross test at the board level generally represent a subset of the device-test vectors that the ASIC vendor applies during production. Therefore, to simplify board-test generation, system manufacturers must have access to those device-level tests. If not, board designers or test engineers have to recreate those vectors, another example of that popular game "reinventing the wheel."

Limiting board-level device tests to looking only for gross failures applies to other complex devices as well. Sophisticated microprocessors, for example, should be treated in this way. The only alternative would be to test them thoroughly on the boards, an unpromising prospect. In-circuit testers, which make contact with individual devices through a bed-of-nails fixture, find achieving necessary test speeds difficult or impossible. Capacitance contributed by the fixture nails compromises signal integrity.

At functional test, which examines the board as a unit from the edge connector or other accessible point, intervening circuitry between the test's input and device pins, and between device outputs and the board's access point, enormously complicate the task of creating comprehensive test programs.

Even design-for-testability techniques do not reduce such efforts by much. Boundary scan, which allows access to internal logic segments from the board's edge, provides only serial signals. For long tests, this approach is painfully slow. A good device self-test helps, but if the self-test exists, why not execute it before expending the labor to install the device on the board and risking the added expense of removing a bad device later? The most cost-effective solution remains ensuring, through vendor testing or other means, that board assembly proceeds with good components only.

1.6.3 Board Characteristics

What do the boards look like? How large are they? A 5″ × 8″ board presents different problems from one that is 15″ × 18″. Similarly, does the board have 600 nodes or 2000? Some test equipment cannot handle the larger boards or higher pin counts. With a low-pincount board and a high-pincount tester, it may be convenient to test several small boards of one or more board types on a single bed-of-nails fixture.

Are there pads available for probing by a bed-of-nails during in-circuit testing or by a technician for fault isolation during functional test? How many layers does the board contain? Have bare boards been tested? If not, can inner board layers be repaired if a fault emerges? Are there embedded components (resistors and capacitors) that require testing within the board itself?

What are the average lot sizes? Lots of 100 instrument boards suggest a very different strategy from disk-drive-controller board lots of 25,000 or locomotive-engine board lots of one.

What is the product mix? How often does the test operation change board types? One prominent disk-drive manufacturer began production of one product in a U.S.-based facility, devoting one line to each board type and therefore requiring no changeover. When he transferred the product to Malaysia, he discovered an entirely different philosophy. There, the entire facility turned out a single board type until it had produced a month's supply, about three days. Operators then changed over to another product for a month's production, and so on. A test strategy that worked in one location proved awkward in the other.

How often does the product change, and how does it change? Although disk drive technology changes very frequently, the controller board's physical layout changes more than its function. PC motherboard function also does not change much (except for speed and memory capacity) from one generation to the next. Large portions of a functional test would remain the same for several generations, whereas an in-circuit test or inspection step, which relies more on the layout and on the integration level of individual devices, would change much more.

How much of the board logic is accessible through the edge connectors? Are there test connectors? (Test connectors are special connectors that are theoretically transparent to the circuit except when matching connectors are attached during testing. The operative words are "theoretically transparent." Designers hate them because they take up real estate, restrain the design, and may compromise circuit performance.)

What is the mix of surface-mounted and through-hole components? Does board assembly call for components on both sides?

Is the board mechanically stable? That is, are logic-node locations for bed-of-nails testing known? Or are they likely to change many times in the early life of the board? If the board is not mechanically stable, creating a bed-of-nails test in a timely manner will be difficult or impossible.

Is the board *thermally* stable? Multilayer boards consist of layers made from a variety of metallic and nonmetallic compounds. Coefficients of thermal expansion may not match from layer to layer. Inside systems, during normal operation, many boards experience temperature gradients, adding twisting moments and other mechanical stress.

For example, it has been said that air movement inside some early personal computers was about comparable to the air movement produced by the wings of a dead butterfly. Therefore, a motherboard was often warmer near the power supply than away from it. Such "hot spots" can cause the board to deform.

Although deformation is rarely great enough to bother leaded components, which "give," solder around surface-mounted components may crack or the components themselves may pop into an upright position, severing the connections at one end, a phenomenon known as "tombstoning," as Figure 1-7 illustrates. A board that is susceptible to tombstoning may require either a system burn-in or some other kind of an elevated-temperature test to ensure reliability in the field.

A conventional board test may not suffice to determine susceptibility to these thermal problems. Most board tests occur in open air or some other uniform-temperature environment, in which an effect that results from a temperature *gradient* will not manifest. These problems must therefore be detected using another method—such as solder-joint integrity predictions during x-ray inspection—or during subsequent system-level test, where the board experiences environmental conditions identical to those in the final product.

Figure 1-7   Because coefficients of thermal expansion for board materials are different, temperature gradients can cause the boards to deform. Solder around surface-mounted components may crack, or the components themselves may pop into an upright position, severing the connections at one end, a phenomenon known as "tombstoning." (Courtesy Teradyne, Inc.)

1.6.4   The Fault Spectrum

Understanding a board's function and application represents only half the battle. A test engineer (and a design engineer, for that matter) must also know how it is likely to fail.

Some manufacturers establish a test strategy first, anticipating that test results will clearly indicate the product's fault spectrum. Unfortunately, constructing a strategy that does not consider expected failure distributions will not likely create the best mix of inspection, test, and process-control techniques.

Estimating the fault spectrum for a new product often involves understanding how that product relates to the technology of its predecessors. For a company that makes copiers, for example, a new copier would be expected to follow old failure patterns. Historical evidence might show that in the first 6 months, a new board exhibits a failure rate of 2.5 percent, declining over the next 6 months to 1.0 percent, where it remains for the rest of the product's life. Test engineers can incorporate any known factors that affect those numbers, such as a new ASIC that might experience a manufacturing-process learning curve and therefore an initial higher-than-normal failure rate, to "fudge" the estimate.

If the new product is a copier/fax machine, estimating its fault behavior will be more difficult. If the copier's scanner stages show a 0.5 percent failure rate, it is reasonable to estimate the scanner-stage failure rate of the copier/fax at 0.5 percent, subject to the same hedging as before.

On the other hand, a copier company would have had less experience with the modem stage of a fax/copier. Estimating its failure rate might simply mean assuming that it will fail at least as often as any other part of the circuit and designing the test strategy accordingly.

When a company introduces a new product that clearly departs from previous product lines, any failure-mechanism estimates will necessarily be less precise. A personal-computer company that branches out to make printers must contend with a wide range of electromechanical problems in addition to any data-processing failure modes that resemble computer failures. A test strategy could base testing of the computer-like functions on previous experience, then concentrate on tactics required to test the product's more unique characteristics. This approach helps avoid unwelcome surprises, where the unfamiliar portion of the circuit creates the most headaches, while allowing the manufacturer to gather information and adjust the production process so that estimating failure modes for the *next* printer product will be much easier.

In addition to deciding where failures will likely occur, a test engineer must try to determine what *types* of failures they are likely to be. If a paste printer has been known to create large numbers of shorts or off-pad solder joints on production boards, it is reasonable to expect that new boards will exhibit these same problems.

The need for familiarity with the manufacturing and test characteristics of a large number of different board types helps to explain the move to contract manufacturing. The diversity of their client base generally means that contractors have experience with problem-solving techniques that individual client companies likely have never seen.

To illustrate how a fault spectrum influences strategic decisions, consider the hypothetical spectrum in Figure 1-8. Assume an overall first-pass yield of 50 percent.

More than half of the failures, and therefore more than 25 percent of the boards through the process, contain shorts. Removing those shorts will improve board quality more than removing any other fault type. Removing them can involve finding them at test and correcting them through board repair or determining the cause of this large number of shorts and altering the process to prevent their creation.

A lot of shorts on a surface-mount board may suggest a problem with the paste-printer calibration. On a through-hole board, the culprit may be (and often is) the wave-solder machine. Missing, wrong, and backwards components generally result from operator errors. Bent leads could be a problem in manual assembly or with automatic-insertion equipment. Unless the production process changes, analog specification, digital-logic, and performance problems are the only faults that will likely change dramatically from one board design to another.

In fact, in this example, shorts, opens, missing components, backwards components, wrong components, and bent leads—the easiest faults to identify

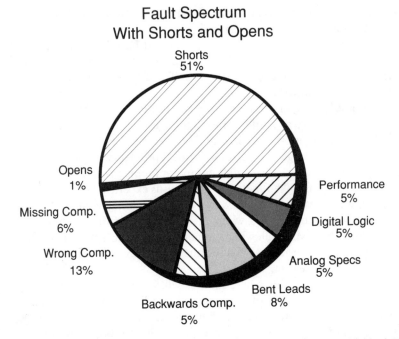

Figure 1-8 A sample fault spectrum. This example assumes a first-pass yield of 50%, with most failures relatively easy to locate.

and remove—comprise 84 percent of the total faults. Without these problems, the production process would yield 92 percent good boards. (Remember, 50 percent of the boards were good to start with.) Therefore, a successful test strategy might employ a less-expensive tester or inspection system as a prescreener to uncover these problems, dedicating more sophisticated equipment to finding only more complex faults.

Suppose that the paste printer, wave-solder machine, or whatever caused the large number of shorts and opens is repaired or replaced, so that boards now get through production with no shorts or opens. Figure 1-9 shows the resulting fault spectrum. The first important consequence of this process improvement is not evident from the figure—the total number of failures is now half of what it was before. To manufacture a lot of 100 good boards previously required about 150 board tests. There are 100 boards in the first pass, but 50 must be tested again after repair. Some boards may need an additional cycle through the system if the repair operation created a new fault or if the first test did not detect all existing faults.

Without shorts and opens, about 75 of the 100 boards will pass right away. Producing 100 boards now requires only about 125 tests, reducing necessary test capacity by 16.7 percent.

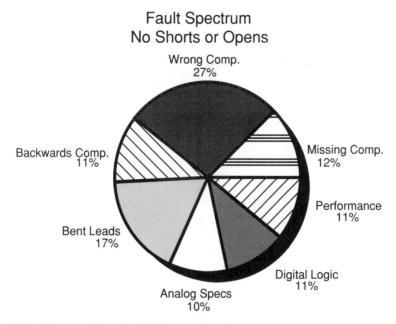

Figure 1-9 The same board as in Figure 1-8. This time, the wave-solder machine, or whatever prime culprit was causing the shorts and opens, has been repaired or replaced.

In addition, eliminating shorts and opens leaves only 55 percent of the failures in the easy-to-detect category. Therefore, it may no longer be worthwhile for the strategy to include both a simple prescreening step and a more complex test step. An economic analysis may show that, after considering tester costs as well as labor and other support costs for the extra machine, allowing the more expensive tester to find even the simpler failures is less expensive. Alternately, it may show that an inspection step can replace a test and thereby reduce the burden on subsequent test steps. The test strategy has changed to reflect changes in the fault spectrum, and the new situation, which produces a higher first-pass yield, includes fewer (or possibly different) test operations.

Figure 1-10 shows an actual electronics manufacturer's cumulative fault spectrum for all of his facility's board types during a recent 1-year period. To improve product quality, a test engineer attacks the most serious problem first, if possible. In this case, bad parts represent the largest single failure category. This result often means either defective parts coming from the vendor or damage to the parts (such as that from electrostatic discharge [ESD]) during handling. Data analysis of board tests will identify those parts most likely to fail. Pinpointing the cause and remedying it, either through vendor action or by modifying handling procedures, improve overall board quality and may change the test strategy's tactic mix.

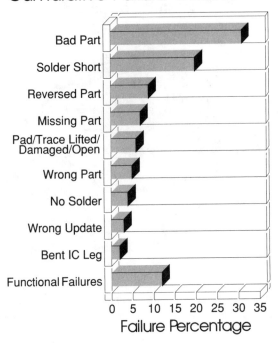

Figure 1-10   Actual electronics manufacturer's cumulative fault spectrum for all of a facility's board types during a recent 1-year period.

Figure 1-11 provides a different view of the same manufacturer's data. This time, the analysis shows failure rates for individual board types. Clearly, types *A* and *B* fail more than three times as often as any of the others. Therefore, reducing failure rates for these two types would have the greatest impact on total product quality. In addition, examining the fault spectrum for each board type might show more process-type failures for types *A* and *B*, while a higher percentage of failures in types *C* through *I* results from design-related causes. Therefore, a strategy that works very well for *A* and *B* would not necessarily represent the optimum choice for the other types.

Looking at these two charts together might suggest that the large number of component failures comes primarily from assemblies *A* and *B*. Perhaps they contain ASICs or other high-failure components that the other assemblies do not. Such reasoning in no way *guarantees* that correlation, but it gives the engineer a starting point for analysis.

The issue of what you are trying to test also depends heavily on who the customer is. In some cases, for example, a government agency may dictate certain aspects of the test strategy, such as choice of equipment or location for field test-

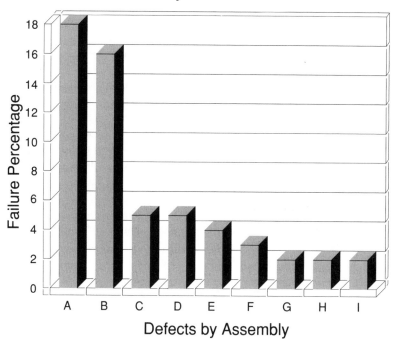

Figure 1-11 A different view of the example in Figure 1-10. This time, the analysis shows the failure rates for individual board types. Clearly, types *A* and *B* fail more than three times as often as the other varieties.

and-repair operations. One maker of locomotive electronics, for example, supplies local transportation departments, one of whom mandated not only spare boards but also test equipment, fixtures, and test programs to allow full-blown testing in the field. This arrangement added considerably to the manufacturing cost of the product, consequently raising its price.

Another company is setting up service facilities around the world, trying to furnish all of them with testers comparable to equipment in the factory. One consequence of proliferating sophisticated and expensive test equipment in the field is the maintenance burden. This burden includes not only keeping the equipment up and running but also distributing machine, fixture, and test-program updates to ensure that everyone is measuring the same parameters in the same way. Otherwise, boards will fail that should pass, and vice versa, and available data for manufacturing-process and design improvements will be less meaningful.

Another drawback to this approach is that oftentimes people conducting tests at field installations are much less familiar with the product and the test program than are their counterparts in the factory. New boards may contain technology that field people have rarely seen. As a result, board testing and repair can take much longer than in the factory, and identifying the actual cause of a subtle failure may be difficult or impossible.

If, as with governments and military contracts, customers demand unusual conditions such as particular test equipment in the field, there is no alternative but to comply. These constraints simply become part of any test-strategy specification.

If, on the other hand, the requirements are part of a *request* and are subject to negotiation, the system manufacturer should try to ascertain the reasons for them and try to address those concerns in a more appropriate way. One alternative to test equipment in the field might involve keeping full sets of spares on-site, isolating a problem to the offending board or assembly module, swapping it for a good one, and sending the faulty copy back to the factory or to a central depot for repair. In the case of the locomotive-electronics manufacturer, within 6 months of first delivery, the customer realized the additional headaches that the requirement for extra test equipment had caused, and abandoned the idea in favor of additional spares.

One large-system manufacturer's customers demanded a product that they could repair themselves to the component level. The manufacturer, therefore, included a lot of built-in diagnostics and troubleshooting techniques. After a few months, however, those same customers decided that the ordeal of stocking extra components and performing tests was too great, so they reverted to board swapping and sending defectives back to the vendor.

Bundling a comprehensive spares kit with system sale is generally far more cost-effective than supplying a tester. Unfortunately, backing away from a strategy that is already in place is sometimes difficult because of all the money and political and emotional energy that went into the original decision. Therefore, companies often stay with a strategy long after it becomes clear that it is not the best alternative.

A different set of concerns arises when the product is a PC, television set, or other relatively expensive piece of consumer equipment that an independent general-purpose service facility will repair. These repair technicians see a wide variety of failures every day, but they rarely have a great deal of expertise on any particular brand or model. Manufacturers of this type of product should pay careful attention to self-tests, service manuals, availability of test nodes for probing, and other conveniences to facilitate fault isolation and repair. Some of these solutions will affect test strategies in the factory, whereas others will relate only to customer-initiated service.

1.6.5 Other Considerations

The question of who performs testing depends heavily on where actual production occurs. Some system manufacturers test their own boards; others farm some or all of the task out to a third party. Manufacturers who produce boards in-house will likely test them as well. Contract manufacturers may or may not test the boards that they build or may perform only some of the necessary tests. Customers may prefer to maintain control over quality by performing all testing or functional testing, or may spot-check the contractor's process by testing lot samples. PC system manufacturers and other makers of "commodity" products often purchase tested boards directly from third parties, performing only burn-in or other environmental stress screening (ESS) and system-level testing before shipment.

Sometimes, contract manufacturers do everything from board assembly to boxing systems and shipping them to distributors and customers. In those cases, the system "manufacturer" (the company whose name goes on the product) must trust that test, repair, and other quality control steps produce high-quality, reliable systems.

At what *level* does test occur? Are bare boards tested before assembly? Do loaded-board tests examine individual components, or do bed-of-nails access limitations require resorting to logic clusters? Does the board contain independent functional units for functional testing, or must a test exercise the entire board at once? Are parts of the board testable only by one technique or the other? Do some boards or parts of boards defy conventional testing altogether, requiring inspection or a comprehensive self-test?

Bare-board testing is not expensive, but it does require investing in equipment, floor space, work-in-process inventory, and people. Most companies buy their bare boards from suppliers who test the boards and ensure their quality. Therefore, board vendors should pay severe penalties if any of their supposedly good boards fails.

Sometimes leaving bare-board test with the vendor is inappropriate. Testing very complex boards that demand specialized test techniques, such as the 50-layer monster described earlier, usually requires work by the board designer—in this case the system manufacturer.

At times a manufacturing-process decision demands an unusual test strategy. In an attempt to minimize environmental damage, one company abandoned freon-based board cleaning, adopting an aqueous technique instead. Unfortunately, deionized water could not dissolve fluxes from the soldering step, so the engineers resorted to organic-acid fluxes, which represent a serious reliability problem if left on the board. Therefore, the company had to create a test strategy that involved testing the boards again after cleaning.

Is it possible to test boards automatically during assembly? This type of test would catch through-hole leads crumpled during automated insertion, for example. Although many manufacturers have considered this approach for a number of years, few, if any, actually do it.

Can the test time meet throughput requirements? High-volume production often demands that a board come off the line every, say, 30 seconds. That means that no holdup in the line can last more than that 30 seconds. The manufacturer must "spend" that time very carefully. Meeting this specification might require resorting to sampling techniques or testing only "trouble spots" if overall yields are high enough to permit that approach.

Is testing necessary at the board level at all? As process yields from board production increase, more and more manufacturers are asking this important question. Foregoing test at the board level means assuming that the boards are good or at least that very few will fail at system test. Another alternative is to test boards and backplanes, but not full systems. That approach assumes that if the boards are good and the backplane is good, the system will work. Most companies today prefer at least some assurance that the system functions as an entity. System test therefore remains part of most strategies.

Is there a burn-in or other ESS step? If so, does test occur before, during, or after it?

Perhaps the most important question—Is test necessary?—is being asked more today than ever before. Some years ago, a Japanese television manufacturer decided to perform no testing until final product turn-on just before shipment. By carefully controlling vendor product quality and the manufacturing process, the company achieved a 96 percent turn-on rate. With that type of quality directly from the process, the most cost-effective strategy would probably be to test only those products that fail.

An American company trying to implement the same strategy fared less well. They managed a turn-on rate of only 36 percent. Further efforts on process and vendor control got yields up to about 74 percent, but either result would mandate continuing to test every board.

A few years later, a Japanese printer manufacturer decided to concentrate on process and vendor control, shipping products without testing them at all—not even the turn-on test. The number of DOAs ("dead on arrivals") that resulted forced the company to reinstitute a more traditional strategy.

How will a test gain access to the board's components? Does it employ a bed-of-nails? Is access available only through an edge connector, or have the designers provided a test connector as well? Are the components and boards set up to take advantage of boundary-scan techniques, allowing access to internal nodes from the board edge?

Some in-circuit–type testers aimed at low-volume applications use clips and probes for access to standard digital parts. An operator clips an appropriate connector to one of the devices, then executes a test for that device, moves the clip or chooses a different clip for the next device, and executes the next test.

Because clips are much less expensive than conventional bed-of-nails fixtures, this approach can be cost-effective. It is also relatively common for diagnosing field failures. For this technique to work reliably, the board's population density must be fairly low, and the devices have to be leaded to permit clips to make sufficient contact.

Does the product include a self-test? Many electronic products today automatically test critical core functions on power-up. Incorporating this self-test to supplant physical node access and reduce test-generation efforts can significantly simplify any test strategy.

Self-tests often cover more than a third of possible failure mechanisms. Some companies have achieved more than 80 percent fault coverage during self-test. Unfortunately, test strategies often fail to take advantage of the self-test other than to verify that it is present and that it works. Exercising the self-test during production and not looking again for those same faults downstream can simplify program generation and fixture construction considerably.

For a number of years, boards and components have been shrinking, while the number of components per board has soared. Surface mounting, small-outline ICs, and other space-saving solutions have permitted cramming huge amounts of circuitry into ever-smaller areas. With ball-grid arrays (BGAs), flip-chips, and other technologies, nodes reside underneath the components. Where conventional wisdom once demanded restricting components to one board side and ensuring that all nodes feed through to the other side for testing, new products often cannot afford such luxuries. For these products, bed-of-nails access is becoming more difficult. Therefore, in-circuit test can verify only component clusters rather than single parts. Some manufacturers are returning to functional test as a primary tool because bed-of-nails access is nearly impossible. Lack of access has also fueled renewed interest in inspection as a viable tactic in a "test" strategy.

At the same time, some products are proceeding in the opposite direction. Disk drives, for example, have shrunk to the point where laptop computers weighing barely 5 pounds can have multigigabyte hard-disk capacities, with no end in sight. To increase reliability and simplify both manufacturing and testing, designers are integrating more functions onto ASICs, reducing the total number of components on the board as well as the density of associated traces. CD-ROM and DVD drives and writers—because of the huge amounts of data they hold and their high data-transfer rates—are exhibiting these same design characteristics.

This development has pushed much quality assurance back to the component level, while providing real estate for the test nodes and through-holes that designers have been resisting. Assuming that even complex devices worked before board assembly, many test engineers are returning to beds-of-nails rather than edge-connector techniques to verify that the board was built correctly, further reducing test-generation costs and schedules.

1.6.6 The How of Testing

What types of equipment does a test strategy include? Some manufacturers choose rack-and-stack test instruments controlled from a host PC or workstation. Others may design custom testers or may select from vendor offerings. Still others may farm out all test responsibilities to third parties, even if they do the manufacturing themselves, although with the expansion of contract manufacturing this practice is far less common than it was just a few years ago.

Test tactics may involve in-circuit–type or functional-type tests. An emulation test allows a board to behave as in the target system, stopping at convenient times to monitor register contents and other specific internal conditions.

Inspection may supplement or replace test steps that look for manufacturing defects. Inspection steps can be manual or automatic, and can occur post-paste, post-placement, or post-reflow. Each inspection method and location provides different information about the process and about board quality.

A "hot-mockup" test takes place in a model of the real system. Hot mockup of a disk-drive controller board, for example, requires a system that is complete except for the board under test. Other than testing an assembled system at the end of the production line, hot mockup is the only way to ensure that a particular board actually performs its intended function.

Strategic choices directly affect how engineers generate test programs. Some techniques, such as emulation, permit only manual generation and require that the person responsible intimately understand the board's function and technology. Shorts-and-opens testers and some forms of inspection can learn their programs from one or more known-good boards, with little human intervention. Programming in-circuit and functional testers ranges from completely manual to completely automatic. Vendor-supplied or manufacturer-developed tools can help with manual steps.

How well a product incorporates design-for-testability and concurrent-engineering principles may expand or limit strategic choices. Inclusion of boundary-scan or self-test circuitry on devices or board segments significantly simplifies test generation and enhances final product quality. Burn-in or ESS in a manufacturing strategy will affect the test steps around it.

Much also depends on the required depth of failure diagnostics. If bad products end up in the garbage, engineers identify failing components but not specific faults except as necessary for feedback into the process. In military-system and other modular applications, testing identifies the failing *module* for replacement and goes no further. On the other hand, if test diagnostics must find the specific component that failed, as is done for process monitoring, test steps and program generation must take that fact into account.

The likelihood that a board device will fail affects the diligence with which a test should verify that it works exactly as expected. A device that rarely fails falls into the same category as an ASIC that used to work. Board-level test need only confirm that it is there, inserted correctly, and alive. Conversely, if devices are suspect, board-level tests must more completely exercise them.

When planning a test strategy, test-engineering managers must know who their engineers are and with what machines, strategies, and technologies they are already familiar. If the people have no experience with functional testing, for example, any strategy that includes that option will require a longer learning curve than one that does not. This constraint does not preclude using a strategy element, but including that element brings with it additional baggage.

Similarly, experience with a particular vendor or machine may help the decision-making process. Programmers familiar with one tester type will have an easier time developing tests for that machine than for something else. By the same token, if the experience is negative, the range of strategic choices can be narrowed by eliminating that vendor or machine from consideration.

Some manufacturers would like to use the same or similar equipment in product development, production, and field service. That approach limits the number of possibilities in the factory. Bed-of-nails test, for example, is rarely practical in the field because of fixture costs and the difficulty of maintaining a complete fixture inventory for old or obsolete products.

One advantage of a common-machine strategy is the freedom to develop test programs only once, rather than for each process step. Also, this solution allows comparing failure data from each stage directly, permitting quality tracking from "cradle to grave." Long-lived, low-volume products particularly benefit from this approach, because field failures provide necessary information about the manufacturing process. With low-volume products, a manufacturer appreciates the traceability of each piece.

In high-volume operations, common data facilitate statistical process control. That is, if analyzing a process indicates that a capacitive network is the most likely circuit section to fail and field data confirm that fact, the test engineer can assume a good understanding of the process. If that network does not actually fail, then the process requires further study.

Of course, any test strategy requires gathering and analyzing test and failure data. Computer-aided test equipment generates a lot of data. Unless those data furnish a better understanding of the product or the process, they are useless.

Test people must remember that no test-strategy decision is cast in stone. Strategies, even strategies that involve considerable investment, can be changed. Also, no law says that a tester must always test to the limits of its abilities. A high-end machine can still perform simpler tests if the occasion warrants it. This option permits running equipment as close as possible to its maximum throughput capacity, even when its most sophisticated capabilities are not needed for a particular process, a particular advantage for a small company with a limited capital budget.

In addition, a manufacturer can change strategies without buying additional equipment by reassigning what already resides within the factory. For example, a product early in its life may require in-circuit testing. As the process matures, a manufacturing defects analysis may suffice. The simpler test is faster, less expensive to develop, and less expensive to conduct. If the in-circuit tester is not needed elsewhere, it can also execute the new test, avoiding the necessity to acquire additional equipment.

Constructing a test strategy requires answering one other very important question: *How much will it cost?* The next section begins to address this critical issue.

1.7 Test-Strategy Costs

Every test-strategy decision brings with it both technical and economic consequences. If a particular test problem permits one and only one solution, either because of technical or managerial constraints, then the cost of that solution may be inconvenient, but it is unavoidable. Most manufacturing situations, however, allow a range of options, with differing quality, scheduling, and cost ramifications. Test managers must examine each alternative and select the one that provides the best balance. One cardinal rule applies: You generally cannot buy less than one of anything.

1.7.1 Cost Components

The total cost of any manufacturing and test strategy comprises a large number of contributing factors. Most of these factors are interrelated, so that saving money in one area often means spending it somewhere else. Design costs represent the most obvious—although by no means the only—example of this phenomenon. Designing a product for manufacturability and testability increases costs at the design stage. The anticipated benefit is downstream costs that are lower by at least enough to compensate.

Moreover, a product that facilitates manufacture and test permits easy process changes to accommodate increasing or decreasing product demand, design updates, or quality improvements. Conversely, designers concerned only with keeping their own costs down condemn the product to delays in manufacturing, quality problems, and higher-than-necessary overall costs.

Translating a theoretical circuit design into a practical system directly affects purchasing and inventory costs. Consider, for example, a design that calls for a number of resistor types, including 2-kΩ 2 percent resistors, as well as pull-up resistors for some of the digital signal lines. For this application, pull-ups can range from about 1 kΩ to 5 kΩ. Nominal resistances can be off by 50 percent or more and the circuit will still function adequately. The less-precise parts undoubtedly cost less than the 2 percent versions. However, adding a part number to purchasing and inventory tracking and allocating additional inventory space often cost more than simply specifying the already stocked component for the less demanding board position as well.

This simple tactic also has test implications. Just because a parts list calls for a 2 percent pull-up resistor does not mean that the board test should necessarily flag any part that fails to meet that specification. Test programmers must consider the board function. Testing a pull-up resistor at 2 percent will produce more failures than testing it at, say, 25 percent. Boards will fall out of the production cycle for diagnosis, repair, and retest, increasing test costs. But because the pull-up application does not demand 2 percent precision, the higher costs will not improve actual board quality at all.

A manufacturing engineer's job is to build a functioning product at the lowest possible cost. Design engineers and test engineers must ensure that the product works. Creating a test specification blindly from a purchasing or manufacturing document ignores differences in the responsibilities of each group. Developing test programs in this way is often less expensive than painstakingly modifying automatically (or manually) generated programs to accommodate practical differences from those specifications, but test costs will be higher. Again, the aim is to reduce the total cost burden, regardless of the distribution among individual departments.

Startup costs for any new-product introduction include the cost of idle manufacturing capacity during retooling and ramp-up, the cost of simulations and physical prototypes, and the salaries of engineers, supervisors, and other non-routine personnel who participate in this project phase. A product that is relatively

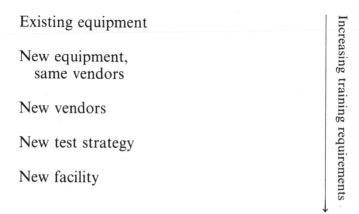

Training-cost impact of
strategy decisions

Existing equipment

New equipment,
    same vendors

New vendors

New test strategy

New facility

Increasing training requirements

Figure 1-12   Training costs are inversely related to everyone's familiarity with various test-strategy components.

easy to manufacture and test will require less ramp-up time and less attention from management people, resulting in lower startup costs.

Initiating manufacturing incurs costs for parts, labor, holding work-in-process inventory, and people or equipment to move components, boards, and systems from station to station through the process. Test costs include costs for testing components at the vendor, at incoming inspection, or by a third party, as well as costs for board test and system test. Burn-in and ESS introduce their own costs, influencing cost distributions at board and system levels.

One major cost component when introducing a new test strategy is training. Operators need to know how to run the equipment, programmers must understand the nuances and peculiarities that characterize any tester, and managers must become familiar with the anticipated personality of the selected test operation as well as the nature and value of resultant data. Training costs are inversely related to everyone's familiarity with various test-strategy components. Relying on existing equipment incurs the lowest training costs, establishing a new test facility, the highest, as Figure 1-12 shows.

Another factor that can significantly increase training costs is personnel turnover. High turnover demands extra effort to train new people as they come on board. Training new people takes longer than training current people, even on an entirely new project or test strategy, because current employees are already familiar with the management personnel, management style, and other environmental aspects of the corporate personality.

Involving manufacturing people in test-strategy decision-making and then training them adequately to implement the agreed-on approaches make them more productive. Increased productivity also generally translates into greater job satisfaction, thereby lowering turnover and reducing costs still further.

Every piece of test and test-related equipment—from PCs or workstation hosts to individual instruments, to benchtop testers, to freestanding "monolithic" testers, to fixtures and other interface hardware—requires maintenance. Maintenance includes instrument calibration and other scheduled activities, pin replacement on bed-of-nails fixtures, and unscheduled repairs of equipment failures. Choice of specific equipment and classes of equipment can significantly affect these costs.

For example, most companies can quickly locate a PC to replace a failed PC host in a tester or rack-and-stack system that conducts measurement and test functions over an IEEE-488 interface. Similarly, a big tester that can pinpoint a faulty switching card after a failure permits quick board swapping and a minimum of manufacturing downtime.

On the other hand, more time-consuming equipment repairs increase costs not only for the repairs themselves but also for lost production while the line is shut down. In-house–designed test equipment often falls into this category because of inadequate or outdated documentation or because tester designers have moved on to other projects and are no longer available for consultation.

Board-design and test-strategy characteristics directly affect the cost of diagnosing faulty boards to identify exact causes of failure and then repairing them. Many strategies will fail some good boards, creating false failures, which increase testing and diagnosis-and-repair costs. More forgiving test-strategy elements may pass bad boards to the next test stage, where finding failures is more expensive ("escapes"). Failing boards may defy diagnosis, leading them to pass through the test process several times, only to end up on some bone pile.

Board repair costs include the cost of concomitant failures—secondary failures that result from repair activities or that occur when powering up a board containing a specific fault causes another device to fail as well.

Test-strategy flexibility and design stability can influence costs related to product upgrades. Test devices that connect to a board via the edge connector or test connector or through a microprocessor socket are less affected by many changes, because these board characteristics are likely to remain relatively constant. Board layout, on the other hand, frequently changes as designers replace off-the-shelf components with ASICs or with more capable or more reliable alternatives. A strategy that includes a bed-of-nails works better with a more stable product, because altering the fixture to accommodate design changes can be quite expensive.

Other test-operation costs include indirect labor costs, including supervisors' and managers' salaries. Facilities costs cover floor space, security, utilities, and other fixed expenses.

Note that reducing the amount of floor space devoted to one product does not necessarily improve the financial burden on the company as a whole. Factories are not elastic. Abandoned space does not disappear. Therefore, unless another project moves into the now empty space, fixed facilities costs per project remain unchanged.

On the other hand, freeing up floor space can prevent overestimating a factory's percent utilization. Managers who realize that a portion of the factory lies idle can plan program or project expansion without budgeting for new facilities.

The impact of a factory's test strategy on product costs does not end with shipment to customers. Field service and warranty repair generally cost far more than eliminating those same faults during production. In addition, products that do not perform to customers' expectations can erode future sales. One reason for many companies' interest in environmental stress screening, despite its logistical and economic consequences, is to remove problems in the factory that otherwise would occur in the field. Manufacturers of so-called "ruggedized PCs" differentiate their products from the competition by touting ESS and other reliability programs in their manufacturing process, which in turn permits premium pricing.

Field operations must create their own test strategies and methods. As stated, the closer these methods are to their siblings in the factory, the lower the test-development costs at that stage. A great deal also depends on field-service logistics. Repair of expensive but fairly standard consumer products, such as PCs, often involves swapping defective boards or modules and sending them back to the factory for diagnosis and repair. Companies that rely on contract manufacturing and test may merely return any bad boards from the field to the board vendor.

International manufacturing companies often maintain repair depots in individual countries to minimize customs complications. These facilities are not as elaborate as comparable operations in the factory, but they offer more capability than a field engineer could carry to a customer site.

Field-test-strategy choices also determine costs to customers for keeping inventories of replacement parts. Manufacturers of large systems may require customers to stock a specific mix of spare parts as a sale or service-contract condition.

Field-test strategies differ from their production counterparts in one important respect. Factory testing must determine whether the product works. Products returned from the field worked at one time but do not work anymore. Therefore, even if the equipment is similar to the factory's arsenal, the failure diagnostics will likely be somewhat different.

1.7.2 Committed vs. Out-of-Pocket Costs

Most discussions of life-cycle costs assume flexibility until someone actually spends the money. One of the cornerstone principles of concurrent engineering is that *designers* commit most of the money to develop and manufacture a product before it ever leaves their hands—in fact, before there is an actual product at all.

Figure 1-13 illustrates a study by British Aerospace in the U.K. that demonstrates this apparent dichotomy. The lower curve indicates the timing of money

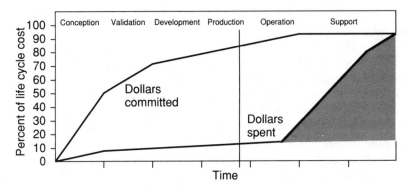

Figure 1-13 From a study by British Aerospace in the UK. The lower curve indicates actual timing of money spent during product development and production ramp-up. The upper curve shows the number of dollars committed by design decisions during product development. (Prang, Joe. 1992. "Controlling Life-Cycle Costs Through Concurrent Engineering," ATE & Instrumentation Conference, Miller-Freeman Trade-Show Division, Dallas, Texas.)

spent during product development and production ramp-up. Notice that, as expected, the development stage is relatively inexpensive. Designers rely on computer equipment and simulations, whereas manufacturing processes require bare boards and components, assembly equipment, test equipment, handlers, and people.

Unfortunately, looking only at when money changes hands does not tell the whole story. Every design decision commits dollars that manufacturing and test operations and other downstream processes must spend. By the end of the validation phase, designers have already committed 70 percent of project expenditures, 85 percent by the time engineering people release a product for manufacturing. The difference between the dollars committed and dollars spent are the so-called "hidden costs" of the design process. The magnitude of this gap emphasizes the value of building design-for-manufacturability, design-for-testability, and other concurrent-engineering principles into a product from its conception.

Chapter 10 will discuss test economics in much more detail.

1.8 Project Scope

Considerations for building a successful test strategy depend heavily on the object of the exercise. Setting up a new facility, for example, generally means that the range of available strategic choices is virtually unlimited. The only criterion for vendor, equipment, and test-method evaluation is to achieve the most economical, highest-quality product possible within available budget constraints.

Establishing a new product line in an existing facility can also permit a broad spectrum of options, but floor space may be limited. In addition, because personnel may already have expertise with particular tester types, managers may be reluctant to experiment with new vendors or test technologies. Managers may also feel pressured to include surplus test equipment from maturing, waning, or discontinued manufacturing lines rather than make new capital purchases.

Adding a product to an existing product line may require conforming to an established test strategy. Although slight tactical modifications may be possible, in this circumstance the range of alternatives will likely be quite limited. Product test may have to use excess capacity from existing equipment. Any new equipment may have to conform to what is already on the factory floor. Test engineers will have to justify carefully any requested changes using economic or quality criteria or both.

Enhancing existing processes rather than introducing new products usually occurs either because too many or too few products on the factory floor are failing. Adding inspection or manufacturing-defects analysis, for example, may minimize the number of simple failures that an in-circuit tester must catch. Similarly, introducing a functional tester may reduce or eliminate failures at system test and can minimize the chance that a faulty board will escape into the field.

As a product matures, a manufacturing process may improve to the point where early test steps, such as manufacturing-defects analysis, no longer find many problems. An economic evaluation may show that the number of failures identified at that step does not justify equipment depreciation, labor, facilities, and other costs. Transferring the equipment to a new product or even to another facility might prove more cost-effective.

Based on the foregoing, a test strategy can include any of the steps outlined in Figure 1-14. Few manufacturing operations would include them all. Incoming inspection or vendor test verify bare boards or components. Prescreening can mean an automatic shorts-and-opens test, or it could indicate some form of inspection or manual test. The term simply indicates an effort to get out obvious problems before the next test step, even if that step is as simple as manufacturing-defects

Possible test steps

*Incoming inspection or vendor test
*Prescreen
*In-circuit or manufacturing-defects analyzer
*Board-level burn-in or other ESS
*Functional or performance
*Hot mockup
*System burn-in
*System test
*Field test

Figure 1-14   A test strategy can include any of these steps. Few manufacturing operations would include them all.

analysis. Visual inspection on a through-hole board, for example, might spot solder splashes from a poor wave-solder process or components that are obviously missing or backwards.

Determining the success of any test strategy requires examining the effectiveness of each step, then making adjustments as required. Not long ago, a locomotive manufacturer discovered that, after assembling his system from tested and supposedly good boards and beginning system-level test, the locomotive horn began to blast, and he could not find a way (other than powering down the system) to turn it off.

The problem turned out to be a faulty field-effect transistor (FET). Although an in-circuit test for that part was feasible, the program running at the time included no such test. The automatic program generator did not know how to turn the FET on and off, so it did not create the test. The overall strategy was adequate, but a test engineer had to add the FET test manually to achieve sufficient comprehensiveness. In that case, the tester could not attain the voltage levels necessary to turn the part on and off, necessitating extra circuitry in the fixture. Because the product was very expensive, almost any effort to improve fault coverage was justifiable.

Again, knowing actual or projected defect levels and likely defects helps in planning an effective strategy. Also, any strategy must be flexible enough to accommodate new information as painlessly as possible.

1.9   Statistical Process Control

Statistical process control (SPC) is a strategy for reducing variability in products, delivery times, manufacturing methods, materials, people's attitudes, equipment and its use, and test and maintenance. It consists of answering the following questions:

- Can we do the job correctly?
- Is the process correct?
- Is the product correct?
- Can we do the job better?

Electronics manufacturers who diligently monitor their processes with those questions in mind will see fewer and fewer failures as the product matures. Eventually, a well-controlled process will produce no failures at all.

At times, however, especially in high-volume processes, even zero defects is not enough. Manufacturing engineers need to anticipate problems from the production cycle before they occur to avoid the considerable time, expense, and schedule disruption of having to fix large numbers of defective products or the inconvenience of product parameters that "drift" in the field to the point where the products fail.

Floppy disk drives provide a perfect example. During disk formatting, the heads of a floppy drive lay down tracks in a predetermined pattern so that they can later write and read data reliably. Specifications include a target head position for each track, along with "slop"—a tolerance—to prevent crosstalk between tracks and head drift as the drive ages from making previously written information unreadable.

Production testing must ensure not only that a particular drive can read what it writes, but also that it can read a purchased disk or a disk created on another drive in the same or another computer. Therefore, test tolerances must be considerably narrower than the tolerances permitted during normal drive operation.

Similarly, whatever tolerances permit a drive to pass during testing, manufacturing and test engineers must know whether drives emerge from the production process sufficiently close to tolerance limits that further drift would cause them to fail during additional testing or after a period of life with customers. CD burners, when dealing with read-write disks (which use a totally different data format from CD-ROMs) can experience these same problems.

Once a process achieves zero failures, manufacturers set up two tiers of test tolerances. Exceeding the narrower set indicates a developing process problem, which can be addressed before production-line units actually begin failing.

Statistical process control employs various tools to help describe the process and its current condition. Flowcharts, as in Figure 1-15, outline the current process so that engineers can more easily examine each step to determine whether it is in or out of control. Many industry experts suggest that not sufficiently understanding a facility's current manufacturing process is the biggest single impediment to implementing SPC and concurrent-engineering principles.

Analyzing test results using histograms, as in Figure 1-16, can show the process condition and warn of impending problems. Figure 1-16a shows an ideal

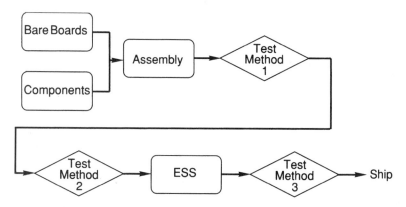

Figure 1-15   SPC flowcharts outline the current process so that engineers can more easily examine each step to determine whether it is in or out of control.

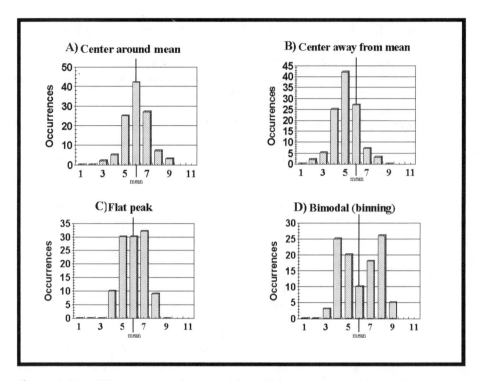

Figure 1-16   Histograms can show process conditions and warn of impending problems. (a) Measurements center tightly around the nominal, with no product failures and no results outside narrow control limits. (b) Some process drift. A few values now lie outside the in-control boundaries. All products still pass, but the process may require recalibration. (c) Values away from the nominal are about as likely as values precisely on target. Again, this result may signal a process analysis and recalibration. (d) Binning, where few values fall at the mean.

situation, where the mean of a set of measurements coincides with the nominal value in the specification. The components in Figure 1-16b, where the measurement mean is displaced from the nominal value, will likely produce more failures. Figure 1-16c shows a flat peak around the mean. Assuming that the tolerance is wider than the range of the three tall bars, no more boards will fail than in case (a). Figure 1-16d shows a common result from analog components. Customers who buy, say, 10 percent resistors will receive few resistors with tolerances much lower than 10 percent. Those parts go to customers paying a premium for 5 percent or 2 percent parts. Parts vendors routinely test their parts and "bin" them for sale according to the test results.

Pareto charts, such as Figures 1-10 and 1-11, illustrate the types of problems that occur most often. They indicate where corrective efforts will bear the most fruit. Looking at Figure 1-10, a manufacturing engineer would logically conclude that eliminating bad parts would generate the largest improvement in overall product quality. Similarly, finding out why assemblies *A* and *B* in Figure 1-11 fail

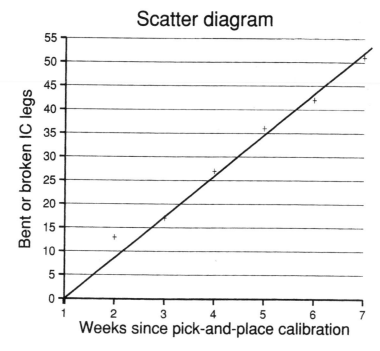

Figure 1-17 A scatter diagram can demonstrate the interdependence among process variables. This example charts the relationship between the number of bent or broken IC legs and the length of time since the last pick-and-place machine maintenance. (Oakland, John S., and Roy F. Followell. 1990. *Statistical Process Control—A Practical Guide*, 2nd ed., Heinemann-Newnes, Oxford.)

so much more often than the others will provide the best quality improvement for the expended effort.

A scatter diagram like Figure 1-17 can demonstrate interdependence among process variables. This example charts the relationship between the number of bent or broken IC legs on through-hole boards against the length of time since the last scheduled or unscheduled maintenance on pick-and-place machines. Manufacturing and test engineers decide how many of this type of fault they will tolerate. Based on that fact and the diagram, they decide on an appropriate preventive-maintenance and calibration schedule.

Control charts, such as Figure 1-18, show process behavior over time. Unlike histograms, which group like occurrences together, control charts indicate each event, allowing engineers to observe whether process parameters are drifting out of control.

In both Figure 1-18a and Figure 1-18b, the inner dashed lines represent limits outside of which the process may be getting out of control but the product still works. The outer lines delineate product failures. Engineers must decide for each process how many failures or out-of-control values constitute sufficient reason to

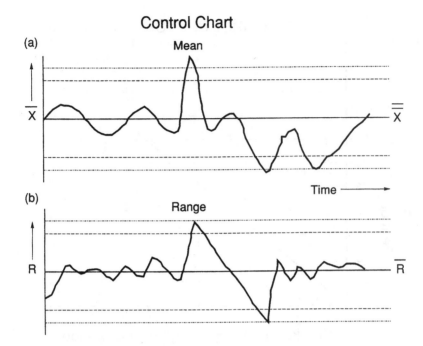

Figure 1-18   Control charts show process behavior over time. (a) Lot mean values over time, comparing them to a universal mean or process specification. (b) The range of individual measurements for expensive or low-volume products. (Oakland, John S., and Roy F. Followell. 1990. *Statistical Process Control—A Practical Guide*, 2nd ed., Heinemann-Newnes, Oxford.)

calibrate, adjust, or change the process. Requirements for SPC will also determine some of the factors affecting test-strategy choices. Figure 1-19 presents a sample technique for applying SPC to developing or improving an existing manufacturing process.

1.10 Summary

A test strategy consists of all the steps that occur at any process stage that help to ensure a quality product. Planning a strategy begins at product inception and does not end until the last piece is retired. Many important test-strategy decisions relate only indirectly to test activities.

Creating a successful strategy requires thoroughly understanding the product, the manufacturing process, and the intended customer. Choice of specific strategic elements depends on many interrelated factors, including who will perform manufacturing and test, expected board-failure rates, likely fault spectrum, and company size.

Throughout the planning stages, test engineers and engineering managers must balance technical requirements with expected costs. A project's costs depend

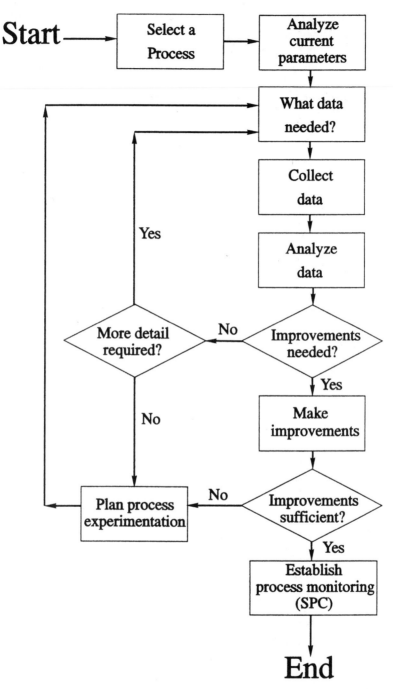

Figure 1-19 A sample technique for developing or improving a manufacturing process. (Oakland, John S., and Roy F. Followell. 1990. *Statistical Process Control—A Practical Guide*, 2nd ed., Heinemann-Newnes, Oxford.)

heavily on design and manufacturing-process decisions. Departments must work together to get the best results. Very often, the scenario carrying the lowest overall costs is different from the one that produces the lowest cost for a particular department. This "we're all in this together" attitude lies at the heart of the concept of concurrent engineering. Tools such as SPC help monitor the process and the test-and-manufacturing strategy to meet everyone's goals.

CHAPTER **2**

Test Methods

Chapter 1 discussed many of the decision parameters that constitute an effective board-test strategy. This chapter describes various test techniques, examining their capabilities, advantages, and disadvantages.

2.1 The Order-of-Magnitude Rule

An accepted test-industry aphorism states that finding a fault at any stage in the manufacturing process costs 10 times what it costs to find that same fault at the preceding stage, as Figure 2-1 illustrates. Although some may dispute the factor of 10 (a recent study by Agilent Technologies puts the multiplier closer to 6) and the economics of testing at the component level, few would debate the principle involved. In fact, for errors that survive to the field, 10 times the cost may be optimistic. Field failures of large systems, for example, require the attention of field engineers, incurring time and travel expenses, and may compromise customer goodwill.

The single biggest cost contributor at every test level is troubleshooting time. Uncovering design problems and untestable circuit areas before production begins can prevent many errors altogether. Similarly, analyzing failures that do occur and feeding the resulting information back into the process can minimize or eliminate future occurrence of those failures.

A prescreen-type tester such as a manufacturing-defects analyzer (MDA) can find shorts or other simple faults much more easily than a functional tester can. In addition, because the functional level generally requires the most expensive and time-consuming test-program development, eliminating a fault class at the earlier stage may obviate the need to create that portion of the functional test program altogether. Equipment and overhead for people and facilities at each stage also tend to be more expensive than at the preceding stage.

As an example, consider an automated assembly-board-manufacturing process that includes a complex soldered part costing $10. A board test reveals that the part is bad. Repair consists of removing the part, manually inserting its replacement, soldering, and cleaning, perhaps 30 minutes' work. At a burdened

Order–of–Magnitude Rule

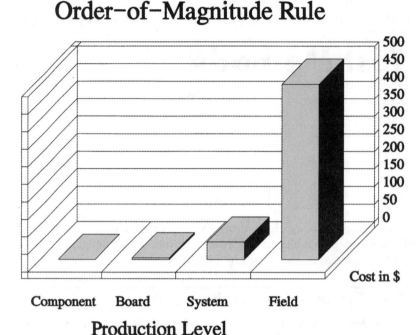

Production Level

Figure 2-1   The order-of-magnitude rule states that finding a fault at any stage in the manufacturing process costs 10 times what it costs to find that same fault at the preceding stage.

labor rate (including benefits and other costs) of $50, the repair costs $35 for parts and labor.

If the board passes until system-level test, the repair process is more complicated. An operator or technician must first identify the bad board, which requires partially disassembling the system and either swapping boards one at a time with known-good versions and rerunning tests or taking measurements at specific system locations with bench instruments.

Diagnosing and replacing the bad component can occur offline at a repair station or in the system itself, depending on the nature and maturity of the product and on the manufacturing process. In any case, the 30-minute repair has now ballooned to two or three hours, and the cost, even if labor rates are the same, has increased to $110 to $160. Since many organizations pay system-test technicians more than technicians supporting board test, actual costs will likely be even higher.

In addition, if a large number of boards fall out at system test, that process step will require more people and equipment. Increased space requirements are not free, and the extra equipment needs maintenance. Hiring, management, and other peripheral costs increase as well. Again, costs are higher still if problems escape into the field.

In many companies, a rigidly departmentalized organizational structure aggravates these cost escalations by hiding costs of insufficient test comprehensiveness at a particular manufacturing stage. Each department minimizes its own costs, passing problems on to the next stage until they fall out at system test or, worse, at warranty repair. The classic example is the increased cost of design-for-testability, including engineering time, additional board components, and extra testing for testability circuitry. Design activities cost more, assembly may cost more, but test costs are much lower. Designers often contend that their extra work benefits other departments to the detriment of their own. Adding inspection to pre-screen traditional test introduces costs as well. Again, a department that merely looks out for its own interests rather than considering *overall* costs will not adopt the extra step.

Combating this attitude requires attacking it from two directions. Managerially, sharing any extra costs or cost reductions incurred at a particular stage among all involved departmental budgets encourages cooperation. After all, the idea is that total benefits will exceed total costs. (Otherwise, why bother?)

Addressing the cultural barriers between designers or manufacturing people and test people is both more difficult and more important. Historically, design engineers have regarded test responsibilities as beneath them. They do the "important" work of creating the products, and someone else has to figure out how to make them reliably. This cavalier "over-the-wall" attitude ("throw the design over the wall and let manufacturing and test people deal with it") often begins at our most venerable engineering educational institutions, where students learn how to design but only vaguely understand that without cost-effective, reliable manufacturing and test operations, the cleverest, most innovative product design cannot succeed in the real world.

Fortunately, like the gradual acceptance of concurrent-engineering principles, there is light at the end of the educational tunnel. People such as Ken Rose at Rensselaer Polytechnic Institute in Troy, New York, are actively encouraging their students to recognize the practical-application aspects of their work. For many managers of engineers already in industry, however, cultivating a "we're all in this together" spirit of cooperation remains a challenge. Averting test people's historical "hands-off" reaction to inspection equipment requires similar diligence.

2.2 A Brief (Somewhat Apocryphal) History of Test

Early circuit boards consisted of a handful of discrete components distributed at low density to minimize heat dissipation. Testing consisted primarily of examining the board visually and perhaps measuring a few connections using an ohmmeter or other simple instrument. Final confirmation of board performance occurred only after system assembly.

Development of the transistor and the integrated circuit in the late 1950s precipitated the first great explosion of circuit complexity and component density, because these new forms produced much less heat than their vacuum-tube predecessors. In fact, IBM's first transistorized computer, introduced in 1955, reduced power consumption by a whopping 95 percent! Increased board functionality led to a proliferation of new designs, wider applications, and much higher production volumes. As designers took advantage of new technologies, boards took on an increasingly digital character. Manufacturing engineers determined that if they could inject signals into digital logic through the edge-connector fingers that connected the board to its system and measure output signals at the edge fingers, they could verify digital behavior. This realization marked the birth of functional test.

Unfortunately, functionally testing analog components remained a tedious task for a technician with an array of instruments. Then, a group of engineers at General Electric (GE) in Schenectady, New York, developed an operational-amplifier-based measurement technique that allowed probing individual analog board components through a so-called "bed-of-nails," verifying their existence and specifications independent of surrounding circuitry. When GE declined to pursue the approach, known as "guarding," the engineers left to form a company called Systomation in the late 1960s. Systomation incorporated guarding into what became the first true "in-circuit" testers.

In-circuit testing addresses three major drawbacks of the functional approach. First, because the test examines one component at a time, a failing test automatically identifies the faulty part, virtually eliminating time-consuming fault diagnosis. Second, in-circuit testing presents a convenient analog solution. Many boards at the time were almost entirely either analog or digital, so that either in-circuit or functional testing provided sufficient fault coverage. Third, an in-circuit tester can identify several failures in one pass, whereas a functional test can generally find only one fault at a time.

As digital board complexity increased, however, creating input patterns and expected responses for adequate functional testing moved from the difficult to the nearly impossible. Automated tools ranged from primitive to nonexistent, with calculations of fault coverage being equally advanced.

Then, in the mid-1970s, an engineer named Roger Boatman working for Testline in Florida developed an in-circuit approach for digital circuits. He proposed injecting signals into device inputs through the bed-of-nails, overwhelming any signals originating elsewhere on the board. Measuring at device outputs again produced results that ignored surrounding circuitry.

Suddenly, test-program development became much simpler. In-circuit testers could address anything except true functional and design failures, and most came equipped with automatic program generators (APGs) that constructed high-quality first-pass programs from a library of component tests. Results on many boards were so good that some manufacturers eliminated functional test, assembling systems directly from the in-circuit step. Because in-circuit testers were significantly less expensive than functional varieties, this strategy reduced test costs considerably.

As circuit logic shrank still further, however, designers incorporated more and more once-independent functions onto a few standard large-scale integration (LSI) and very-large-scale-integration (VLSI) devices. These parts were both complex and expensive. Most board manufacturers performed incoming inspection, believing the order-of-magnitude rule that finding bad parts there would cost much less than finding them at board test.

For as long as the number of VLSI types remained small, device manufacturers, tester manufacturers, or a combination of the two created test-program libraries to exercise the devices during board test. Custom-designed parts presented more of a problem, but high costs and lead times of 15 months or longer discouraged their use.

More recently, computer-aided tools have emerged that facilitate both the design and manufacture of custom logic. Lead times from conception to first silicon have dropped to a few weeks. Manufacturers replace collections of jellybean parts with application-specific integrated circuits (ASICs). These devices improve product reliability, reduce costs and power consumption, and open up board real estate to allow shrinking products or adding functionality.

In many applications, most parts no longer attach to the board with leads that go through to the underside, mounting instead directly onto pads on the board surface. Adding to the confusion is the proliferation of ball-grid arrays (BGAs), flip chips, and similar parts. In the interest of saving space, device manufacturers have placed all nodes *under* the components themselves, so covered nodes have progressed from a rarity to a major concern. Ever-increasing device complexity also makes anything resembling a comprehensive device test at the board level impractical, at best. Today's microprocessors, for example, cram millions of transistors onto pieces of silicon the size of pocket change.

Board manufacturers have found several solutions. To deal with nodes on both board sides, some use "clamshell" beds-of-nails, which contact both board sides simultaneously. Clamshells, however, will not help to test BGAs and other hidden-node parts. Other approaches create in-circuit tests for circuit clusters, where nodes are available, rather than for single devices. Although this method permits simpler test programming than functionally testing the entire board does, the nonstandard nature of most board clusters generally defies automatic program generation. To cope with the challenges, strict design-for-testability guidelines might require that design engineers include test nodes, confine components to one board side, and adopt other constraints. For analog circuitry, new techniques have emerged that allow diagnosing failures with less than full access.

As manufacturing processes improved, in-circuit–type tests often uncovered few defects, and a larger proportion fell into the category of "functional" failures. As a result, many manufacturers returned to functional testing as their primary test tactic of choice to provide comprehensive verification that the board works as designers intended without demanding bed-of-nails access through nodes that do not exist.

Unfortunately, proponents of this "new" strategy had to contend with the same problems that led to its fall from grace in the first place. Functional testers can be expensive. Test programming remains expensive and time-consuming, and fault diagnostics can be very slow. Complicating matters further is the fact that electronic products' selling prices have dropped precipitously over the years, while the pace of product change continues to accelerate. Meanwhile, competition continues to heat up and the time-to-market factor becomes ever more critical. A test that might have sufficed for earlier product generations might now be too expensive or delay product introduction intolerably.

The evolution of test has provided a plethora of methods, but none represents a panacea. Thoroughly understanding each approach, including advantages and disadvantages, permits test managers to combine tactics to construct the best strategy for each situation.

2.3 Test Options

There are two basic approaches to ensuring a high-quality electronic product. First, a manufacturer can design and build products correctly the first time, monitoring processes at every step. This technique, popular with managers because it saves both time and money, includes design verification, extensive process and product simulation, and statistical process control. The alternative is to build the product, test it, and repair it. As Figure 2-2 shows, test-and-repair is more cost-effective for simpler products, but its cost increases with product complexity faster than testing the process does. At some point, process testing becomes less expensive. The exact crossover point varies from product to product and from manufacturer to manufacturer.

Most companies today follow a strategy that falls somewhere between the two extremes. Relying heavily on product test does not preclude being aware of process problems, such as bad solder joints or vendor parts. Similarly, products from even the most strictly controlled process require some testing to ensure that

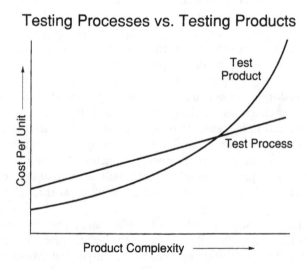

Figure 2-2   Test-and-repair is cost-effective for simple products, but its cost increases with product complexity faster than testing the process does.

Board-Test Categories

* Shorts and opens
* Manufacturing defects analyzer
* In-circuit
* Functional
* Combinational
* "Hot mockup"
* Emulation
* System

Figure 2-3 Electronics manufacturers select test strategies that include one or more of these test techniques.

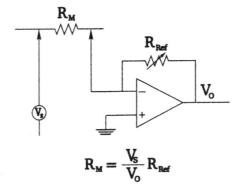

$$R_M = \frac{V_S}{V_O} R_{Ref}$$

Figure 2-4 The "apply voltage, measure current" in-circuit measurement technique.

the process remains in control and that no random problems get through the monitoring steps.

Electronics manufacturers select test strategies that include one or more of the techniques listed in Figure 2-3. The following sections will explore each test technique in detail. The next chapter will address the issue of inspection.

2.3.1 Analog Measurements

Bed-of-nails-based automatic test equipment generally performs analog measurements using one or both of the operational-amplifier configurations shown in Figures 2-4 and 2-5.

In each case, the operational amplifier, a high-gain difference device featuring high-impedance inputs and a low-impedance output, serves as a current-to-voltage converter. Both versions tie the positive input to ground. Because (ideal) op-amp design requires that the two inputs be at the same potential, the amplifier is stable only when the negative input is at virtual ground and no current flows at that point.

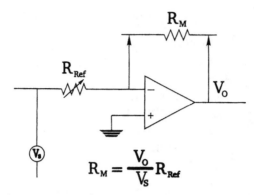

$$R_M = \frac{V_o}{V_S} R_{Ref}$$

Figure 2-5   The "apply current, measure voltage" in-circuit measurement configuration.

Therefore, the current through the two resistors must be identical, and the value of the unknown resistor is proportional to the value of the output voltage V_o.

In Figure 2-4, the measurement injects a known voltage V_S. Measuring the current through R_{Ref} in the feedback loop determines the unknown resistance R_M. This version is often called "apply voltage, measure current." In Figure 2-5, the position of the two resistors is reversed, so that the applied *current* is known. Again, the value of the unknown resistor depends on the ratio of the output voltage to V_S, but the factor is inverted. An AC voltage source allows this technique to measure capacitances and inductances.

2.3.2   Shorts-and-Opens Testers

The simplest application of the analog measurement technique is to identify unwanted shorts and opens on either bare or loaded boards. Shorts-and-opens testers gain access to board nodes through a bed-of-nails. A coarse measurement determines the resistance between two nodes that should not be connected, calling anything less than some small value a short. Similarly, for two points that should connect, any resistance higher than some small value constitutes an open.

Some shorts-and-opens testers handle only one threshold at a time, so that crossover from a short to an open occurs at a single point. Other testers permit two crossovers, so that a short is anything less than, say, $10\,\Omega$, but flagging two points as open might require a resistance greater than $50\,\Omega$. This "dual-threshold" capability prevents a tester from identifying two points connected by a low-resistance component as shorted.

In addition, crossover thresholds are generally adjustable. By setting open thresholds high enough, a test can detect the presence of a resistor or diode, although not its precise value.

Purchase prices for these testers are quite low, generally less than $50,000. Testing is fast and accurate within the limits of their mission. Also, test-program generation usually means "merely" learning correct responses from a good board. Therefore, board manufacturers can position these testers to prescreen more expensive and more difficult-to-program in-circuit and functional machines.

Unfortunately, the convenience of self-learn programming depends on the availability of that "good" board early enough in the design/production cycle to permit fixture construction. Therefore, manufacturers often create both fixture drill tapes and test programs simultaneously from computer-aided engineering (CAE) information.

Shorts-and-opens testers, as the name implies, directly detect only shorts and opens. Other approaches, such as manufacturing-defects analyzers, can find many more kinds of failures at only marginally higher cost and greater programming effort. Also, the surface-mount technology on today's boards makes opens much more difficult to identify than shorts. As a result, shorts-and-opens testers have fallen into disfavor for *loaded* boards (bare-board manufacturers still use them), having been replaced by inspection and more sophisticated test alternatives.

In addition, as with all bed-of-nails testers, the fixture itself represents a disadvantage. Beds-of-nails are expensive and difficult to maintain and require mechanically mature board designs. (Mechanically mature means that component sizes and node locations are fairly stable.) They diagnose faults only from nail to nail, rather than from test node to test node. Sections 2.3.5 and 2.3.6 explore fixture issues in more detail.

2.3.3 *Manufacturing-Defects Analyzers*

Like shorts-and-opens testers, manufacturing-defects analyzers (MDAs) can perform gross resistance measurements on bare and loaded boards using the op-amp arrangement shown in Figures 2-4 and 2-5. MDAs actually calculate resistance and impedance values, and can therefore identify many problems that shorts-and-opens testers cannot find. Actual measurement results, however, may not conform to designer specifications, because of surrounding-circuitry effects.

Consider, for example, the resistor triangle in Figure 2-6. Classical calculations for the equivalent resistance in a parallel network,

$$\frac{1}{R_M} = \frac{1}{R_1} + \frac{1}{R_2 + R_3}$$

produce a measured resistance of 6.67 kΩ. Like shorts-and-opens testers, MDAs can learn test programs from a known-good board, so 6.67 kΩ would be the expected-value nominal for this test, despite the fact that R_1 is actually a 10-kΩ device.

An MDA tester might not notice when a resistor is slightly out of tolerance, but a wrong-valued part, such as a 1-kΩ resistor or a 1-MΩ resistor, will fail. Creating an MDA test program from CAE information can actually be more difficult than for some more complex tester types, because program-generation software must consider surrounding circuitry when calculating expected measurement results.

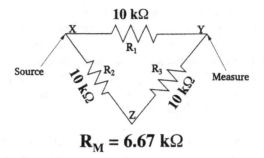

$$R_M = 6.67 \text{ k}\Omega$$

Figure 2-6   In this classic resistor triangle, a measured valued of $6.67\,\text{k}\Omega$ is correct, despite the fact that R_1 is a $10\text{-k}\Omega$ device.

By measuring voltages, currents, and resistances, MDAs can find a host of failures other than gross analog-component problems. A backwards diode, for example, would fail the test, because its reverse leakage would read as forward drop. Resistance measurements detect missing analog or digital components, and voltage measurements find many backwards ICs. In manufacturing operations where such process problems represent the bulk of board failures and where access is available through board nodes, an MDA can identify 80 percent or more of all faults.

Like shorts-and-opens testers, MDAs are fairly inexpensive, generally costing less than $100,000 and often much less. Tests are fast, and self-learn programming minimizes test-programming efforts. Because they provide better fault coverage than shorts-and-opens testers, MDAs serve even better as prescreeners for in-circuit, functional, or "hot-mockup" testing.

Again, these testers suffer because of the bed-of-nails. Contributions from adjacent components severely limit analog accuracy, and there is no real digital-test capability.

2.3.4 In-Circuit Testers

In-circuit testers represent the ultimate in bed-of-nails capability. These sophisticated machines attempt to measure each analog component to its own specifications regardless of surrounding circuitry. Its methods also permit verifying on-board function of individual digital components.

Consider the resistor network in Figure 2-6 if the tester grounds node Z before measuring R_1, as Figure 2-7 shows. Theoretically, no current flows through resistor R_2 or resistor R_3. Therefore, $R_M = R_1 = 10\,\text{k}\Omega$. This process of grounding strategic points in the circuit during testing is called *guarding*, and node Z is known as a *guard point*.

In practice, because the measurement-op-amp's input impedance is not infinite and output impedance is not zero, a small current flows through the guard

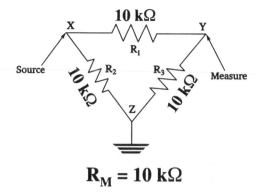

Figure 2-7 Consider the resistor network in Figure 2-6 if the tester grounds node Z before measuring R_1. Theoretically, no current flows through resistor R_2 or resistor R_3.

path. The ratio of measurement-circuit current to guard-path current is known as the *guard ratio*. A simple three-wire in-circuit measurement, as in Figure 2-7, can achieve guard ratios up to about 100.

For high-accuracy situations, and in complex circuits requiring several guard points, assuming that guard-path current is negligible can present a significant problem. Therefore, in-circuit-tester manufacturers introduced a four-wire version, where a guard-point sense wire helps compensate for its current. This arrangement can increase guard ratios by an order of magnitude.

Today, more common six-wire systems address lead resistance in source and measure wires as well. Two additional sense wires add another order of magnitude to guard ratios. This extra accuracy raises tester and fixture costs (naturally) and reduces test flexibility.

Measuring capacitors accurately requires one of two approaches. Both involve measuring voltage across a charged device. In one case, the tester waits for the capacitor to charge completely, measures voltage, and computes a capacitance. For large-value devices, these "settling times" can slow testing considerably. Alternately, the tester measures voltage changes across the device as it charges and extrapolates to the final value. Although more complex, this technique can significantly reduce test times.

In-circuit-tester manufacturers generally provide a library of analog device models. A standard diode test, for example, would contain forward-voltage-drop and reverse-current-leakage measurements. Program-generation software picks tests for actual board components, assigning nominal values and tolerances depending on designers' or manufacturing-engineers' specifications. Analog ICs, especially custom analog ICs, suffer from the same lack of automated programming tools as their complex digital counterparts.

Digital in-circuit testing follows the same philosophy of isolating the device under test from others in the circuit. In this case, the tester injects a pattern of current signals at a component's inputs that are large enough to override any pre-

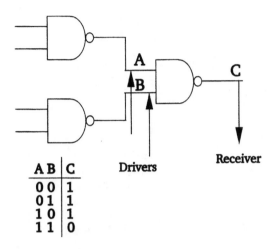

A B	C
0 0	1
0 1	1
1 0	1
1 1	0

Figure 2-8   An in-circuit configuration for testing a two-input NAND gate.

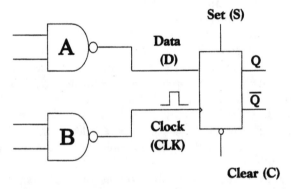

Figure 2-9   A noise glitch that escapes from NAND-gate *B* may look like another clock signal to the flip-flop.

existing logic state, then reads the output pattern. Figure 2-8 shows an in-circuit configuration for testing a two-input NAND gate.

For combinational designs, the in-circuit approach works fairly well. Output states are reasonably stable. Good boards pass, bad boards fail, and test results generally indict faulty components accurately.

Sequential circuits, however, present more of a problem. Consider the simple flip-flop in Figure 2-9.

Most tests assert a state on the *D* line, clock it through, then hold the *D* in the opposite state while measuring the outputs. A noise glitch that escapes from NAND-gate *B* may look like another clock signal to the flip-flop. If *D* is already in the "wrong" state, the outputs will flip and the device will fail. A similar problem may occur from glitches on SET or CLEAR lines.

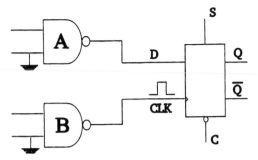

Figure 2-10   Grounding one input on each gate at the flip-flop input in Figure 2-9.

To alleviate this problem, the in-circuit tester can ground one input on each gate at the flip-flop input, as in Figure 2-10, guaranteeing that their outputs (and therefore the inputs to the device under test) remain at 1. This extra step, called *digital guarding*, minimizes the likelihood that any glitches will get through. For standard-voltage devices (TTL, CMOS), an in-circuit tester can overdrive a 1 to a 0 more easily than a 0 to a 1, so digital guarding ensures that test-device inputs are all 1, if possible. For ECL technologies, because of their "high ground," guarding software places device inputs at 0 before testing.

Test and CAE equipment vendors offer test routines for simple "jellybean" digital devices, such as gates, counters, and flip-flops, as well as cursory tests for fairly complex devices, such as microprocessors. Constructing in-circuit test programs for boards whose designs consist primarily of those devices requires merely "knitting" individual device tests together, based on interconnect architecture, and generating analog and digital guard points where necessary.

Many of today's complex boards require a large number of bed-of-nails access pins, sometimes as many as several thousand. Because only a relative handful must be active at any time during testing, tester vendors can minimize the number of actual pin drivers and receivers through a technique known as *multiplexing*. With multiplexing, each real tester pin switches through a matrix to address one of several board nodes. The number represents the *multiplex ratio*. An 8:1 multiplex ratio, for example, means that one tester pin can contact any of eight pins on the board.

The technique reduces in-circuit-tester costs while maximizing the number of accessible board nodes. On the other hand, it introduces a switching step during test execution that can increase test times. In addition (and perhaps more important), it significantly complicates the twin tasks of test-program generation and fixture construction, because for any test, input and output nodes must be on different pins. Aside from simple logistics, accommodation may require much longer fixture wires than with dedicated-pin alternatives, which can compromise test quality and signal integrity.

Most in-circuit testers today employ multiplexing. To cope, many test-program generators assign pins automatically, producing a fixture wiring plan along with the program. Waiting for that information, however, can lengthen project schedules. Revising a fixture to accommodate the frequent changes in board design or layout that often occur late in a development cycle may also prove very difficult. Nevertheless, if the board contains 3000 nodes, a dedicated-pin solution may be prohibitively expensive or unavailable.

In-circuit testers offer numerous advantages. Prices between about $100,000 and $300,000 are generally lower than for high-end functional and some inspection alternatives, and they are generally less expensive to program than functional testers. Test times are fast, and although functional testers feature faster test times for good boards, bad-board test and fault-diagnostic times for in-circuit testers can be substantially lower. In-circuit testers can often verify that even complex boards have been built correctly.

At the same time, however, three forces are combining to make test generation more difficult. First, flip-chips, BGAs, and other surface-mount and hidden node varieties often severely limit bed-of-nails access. In addition, the explosion of very complex devices increases the burden on tester and CAE vendors to create device-test programs. As stated in Chapter 1, these tests are not identical to tests that verify device-design correctness.

Also, as electronic products become both smaller and more complex, hardware designers increasingly rely on ASICs and other custom solutions. Test programs for these parts must be created by device designers, board designers, or test engineers. Because ASIC production runs are orders of magnitude lower than production runs for mass-marketed devices such as microprocessors and memory modules, much less time and money are available for test-program creation. Complicating the problem, device designers often do not have final ASIC versions until very near the release date for the target board or system. Therefore, pressure to complete test programs in time to support preproduction and early production stages means that programs are often incomplete.

Perhaps the biggest drawback to in-circuit test is that it provides no assessment of board performance. Other disadvantages include speed limitations inherent in bed-of-nails technology. Nails add capacitance to boards under test. In-circuit test speeds even approaching speeds of today's lightning-fast technologies may seriously distort stimulus and response signals, as square waves take on distinctly rounded edges.

Traditionally, long distances between tester drivers and receivers often cause problems as well, especially for digital testing. Impedance mismatches between signals in the measurement path can cause racing problems, reflections, ringing, and inappropriate triggering of sequential devices. Fortunately, newer testers pay much more attention to the architecture's drawbacks by drastically limiting wire lengths, often to as little as 1 inch.

Board designers dislike the technique because it generally requires a test point on every board node to allow device access. Overdriving some digital device technologies can demand currents approaching 1 amp! Obtaining such levels from an in-circuit tester is difficult at best. In addition, designers express concern that overdriving components will damage them or at least shorten their useful lives. Therefore, many manufacturers are opting to forego full in-circuit test, preferring to use an MDA or some form of inspection to find manufacturing problems and then proceeding directly to functional test.

One recent development is actually making in-circuit testing easier. As designers integrate more and more functionality onto single devices, real-estate density on some boards is declining.

For example, shrinking boards for desktop computers, telephone switching systems, and similar products realizes few advantages. Even notebook computers cannot shrink beyond a convenient size for keyboard and display. A few years ago, IBM experimented briefly with a smaller-than-normal form factor notebook computer. When the user opened the lid, the keyboard unfolded to conventional size. Unfortunately, reliability problems with the keyboard's mechanical function forced the company to abandon the technology. (Although some palmtop computers feature a foldable keyboard design, their keyboards are much smaller than those of a conventional PC, making the mechanical operation of the folding mechanism considerably simpler.) Since that time, minimum x-y dimensions for notebook computers have remained relatively constant, although manufacturers still try to minimize the z-dimension (profile). In fact, some new models have grown *larger* than their predecessors to accommodate large displays.

Two recent disk-drive generations from one major manufacturer provide another case in point. The earlier-generation board contained hundreds of tiny surface-mounted components on both sides. The later design, which achieved a fourfold capacity increase, featured half a dozen large devices, a few resistor packs, bypass capacitors, and other simple parts, all on one side. Boards destined for such products may include sufficient real estate to accommodate through-hole components, test nodes, and other conveniences.

In-circuit testing of individual devices, however, has become more difficult, especially in applications such as disk-drive test that contain both analog and digital circuitry. At least one in-circuit tester manufacturer permits triggering analog measurements "on-the-fly" during a digital burst, then reading results after the burst is complete. This "analog functional-test module" includes a sampling DC voltmeter, AC voltmeter, AC voltage source, frequency- and time-measurement instrument, and high-frequency multiplexer.

Consider, for example, testing a hard-disk-drive spindle-motor controller such as the TA14674 three-phase motor driver with brake shown in Figure 2-11. This device offers three distinct output levels—a HIGH at 10 V, LOW at 1.3 V, and OFF at 6 V. Durickas (1992) suggests that although a CMOS logic-level test will verify basic functionality, accurately measuring the voltages themselves increases test comprehensiveness and overall product quality.

To conduct such a test, Durickas requires prior testing of certain passive components that surround the controller, then using those components to set important operating parameters for the primary device test. Therefore, his approach necessitates bypassing the controller test if any of the passive components fails.

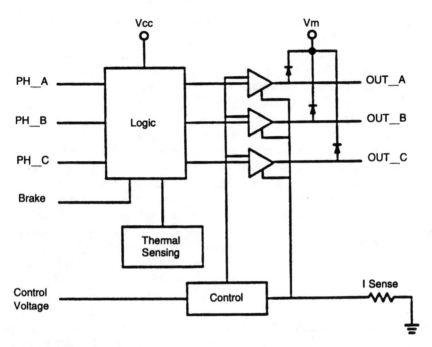

Figure 2-11 A TA14674 three-phase hard-disk-drive spindle-motor controller with brake. (Durickas, Daniel A. 1992. "228X AFTM Applications," GenRad, Concord, Massachusetts.)

The main test consists of four steps. The first provides all valid and invalid logic input states and measures digital outputs at CMOS logic levels—$V_{OH} = 9.5\,V$, $V_{OL} = 1.75\,V$. A pass initiates three additional tests, one for each output. Each output test requires six analog voltage measurements, as Figure 2-12 shows. This hybrid test capability minimizes the number of boards that pass in-circuit test only to fall out at the next test station, in this case usually a hot-mockup.

2.3.5 Bed-of-Nails Fixtures

As indicated earlier, beds-of-nails represent a disadvantage for any test method that must employ them. Nevertheless, the technique can be the only solution to a test problem.

All bed-of-nails fixtures conform to the same basic design. At the base of each fixture is a receiver plate, which brings signals to and from the tester, usually on a 100-mil (0.100-inch) grid. Wires connect appropriate receiver pins to spring-loaded probes that contact the board under test through a platen that is drilled to precisely match the board's electrical nodes and other test points. Each receiver pin corresponds to one and only one board node.

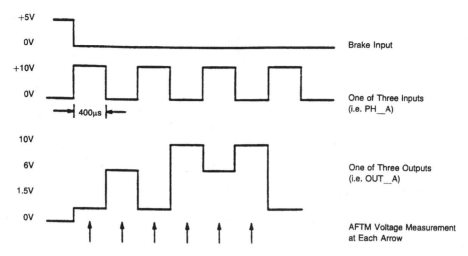

Figure 2-12 An in-circuit test for the component in Figure 2-11. Testing each output requires six analog voltage measurements. (Durickas, Daniel A. 1992. "228X AFTM Applications," GenRad, Concord, Massachusetts.)

Vacuum or mechanical force pulls the board under test down onto the fixture, pushing the probes out through the platen to make contact. Spring-probe compression exerts about 4 to 8 oz. of force at the test point, ensuring a clean electrical connection.

Some fixtures forego the "rat's nest" of wires from receiver to probes in favor of a printed-circuit board. This choice improves speed and reliability while minimizing crosstalk and other types of interference. Such "wireless fixtures" are less flexible than their more traditional counterparts. In addition, lead times to manufacture the circuit board may make delivering the fixture in time to meet early production targets more difficult.

There are four basic types of bed-of-nails fixtures. A *conventional*, or *dedicated*, fixture generally permits testing only one board type, although a large fixture sometimes accommodates several related small boards.

A *universal* fixture includes enough pins on a standard grid to accommodate an entire family of boards. The specific board under test rests on a "personality plate" that masks some pins and passes others. Universal fixtures generally cost two to three times as much as their conventional counterparts. Pins also require more frequent replacement because although testing a particular board may involve only some pins, all pins are being compressed. Therefore, if a pin permits 1,000,000 repetitions, that number must include occasions where the pin merely pushes against the bottom of the personality plate. The high pin population on universal fixtures makes troubleshooting faulty pins and other maintenance much more difficult than in the conventional type.

Surface-mount fixtures also attach to a standard 100-mil receiver grid. In this case, however, a series of translator pins pass through a center fixture platen so

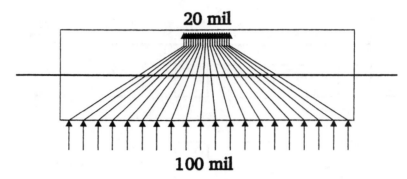

Figure 2-13   In surface-mount fixtures, a series of translator pins pass through a center fixture platen so that at the board the pins can be much closer together.

that at the board, the pins can be much closer together, as Figure 2-13 shows. The principle here is that even on heavily surface-mount boards, not all nodes require very close centers. The center platen minimizes pin bending and therefore improves pointing accuracy.

This type of fixture is about 20 to 50 percent more expensive than a conventional type. In addition, if too many areas on the board contain close centers, the number of available tester pins may be insufficient, or wires from the receiver may be too long to permit reliable testing.

Each of these fixture designs assumes that the tester requires access to only one board side at a time. This approach means either that all test nodes reside on one side (usually the "solder" side) or that board design permits dividing the test into independent sections for each side. This latter approach, of course, means that each board requires two fixtures and two tests. In many manufacturing operations, however, capacity is not at issue, so that testing boards twice per pass does not present a problem.

Testing both board sides simultaneously requires a *clamshell* fixture that contains two sets of probes and closes onto the board like a waffle iron. Clamshell fixtures are extremely expensive. A test engineer from a large computer company reported that one of his fixtures cost nearly $100,000. To be fair, the board was 18″ × 18″ and contained several thousand nodes. Nevertheless, fixture costs that approach or exceed tester costs may necessitate seeking another answer.

Clamshell-fixture wires also tend to be much longer than wires on a conventional fixture. Wire lengths from the receiver to the top and bottom pins may be very different, so that signal speeds do not match. If a pin or wire breaks, finding and fixing it can present quite a challenge. The rat's nest in one of these fixtures makes its conventional counterpart look friendly by comparison.

The accuracy, repeatability, and reliability of a clamshell fixture's top side are poorer than those of a conventional solution. Access to nodes and components on the board for probing or pot adjustment is generally impossible. In addition, the top side of a clamshell fixture often contacts component legs directly. Pressure from nails may make an electrical connection when none really exists on the board, so the test may not notice broken or cold-solder joints.

Although clamshell fixtures present serious drawbacks, they often provide the only viable way to access a board for in-circuit or MDA test. Because of their complexity, they bring to mind a dog walking on its hind legs: It does not do the job very well, but, considering the circumstances, you have to be amazed that it can do the job at all.

Bed-of-nails fixtures permit testing only from nail to nail, not necessarily from node to node. Consider a small-outline IC (SOIC) or other surface-mounted digital component. Actual electrical nodes may be too small for access. For BGAs and flip-chips, nodes reside underneath. In any case, pads are often some distance from the actual component under test. Therefore, a trace fault between the component and the test pad will show up as a component fault, whereas the continuity test between that pad and the next component or pad will pass.

Service groups rarely use bed-of-nails systems. Fixtures are expensive and difficult to maintain at correct revision levels when boards returning from the field may come from various product versions, some of which the factory may not even make anymore. Necessary fixture storage space for the number of board types handled by a service depot would generally be excessive.

Alternatives include foregoing the factory bed-of-nails test altogether to permit a common test strategy in the factory and the field. Some vendors offer service testers that can mimic the behavior of their bed-of-nails products without the need for conventional fixtures. Operation involves scanners, *x-y* probers, or clips and probes. These testers can be less expensive than their factory-bound counterparts. Software and test programs are compatible or can be converted between the two machine types.

2.3.6 *Bed-of-Nails Probe Considerations*

Figure 2-14 shows a typical bed-of-nails probe construction, including a plunger, barrel, and spring. This design allows variations in target height. Spring-probe tips must pierce board-surface contamination to permit a low-resistance connection. Therefore, probe manufacturers provide a plethora of tip styles for different applications. Figure 2-15 presents some common types. To avoid the confusion of including several tip styles on a single fixture, test engineers usually choose one that provides the best compromise for a particular board.

Chisel tips easily penetrate oxide contamination on solder pads and plated-through holes, the most common targets. Other solutions include *stars*, *tulips*, and *tapered crowns*. These tips are self-cleaning, meaning that each probe cycle wipes away flux residue, solder resist, and other contaminants, increasing test accuracy and extending probe life. Many board manufacturers, especially in Japan, choose star tips for plated-through-hole applications, such as bare-board testing. Pene-

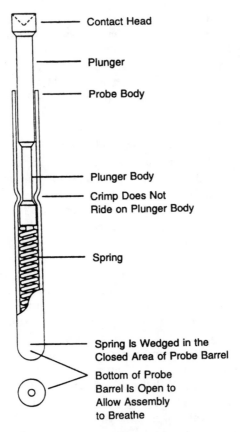

— Contact Head

— Plunger

— Probe Body

— Plunger Body

— Crimp Does Not
Ride on Plunger Body

— Spring

— Spring Is Wedged in the
Closed Area of Probe Barrel

— Bottom of Probe
Barrel Is Open to
Allow Assembly
to Breathe

Figure 2-14 Construction for a typical bed-of-nails probe. (Reprinted with the permission of Interconnect Devices, Kansas City, Kansas.)

trating an oxide layer with a star tip requires higher spring force than with other designs, but the star provides more contact points.

Other probe tips offer advantages, as well. *Concave* tips accommodate long leads and wire-wrap posts. A tendency to accumulate debris, however, makes these tips most effective in a "pins down" position.

Spear tips, which are usually self-cleaning unless they contain a flat spot or large radius, can pierce through surface films on solder pads. As a general solution, however, Mawby (1989) cautions against probing holes larger than half the diameter of the spear-tip plunger.

Flat tips and *spherical-radius* tips work well with gold-plated (and therefore uncorroded) card-edge fingers. *Convex* designs apply to flat terminals, buses, and other solid-node types. These tips are not self-cleaning when probing vias.

Flex tips can pierce conformal coatings, as on many military boards. Of course, these boards require recoating after test. *Serrated* tips work well for access to translator pins or on terminal strips that contain some surface contamination.

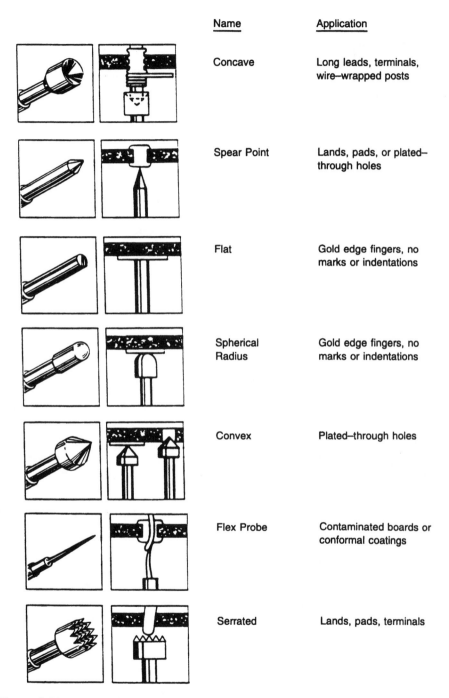

Name	Application
Concave	Long leads, terminals, wire–wrapped posts
Spear Point	Lands, pads, or plated–through holes
Flat	Gold edge fingers, no marks or indentations
Spherical Radius	Gold edge fingers, no marks or indentations
Convex	Plated–through holes
Flex Probe	Contaminated boards or conformal coatings
Serrated	Lands, pads, terminals

Figure 2-15 A selection of available probe tip styles. (Reprinted with the permission of Interconnect Devices, Kansas City, Kansas.)

	Name	Application
	Star	Plated–through holes, lands, pads; self–cleaning
	3– or 4–sided Chisel	Lands, pads, leads, holes; self–cleaing
	0.040 Crown	Leads, pads, lands, holes; self–cleaning
	0.050 Crown	Leads, pads, lands; self–cleaning
	Tapered Crown	Leads, pads, lands, holes; self–cleaning
	Tulip	Long leads, terminals, wire–wrapped parts; self–cleaning

Figure 2-15   (*continued*)

Success for a particular tip style depends on both its geometry and the accompanying spring force. Too little force may prevent good electrical contact. Too high a force may damage delicate board pads and other features.

Some probes rotate as they compress. This action helps the probe to pierce through conformal coatings, corrosion, and other surface contamination to ensure good contact.

Test engineers must carefully distribute pins across a board's surface to avoid uneven pressure during testing. Such pressure can cause bending moments, reducing the pointing accuracy of pins near the board edge and possibly breaking solder bonds or tombstoning surface-mounted components.

Ensuring high-quality circuit resistance measurements and digital tests requires low interface resistance and good node-to-node isolation. According to Mawby, fixture-loop resistance—including switching circuitry, wires, and probes—ranges between 1Ω and 5Ω. Although its magnitude is not much of an impediment for shorts-and-opens and other self-learned tests that compensate for it automatically, node-to-node variations must remain much less than shorts-and-opens thresholds. In-circuit testers must subtract out this so-called "tare" resistance to achieve necessary measurement accuracies.

A low *contact resistance*, usually specified as less than 500 mΩ, also ensures maximum test speed and accuracy. Contact resistances tend to increase as probes age, however, primarily from corrosion and other contamination. Cleaning helps minimize the problem but cannot eliminate it completely.

Pointing accuracy specifies how closely a pin can hit its target. Repeatability indicates the consistency from cycle to cycle. Keller and Cook (1985) estimated an accuracy (1 standard deviation, or σ) of ±4 mils for 100-mil probes and ±5 mils for 50-mil probes. Of course, if pins bend as they age, these figures will get worse. The researchers recommend surface-mount-board test pads with diameters of 3σ or 4σ. At 3σ, approximately 3 probes in 1000 would miss their targets. At 4σ, less than 1 in 10,000 would miss. Based on these numbers, 100-mil probe pads should be 24 mils or 32 mils in diameter, and 50-mil pads should be 30 mils or 40 mils. Designers may justifiably balk at consuming so much real estate "merely" for test pads.

Mawby refers to new probe designs that achieve accuracies of ±1.5 mils for 50-mil surface-mount boards. These probes need test pads only 9 mils or 12 mils in diameter to meet 3σ and 4σ requirements. St. Onge (1993) also explores hitting small targets with a conventional bed-of-nails.

Unfortunately, probes do not last forever. Standard models are rated for about 1,000,000 cycles if there is no sideloading. Probe life is approximately proportional to cross-sectional area. Therefore, if a 50-mil-pin barrel diameter is half of the barrel diameter for a 100-mil pin, it will likely last only one-quarter as long. Because pin replacement is a maintenance headache, this shortened life constitutes a major concern. Some probe manufacturers have introduced small-center models with larger-than-normal barrel diameters specifically to address the life issue. Although they do not last as long as their 100-mil siblings, these models can last twice as long as other small-center designs. To hit targets smaller than 50 mils, some probes adopt the traditional design but completely eliminate the receptacle, mounting directly to thin mounting plates. Because small pins are more fragile and can be less accurate, a test fixture should include 100-mil pins wherever possible, resorting to 50-mil and smaller ones only where necessary.

2.3.7 Opens Testing

As stated earlier, the proliferation of surface-mount technologies has aggravated the problem of opens detection to the point where it is now often the most difficult manufacturing fault to detect. Some such problems, which defy electrical testing altogether and encourage some kind of inspection, will be addressed in the next chapter.

Techniques have emerged to detect many opens—assuming (and this assumption is becoming ever more of a constraint) that you have bed-of-nails access to the board nodes. They perform measurements on unpowered boards, and often rely on clamp diodes that reside inside the IC between I/O pins and ground or on "parasitic diodes" formed by the junction between the pin and the substrate silicon.

The common techniques can be broadly divided into two groups. *Parametric process testing* measures voltage or current on the diodes directly or by forming them into transistors. This approach requires no special hardware beyond the fixture itself.

The simplest version applies voltage to the diode through the bed-of-nails, then measures the current. An open circuit generates no current and no forward voltage. Unfortunately, this method will miss faults on parallel paths.

One variation biases the diode input (emitter) and output (collector) relative to ground, then compares collector current on groups of pins. Proponents contend that this approach proves less sensitive to device-vendor differences than the more conventional alternative. It can detect opens right to the failing pin, as well as misoriented devices and incorrect device types. It will sometimes identify a device from the wrong logic family, and may find resistive joints and static damage, depending on their severity.

On the downside, program debugging for this method requires "walking around" a reference board and opening solder joints. It also may not see differences between simple devices with the same pinouts but different functions. It cannot detect faults on power or ground buses, but that limitation is also true with the other techniques.

In *capacitive testing*, a spring-mounted metal plate on top of the IC package forms one side of a capacitor. The IC leadframe forms the other side, with the package material the dielectric. The tester applies an AC signal sequentially to the device pins through the bed of nails. The probe-assembly buffer senses the current or the voltage to determine the capacitance. In this case, the measurement circuits see only the pins, not the bondwire and internal diodes, detecting only opens between the IC pins and the board surface. It can, therefore, examine the connectivity of mechanical devices such as connectors and sockets, as well as ICs. Because of the extra hardware required, this technique increases bed-of-nails fixture costs.

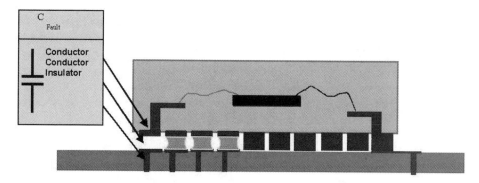

Figure 2-16   Anatomy of an open-solder capacitor. (Courtesy Agilent Technologies.)

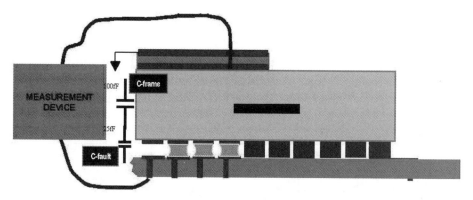

Figure 2-17   The measurement system places a conductor over the lead frame, forming an additional capacitor of about 100 fF. (Courtesy Agilent Technologies.)

Figure 2-16 shows the capacitor formed by an open IC circuit. The leadframe forms one plate, the pad and trace on the PCB the other. The lack of solder in between (air) forms the dielectric, creating a capacitor of about 25 fF. The measurement system places a conductor over the leadframe, as in Figure 2-17, forming an additional capacitor of about 100 fF. An open circuit would produce these two capacitances in series, for an equivalent capacitance of 20 fF. In a good joint, the measurement would see only the 100 fF test capacitor.

This theory also applies to testing other components with internal conductors, such as connectors and sockets, as Figure 2-18 shows. Testing sockets ensures proper operation before loading expensive ICs onto the board at the end of the assembly process. The measurement system grounds pins in the vicinity of the pin under test. The resulting capacitance is often higher than for an IC, which may cause the capacitance of a solder open to be higher as well. Figure 2-19 shows the same principle applied to a switch.

You can probe even right-angle connectors using this technique. In that case, however, you do have to create a bit of custom fixturing. Figure 2-20 shows one

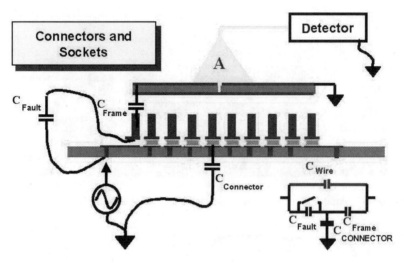

Figure 2-18 Measuring opens in connectors and sockets. (Courtesy Agilent Technologies.)

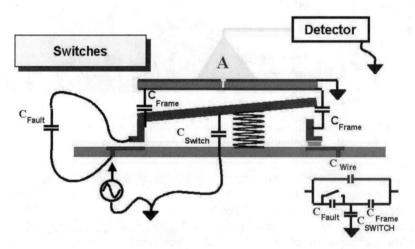

Figure 2-19 Applying the same principle to a switch. (Courtesy Agilent Technologies.)

possibility. The capacitive leadframe is mounted on a sliding mechanism called a *probe carriage*. The spring-loaded probe carriage retracts the leadframe probe when the fixture is open, to allow the operator to more easily load and unload the board under test. In the closed fixture, the cone pushes on a roller bearing that is part of the carriage, moving it and the leadframe probe to the connector.

Some manufacturers have expressed concern that the force of the capacitive probe will close the solder gap, creating a connection where none really exists. As Figure 2-21 shows, the force exerted is far less than would be required to do this.

RIGHT ANGLE CONNECTOR

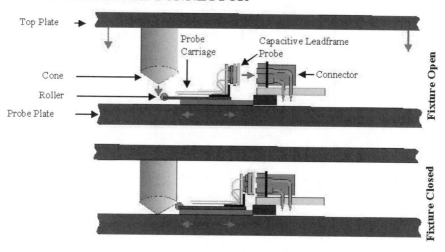

Figure 2-20   Custom fixturing for testing a right-angle connector. (Courtesy Agilent Technologies.)

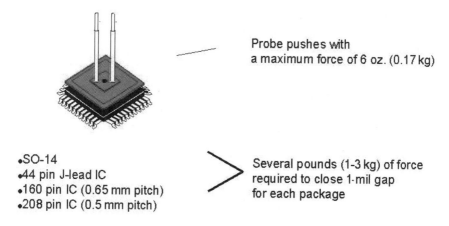

Probe pushes with
a maximum force of 6 oz. (0.17 kg)

• SO-14
• 44 pin J-lead IC
• 160 pin IC (0.65 mm pitch)
• 208 pin IC (0.5 mm pitch)

Several pounds (1-3 kg) of force
required to close 1-mil gap
for each package

Figure 2-21   The force of the probe is far less than that necessary to close the solder gap.

2.3.8   *Other Access Issues*

With bed-of-nails access becoming more difficult, companies often rely more on functional or cluster testing to verify digital circuitry. Because analog circuits do not lend themselves easily to that approach, it has become necessary to find a viable alternative. With an average of two to three analog components on every board node, every node that defies probing reduces the number of testable components by that same two or three. Pads have shrunk to only a few mils, and center-to-center

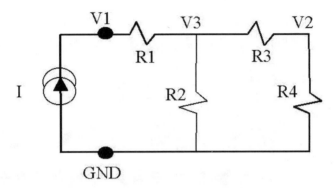

Figure 2-22   Circuit diagram for the limited-access example. (From *Proceedings of Nepcon West*, 1998. Courtesy Agilent Technologies.)

probe distances have fallen as well. So-called "no-clean" processes require higher probing forces to pierce any contaminants on the node, which increases stress on the board during bed-of-nails test. In fact, 2800 12-oz. probes exert a *ton* of force. Clearly, less access may occur even where nodes are theoretically available.

McDermid (1998) proposes a technique for maximizing test diagnostics with as little as 50 percent nodal access. He begins with an unpowered measurement, using a small stimulus voltage to break the circuit into smaller pieces. In this situation, device impedances are sufficient to appear to the tester as open circuits. Clusters of analog components are connected by either zero or one node. Typically, these clusters are small and isolated from one another. We assume no more than one failing node per cluster.

Consider the circuit in Figure 2-22. I is the system stimulus. When circuit components are at nominal values, the voltages are defined as nominal as well. Varying component values within tolerance limits produces voltages that fall into a scatter diagram, such as the one in Figure 2-23. If R1 or R3 fail, the scatter diagram looks like the one in Figure 2-24. If nodes are available for only V1 and V2, you see the two-dimensional shadow depicted, and shown in more detail in Figure 2-25. If only V1 and V3 permit access, the shadow looks like Figure 2-26. In this view, you cannot tell which resistor has failed, demonstrating the importance of selecting test points carefully. Figure 2-27 presents actual results from this technique.

2.3.9 Functional Testers

Functional testers exercise the board, as a whole or in sections, through its edge connector or a test connector. The tester applies a signal pattern that resembles the board's normal operation, then examines output pins to ensure a valid response. Draye (1992) refers to this type of test as "general-purpose digital input/output measurement and stimulus." Analog capability generally consists of a range of instruments that provide analog stimuli or measurements in concert with the board's digital operation.

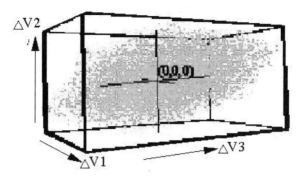

Figure 2-23 Varying component values within tolerance limits produces voltages that fall into a scatter diagram. (From *Proceedings of Nepcon West*, 1998. Courtesy Agilent Technologies.)

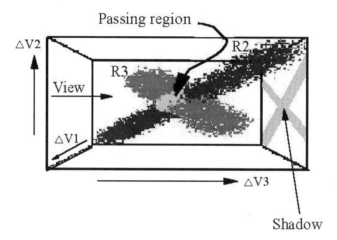

Figure 2-24 The scatter diagram if R1 or R3 fails. (From *Proceedings of Nepcon West*, 1998. Courtesy Agilent Technologies.)

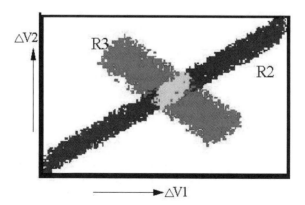

Figure 2-25 Looking at the scatter-diagram "shadow" if nodes are available only on V1 and V2. (From *Proceedings of Nepcon West*, 1998. Courtesy Agilent Technologies.)

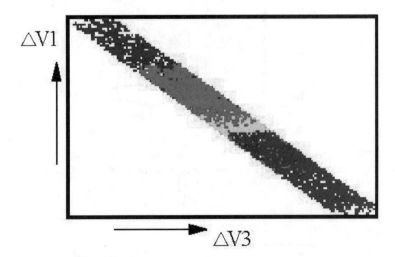

Figure 2-26 The shadow with nodes only on V1 and V3. (From *Proceedings of Nepcon West*, 1998. Courtesy Agilent Technologies.)

#	Nodes	Accessible nodes	Resistors	Inductors	Capacitors
1	6	3	6	0	3
2	17	4	16	1	11
3	21	9	14	2	20

#	# distinct groups	# indistinct groups	Max # components per indistinct group	APG time (seconds)
1	9	0	--	3.3
2	1	3	5	7.3
3	11	6	3	125

Figure 2-27 Actual test results with limited access. (From *Proceedings of Nepcon West*, 1998. Courtesy Agilent Technologies.)

Some complex boards also require a "modified" bed-of-nails to supplement edge and test connectors. Differences between in-circuit and functional beds-of-nails include the number of nails and their purpose. Whereas an in-circuit bed-of-nails provides a nail for every circuit node, functional versions include nails only at critical nodes that defy observation from the edge connector. Relying on only a few nails avoids loading the circuit with excess capacitance, which would reduce maximum reliable test speeds.

A functional-test bed-of-nails cannot inject signals. It provides observation points only. Reducing functional-test logic depth, however, simplifies test-program generation considerably.

Functional beds-of-nails remain unpopular with most board manufacturers because of fixture costs, scheduling pressures, and nail capacitances. These concerns have spurred the growth of boundary-scan designs (see Chapter 5) as an alternative for internal-logic access.

An MDA or an in-circuit test measures the success of the manufacturing process. Functional testing verifies board performance, mimicking its behavior in the target system. Because this test tactic addresses the circuit's overall function, it can apply equally well to testing large circuit modules or hybrids and to system testing.

Functional test can occur at full speed, thereby uncovering racing and other signal-contention problems that escape static or slower-speed tests. This test also verifies the design itself, as well as how well the design has translated to the real world. Test times for passing boards are the fastest of any available technique, and failure analysis can indicate the correct fault, almost regardless of board real-estate population density.

Functional test is traditionally the most expensive technique. Also, *automatic* functional testing is still basically a digital phenomenon. Programming is difficult and expensive and traditionally involves a complex cycle of automatic and manual steps. Analog automatic program generation is nearly nonexistent.

Most functional testers work best at determining whether a board is good or bad. Pinpointing the cause of a failure can take much longer than the test itself. Diagnostic times of hours are not unheard of. Many companies have "bone piles" of boards that have failed functional test, but where the cause remains unknown.

A bed-of-nails tester can identify all failures from a particular category (shorts, analog, digital) in one test pass. For a functional test, any failure requires repair before the test can proceed. Therefore, a test strategy that eliminates bed-of-nails techniques must achieve a very high first-pass yield with rarely more than one fault per board to avoid multiple functional-test cycles.

Solutions that address these issues are emerging. Some new benchtop functional testers are *much* less expensive than their larger siblings. A traditional functional tester can cost hundreds of thousands of dollars, whereas benchtop prices begin at less than $100,000. Larger testers still perform better in applications requiring high throughput and very high yields, but for many manufacturers, small functional testers can offer a cost-effective alternative.

2.3.10 Functional Tester Architectures

Digital functional testing comes in several forms. Differences involve one or more of certain test parameters:

- *Test patterns*: Logical sequences used to test the board
- *Timing*: Determines when the tester should drive, when it should receive, and how much time should elapse between those two events

- *Levels*: Voltage and current values assigned to logic values in the pattern data. Levels may vary over the board, depending on the mix of device technologies.
- *Flow control*: Program tools that specify loops, waits, jumps, and other sequence modifiers

The simplest type of digital functional "tester" is an *I/O port*. It offers a limited number of I/O channels for a board containing a single logic family. The I/O port offers a low-cost solution for examining a few digital channels. However, it is slow and provides little control over timing or logic levels during test, severely limiting its capability to verify circuits at-speed or to-spec.

Emulators exploit the fact that many digital boards feature bus-structured operation and resemble one another functionally. A somewhat general, hardware-intensive test can verify those common functions, reducing overall test-development effort. Emulation replaces a free-running part of the board's logic with a test pod. It then mimics the board's behavior in the target system, stopping at convenient points to examine registers and other hardware states. Figure 2-28 shows a simplified block diagram of a typical emulation tester.

Emulation is perhaps the least well-understood test technique. One problem is that many sources refer to it as *in-circuit emulation*, yet it has nothing to do with in-circuit testing. Calling it *performance testing*, as other sources do, better describes its operation.

There are three basic types of emulation. Most familiar is *microprocessor emulation*, where a test pod attaches to the microprocessor leads or plugs into an empty microprocessor socket. On boards with more than one processor, the test must replace all of them, either serially or simultaneously. A successful test requires that the board's power and clock inputs, reset, data-ready, and nonmaskable interrupts function correctly.

Memory emulation replaces RAM or ROM circuitry on the board under test, then executes a stored program through the existing microprocessor and surrounding logic, including clock, address and data buses, address decoder, and RAM. Because the microprocessor remains part of the circuit, this variation has advantages over microprocessor emulation for production test. Also, the tester does not require a separate pod for each microprocessor, only one for each memory architecture, reducing hardware acquisition and development costs.

Bus-timing emulation does not actually require the presence of a microprocessor on the board at all. It treats the processor as a "black box" that simply communicates with the rest of the board logic over I/O lines. The bus can be on the board or at the edge connector. The technique executes MEMORY READ, MEMORY WRITE, INPUT, OUTPUT, FETCH, and INTERRUPT functions from the processor (wherever it is) to assess board performance.

Bus emulators contain a basic I/O port, but include local memory behind the port for caching digital test patterns as well, giving considerably more control over speed and timing. Emulators offer three types of channels: address field, data field, and timing or control channels. Logic levels are fixed, and the system controls timing

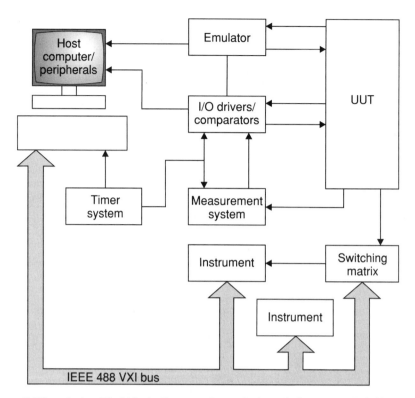

Figure 2-28 A simplified block diagram of a typical emulation tester. (Scheiber, Stephen F. 1990. "The New Face of Functional Testing," *Test & Measurement World,* Newton, Massachusetts. Eisler, Ian. 1990. "Requirements for a Performance-Tester Guided-Probe Diagnostic System," ATE & Instrumentation Conference, Miller-Freeman Trade-Show Division, Dallas, Texas.)

within address and data fields as a group, rather than as individual channels. The emulator loads the bus memory at an arbitrary speed, triggers the stimulus at the tester's fixed clock rate, then collects board responses on the fly in response memory. Unloading the memory at a convenient speed provides the test results.

Bus-timing emulation is an ideal technique for testing boards that do not contain microprocessors but are destined for microprocessor-based systems. The classic example is a personal-computer expansion board or a notebook computer's credit-card-sized PCMCIA board. The PC's I/O bus is well defined. The tester latches onto the bus and executes a series of functions that are similar to what the board experiences inside the PC. The test can even include erroneous input signals to check the board's error-detection capabilities. Tools include noise generators, voltage-offset injectors, and similar hardware.

Emulation testers are generally quite inexpensive. In fact, it is possible to execute a bus-emulation test for some products without an actual tester. A conventional PC contains most of the necessary hardware. An expansion board and proper software may suffice to create a "test system."

Program development may also be less expensive and less time-consuming than development for a more-elaborate functional test. A microprocessor-emulation test resembles a self-test for the corresponding system. Test engineers can avoid creating a software behavioral model of a complex IC, choosing instead to emulate it with a hardware pod.

Emulation tests boards in their "natural" state. That is, it does not apply every conceivable input combination, restricting the input set to those states that the target product will experience.

Emulation can also find obscure faults by "performance analysis." It can identify software problems by tracking how long the program executes in each area of memory. In some cases, software contains vestigial routines that do not execute at all. Recent revisions may have supplanted these routines, yet no one has deleted them. They take up memory space, and their mere presence complicates debugging and troubleshooting the software to no purpose. In other cases, a routine should execute but does not.

For example, an unused routine may provide a wait state between two events. Unless the system usually or always fails without that delay, the fact that the software lies idle goes unnoticed. If the emulation test knows to look for program execution in that section of memory and it never gets there, the test will fail. This technique will also notice if the software remains too long in a section of memory or not long enough (such as an n-cycle loop that executes only once).

Emulation permits finding certain failures that are difficult or impossible to detect in any other way. One such example is *single-bit-stack creep*. A computer system stuffs information into a data stack in bytes or other bit-groups for later retrieval. If noise or some other errant signal adds or deletes one or more bits, all subsequent stack data will violate the specified boundaries, and retrieved data will be garbled. The ability to stop the system to examine hardware registers (such as stacks) will uncover this particularly pernicious problem. As digital-device voltages continue to fall, this kind of fail-safe testing becomes more important.

Disadvantages of this technique include the need for individual emulation modules for each target microprocessor, RAM configuration, or I/O bus. Module availability presents few impediments other than cost if the target configuration is common, such as an ISA or USB PC bus or a Pentium-class microprocessor. Testing custom systems and new designs, however, may have to wait for the pod's completion.

Program generation is nearly always a manual process and requires a programmer with intimate knowledge of circuit behavior. Therefore, test development tends to miss unusual failure mechanisms, because the person most familiar with a board's design is not the person most likely to anticipate an oddball result.

Emulation requires that the target system be microprocessor-based. Creating guided-probe diagnostics is both expensive and time-consuming, especially for memory and bus-timing variations where the microprocessor is not completely

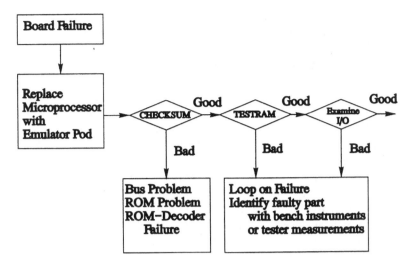

Figure 2-29 A flowchart for finding a failure with a microprocessor-emulation test. (Scheiber, Stephen F. 1990. "The New Face of Functional Testing," *Test & Measurement World*, Newton, Massachusetts.)

in control of the circuit. The technique offers no significant analog testing and no clear way to isolate faults outside of the kernel logic. In addition, there is no easy way to determine fault coverage accurately, so it is difficult to decide when to stop program development. Many test engineers simply establish a set of test parameters and a timetable. When the time is exhausted, the test program is declared complete.

For finding a failure with a microprocessor-emulation test, consider the flowchart in Figure 2-29. Here, the board's internal self-tests have failed, and the microprocessor has either stopped or is running wild. A passive examination of circuit behavior has produced no diagnosable symptoms.

The test operator replaces the board's microprocessor with an emulator pod or clips the pod onto the processor's I/O pins and executes a CHECKSUM test on the ROMs. This test involves reading all ROM locations, adding them up, and comparing the results to the corresponding numbers stored on the ROMs themselves. If the numbers match, the test moves on. Any difference indicates a problem with one of the ROMs, the ROM bus, or the decoder. On a failure, the tester enters a diagnostic loop, providing a sync pulse that allows examining the circuit with a logic analyzer or other external instrument to pinpoint the faulty component.

If the CHECKSUM test passes, the next step is a TESTRAM that verifies read/write memory devices and their surrounding buses. If this test fails and the buses are presenting legal WRITE and CHIP-SELECT signals, the fault lies in one or more devices. Again, looping at a convenient hardware state allows further analysis with bench instruments. Experience has shown that with these symptoms, a RAM decoder or driver is a frequent culprit.

If ROMs, RAMs, and their corresponding buses have passed, the test checks the board's I/O. I/O tests can be simple or complex, depending on the situation. Steps include reading A/D converter output values, reading and setting parallel bits, and examining complex devices such as direct-memory-access (DMA) channels. In some cases, obtaining meaningful data from I/O devices can require programming dozens of internal registers. Therefore, many emulation testers offer *overlay RAM*, which overrides on-board memory during test execution. This approach allows examining board logic, for example by triggering I/O initialization routines, regardless of any memory faults.

Digital word generators resemble the memory-and-state-machine architecture of emulators, but provide a more general-purpose solution. Like the emulator, they store stimulus and response signals in local memory. They also include several types of fixed-logic-level channels, but add some programmable channels as well. Channels again exist as groups, and the tester can control them only as groups. However, you can add or configure signal channels by buying additional modules from the tester manufacturer. This architecture offers more flexibility than the emulators do. Nevertheless, timing flexibility is similarly limited, and programmable timing and test speed is limited to the time necessary to perform one memory-access cycle.

Most sophisticated in the digital-test arsenal is the *performance functional tester*. This system tests the board operation to established specifications, rather than "merely" controlling it. The tester can emulate the board-to-system interface as closely as possible to ensure that the board will work in the final system without actually installing it.

This alternative offers highly flexible digital channels, as well as individually selectable logic levels (to serve several logic families on the same board), timing, and control. To synchronize the test with analog measurements, the tester often includes special circuitry for that purpose. Subject to the tester's own specifications (we will discuss this subject further in Chapter 8), it can precisely place each signal edge. Programs are divided into clock cycles, rather than events, permitting more direct program development through interface with digital simulators.

Conventional stimulus/response functional testing relies on simulation for test-program development. As boards become more complex, however, programmers must trade off test comprehensiveness against simulation time. Design or test engineers who understand board logic often select a subset of possible input patterns to reduce the problem's scope. After obtaining fault-coverage estimates from a simulator using that subset, engineers can carefully select particular patterns from the subset's complement to cover additional faults.

2.3.11 Finding Faults with Functional Testers

Once a board fails functional test, some kind of fault isolation technique must determine the exact reason, either for board repair or for process improvement. Common techniques include manual analysis, guided-fault isolation (GFI), fault dictionaries, and expert systems.

For many electronics manufacturers, especially those of complex, low-volume, and low-failure-rate products (such as test instruments and systems), functional testers do not perform fault isolation at all. Bad boards proceed to a repair bench where a technician, armed with an array of instruments and an understanding of the circuit, isolates the fault manually.

Because manual analysis takes place offline, it maximizes throughput across the tester. Because finding faults on some boards can take hours (or days) even with the tester's help, this throughput savings can be significant.

Manual techniques can be less expensive than tester-bound varieties. Most test operations already own logic analyzers, ohmmeters, digital voltmeters (DVMs), oscilloscopes, and other necessary tools, so the approach keeps capital expenditures to a minimum. Also, this method does not need a formal test program. It merely follows a written procedure developed in cooperation with designers and test engineers. The repair technician's experience allows adjusting this procedure "on the fly" to accommodate unexpected symptoms or analysis results. For the earliest production runs, test engineers, designers, and technicians analyze test results together, constructing the written procedure at the same time. This approach avoids the "chicken-and-egg" problem of trying to anticipate test results before the product exists.

On the downside, manual analysis is generally slow, although an experienced technician may identify many faults more quickly than can an automatic-tester operator armed only with tester-bound tools. The technique also demands considerable technician expertise. Speed and accuracy vary considerably from one technician to the next, and the process may suffer from the "Monday/Friday" syndrome, whereby the same technician may be more or less efficient depending on the day, shift, nearness to lunch or breaks, and other variables.

The semiautomatic fault-finding technique with which functional-test professionals are most familiar is *guided-fault isolation (GFI)*. The functional tester or another computer analyzes data from the test program together with information about good and bad circuits to walk a probe-wielding operator from a faulty output to the first device input that agrees with the expected value. Performing GFI at the tester for sequential circuits allows the tester to trigger input patterns periodically, thereby ensuring that the circuit is in the proper state for probing. The tester can learn GFI logic from a known-good board, or an automatic test-program generator can create it.

Properly applied, GFI accurately locates faulty components. As with manual techniques, it works best in low-volume, high-yield applications, as well as in prototype and early-production stages where techniques requiring more complete information about circuit behavior fare less well.

As with manual techniques, however, GFI is both slow and operator-dependent. It generally occupies tester time, which reduces overall test capacity. Long logic chains and the preponderance of surface-mount technology on today's boards have increased the number of probing errors, which slows the procedure even further. Most GFI software copes with misprobes by instructing the operator to begin again. Software that allows misprobe recovery by starting in the middle of the sequence, as close to the misprobe as possible, reduces diagnostic time considerably. Using automated probe handlers, similar to conventional *x-y* probers, during repair can speed diagnosis and minimize probing errors.

Some manufacturers construct functional tests in sections that test parts of the logic independently. In this way, GFI probing chains are shorter, reducing both time and cost.

As parts, board traces, and connections have shrunk, concern has mounted that physical contact with the board during failure analysis may cause circuit damage. The proliferation of expensive ASICs and other complex components and the possibility that probed boards will fail in the field have increased the demand for less stressful analysis techniques.

Fault dictionaries address some of these concerns. A fault dictionary is merely a database containing faulty input and output combinations and the board faults that cause them. Fault simulators and automatic test-program generators can create these databases as part of their normal operation.

A complete dictionary would contain every conceivable input and output pattern and, therefore, pinpoint all possible failure modes. As a practical matter, manufacturers have experienced mixed success with this technique because few dictionaries are so comprehensive. The analysis may identify a fault exactly, but it more often narrows the investigation to a small board section or a handful of circuit nodes, then reverts to GFI for confirmation and further analysis. Even more often than with GFI techniques, a fault dictionary depends on dividing circuit logic to "narrow the search" for the fault source.

This method is very fast and requires no operator intervention except during supplemental GFI. It is, therefore, appropriate for high-volume applications. Test programming is generally faster than with GFI because the test-program generator does much of the work. For specific faults that it has seen before, the technique is quite accurate.

The need to subdivide board logic represents a design constraint. In addition, not all logic subdivides easily. The method is *deterministic*—that is, an unfamiliar failure pattern generally reverts to GFI. Revising board designs may necessitate manually updating the dictionary, which can be a significant headache during early production.

An interesting technique that has fallen into disuse in the past few years involves *expert systems* (also known as *artificial-intelligence techniques*)—essentially smart dictionaries. Like conventional fault dictionaries, they can examine faulty outputs from specific inputs and identify failures that they have seen before. Unlike their less-flexible counterparts, expert systems can also analyze a never-before-seen output pattern from a particular input and postulate the fault's location, often presenting several possibilities and the probability that each is the culprit. When an operator or technician determines the actual failure cause, he or she informs the tester, which adds that fault to the database.

Expert systems do not require conventional test programming. Instead, the tester executes a set of input vectors on a good board and reads the output patterns. Then, a person repeatedly inserts failures, allowing the tester to execute the vectors each time and read the outputs again. The person then reports the name of the failure, and the machine learns that information. Test engineers can generate input vectors manually or obtain them from a fault simulator or other CAE equipment. For very complex boards, engineers may prefer entire CAE-created test programs.

Teaching instead of programming cuts down on programming time. Users of this technique have reported program-development times measured in days or weeks, a vast improvement over the months that developing some full functional programs requires. In addition, because the program learns from experience, delaying a product's introduction because its test program is incomplete is unnecessary.

Another advantage to teaching the tester about the board is that it permits margin testing and some analog testing. The engineer can submit a number of good boards for testing, declaring that waveforms and other parametric variations are within the normal range. This permits the test to detect timing problems and other faults that do not fit into the conventional "stuck-at" category. Even with only one good board, a test engineer can exercise that board at voltages slightly above and below nominal levels, telling the tester that the responses are also acceptable. In this way, normal board-to-board variations are less likely to fail during production.

Anytime a good board does fail, the test engineer so informs the tester. It then incorporates the information into the database, reducing the likelihood that a similarly performing board will fail in the future.

Software to make an expert system work is extremely sophisticated. The chief proponent of the technique packed it in a few years ago, and as yet no other company has taken up the challenge. I continue to talk about it in the hope that some enterprising test company will resurrect it, like a phoenix from the ashes. Thus far, this personal campaign has not succeeded. Even in its (relative) heyday, most test engineers remained unaware of this option, and therefore could not evaluate its appropriateness for their applications. The learning curve both before and after purchase was not insignificant, but users who tried the approach reported great success.

2.3.12 Two Techniques, One Box

Combinational testers provide in-circuit and functional test capability in a single system. This solution offers many of the advantages of each approach and some of the drawbacks of each.

Board access requires a special bed-of-nails fixture containing two sets of nails. A few long probes permit functional testing at a limited number of critical nodes. All remaining fixture nodes contain shorter pins. Vacuum or mechanical fixture actuation occurs in two stages. The first stage pulls the board down only far enough for contact with the longer pins. Second-stage actuation pulls the board into contact with the shorter pins as well for an in-circuit test. Manufacturers can conduct either test first, depending on the merits of each strategy.

A combinational tester can alleviate headaches that many very densely pop-ulated boards cause. Test programmers subdivide a board's circuitry and create the most effective test on each section. For example, where bed-of-nails access is pos-sible at every node or where there is a lot of analog circuitry, an in-circuit test is generally the best choice. Functional test works better in time-critical areas or where surface mounting or other mechanical board features prevent convenient probing.

Combinational testers can find in-circuit and functional failures in the same process step (although not necessarily in the same pass). Use of one machine instead of two minimizes factory floor space devoted to test operations and may reduce the number of people required. Eliminating one complete board-handling operation (between in-circuit and functional testers) reduces handling-induced fail-ures, such as those from electrostatic discharge (ESD).

On the other hand, taking advantage of the ability to subdivide a board for testing necessitates performing the subdivision during test-program development. This extra analysis step often lengthens programming schedules and increases costs. Combinational testers can also represent the most expensive test alternative.

As with functional testers, some lower-cost solutions are emerging. Smaller than their more expensive siblings, the size of these models limits the amount of test capability that fits inside the box. Speed, accuracy, fault coverage, and throughput capacity are generally lower than with high-end machines. Also, low-end systems do not offer multiplexing of test pins. Therefore, each pin driver and receiver is independent of the others, but the total number of available pins is limited.

2.3.13 Hot-Mockup

The expense and other drawbacks of conventional functional and emulation testing often prohibit their use. Many manufacturers follow in-circuit test with a *hot-mockup*. This approach plugs the board under test into a real system that is complete except for that board, then runs self-tests or other tests specifically designed for this situation.

Disk drives, for example, are electromechanical systems with considerable analog circuitry. Manufacturing occurs in very high volumes, with fast changeover and short product life. Figure 2-30 shows an appropriate disk-drive test strategy.

During hot-mockup test, an operator attaches the board under test to a PC-driven hard-disk assembly using clamps and pogo pins. He or she then executes intense read/modify/write cycles for 5 minutes or more. If the drive fails, the board is bad. One prominent disk-drive manufacturer employs more than 400 such hot-mockups in one Singapore factory, with four per operator on a 5-foot workbench. To change quickly from one board to another, the operator simply pulls two cables, four thumb screws, and four Allen screws.

One fault that only hot-mockup testing can find relates to the way in which disk drives store files wherever there is empty space, often scattering many file pieces across the disk surface. Conventional test scans the disk from the outside in or the

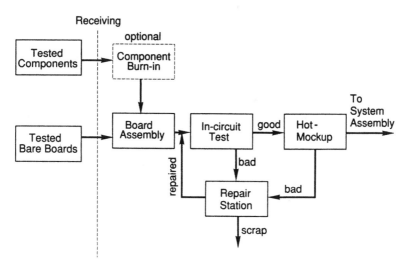

Figure 2-30 A typical disk-drive test strategy.

inside out. A hot-mockup test executes a read cycle on a fragmented file, stops in the middle, moves the heads, and reads the next segment. A data discontinuity between file segments indicates faulty board logic.

Hot-mockup is attractive because, in most cases, the self-tests that it runs already exist. Engineering teams develop self-tests with new products. Therefore, this approach minimizes test's impact on frantic development schedules. The engine, in this case a conventional PC, already exists, as does the target system, so hardware and software development costs are very low.

On the minus side, hot-mockup is very labor-intensive. Results are qualitative, not quantitative, and fault diagnosis depends more on operator experience than on the test itself. Little information is available on fault coverage. Perhaps most inconvenient is the problem of mockup-system wearout. Again referring to the disk-drive manufacturer, consider the logistics of managing 400 sets of test apparatus. Replacing them only when they die is probably the least expensive option, but possibly the most disruptive to the manufacturing cycle. Replacing them all on a schedule, either a few at a time or all at once, minimizes unanticipated failures but increases hardware costs.

Because commercially available functional testers cope better with the challenges of certain products, some manufacturers are again selecting to abandon or considering abandoning hot-mockup in favor of that option.

2.3.14 Architectural Models

Dividing testers into MDAs, in-circuit testers, and so on categorizes them by test method. Within each group, it is possible to separate members by tester architecture into monolithic, rack-and-stack, and hybrid systems.

Monolithic testers are the machines traditionally associated with automatic test equipment (ATE). These are single—often large—boxes from one vendor, containing a computer engine and a collection of stimulus and measurement electronics. Measurement architecture is generally unique to that vendor, although most vendors today use a standard computer engine, such as a PC-type or UNIX workstation, to avoid developing and maintaining computer-bound system software. The vendor defines the machine's overall capability. Customers must meet additional requirements through individual instruments over a standard bus, such as IEEE-488.

One advantage of the monolithic approach is that individual measurement capabilities lack unnecessary features, such as front panels, embedded computer functions, displays, and redundant software, that can make instrument-bound solutions more awkward and more expensive. The vendor generally provides a well-integrated software package and supports it directly, an advantage of "one-stop shopping" for test-program development. The vendor is familiar with every part of the system and how the parts interact, permitting the most effective service and support.

On the other hand, these systems are quite inflexible. If a customer wants a capability that the tester does not already have, one option is IEEE-488 instruments. Expandability is limited to instruments and vendor offerings. A functional tester may contain 256 pins, and the architecture may permit up to 512. The customer can expand to that point, but not beyond. The vendor may offer other features as field upgrades, but only those features are available. Add-on instrument choices are somewhat limited, and they often do not integrate well into system software. Programming the instruments requires one of the "point-and-click" programming tools or—as a last resort—that marvel of cryptic horrors, IEEE-488 language.

Rack-and-stack systems consist of a computer engine, (usually a PC-type), a switching matrix, and an array of instruments, communicating over a common I/O bus such as IEEE-488. Rack-and-stack solutions permit purchasing appropriate instruments from any vendor, as long as they support the communication bus. Therefore, this solution provides the most flexible hardware choices. It may also provide the best-quality hardware. The best switching matrix may come from vendor A, the function generator and spectrum analyzer from vendor B, the waveform analyzer from vendor C, and the logic analyzer from vendor D. Because it permits foregoing capability that the customer does not need, rack-and-stack systems may be less expensive than monolithic or hybrid solutions.

Some vendors will act as consultants, helping customers to assemble rack-and-stack systems from components, regardless of individual-instrument manufacturer. This is a handy service, as long as the vendor is honest about recommending, purchasing, and pricing competitors' products.

For analog and high-frequency applications, rack-and-stack solutions are often more accurate, less expensive, and easier to use than an array of instruments attached to a monolithic tester. In addition, some capabilities are available only as individual instruments. For an application that requires them, there is no other choice.

Disadvantages to this approach include test-program generation, which remains primarily a manual process. A user teaches the "tester" about instruments in the system and loads vendor-supplied or in-house-developed instrument drivers. Automatic software tools then guide the programmer through code creation. Although these tools can analyze information from computer-aided design and other CAE equipment, people must still create and select the final tests.

Bus-based test systems tend to be noisy, limiting measurement precision and possibly compromising functional board performance. Users must generally design their own fixtures or other interfaces between tester and board under test.

Hybrid systems offer some features of monolithic and rack-and-stack alternatives. They consist of an embedded or stand-alone computer engine, again usually a PC-type, and a collection of printed-circuit-board-based instrument modules connected through a standard I/O bus designed specifically for this purpose. The current frontrunner for this arrangement is the VME eXtension for Instrumentation—VXI—along with its more recent siblings, such as MXI and PXI.

VXI provides what the IEEE-488 bus would call the "switching matrix" as a special card cage. The standard specifies how to connect the modules, and the signals that allow them to communicate. This architecture permits I/O speeds of about 10 MHz and as many as 255 individual instruments, all of which may be talkers. Board and system manufacturers are beginning to create in-house-built testers around this design. In addition, traditional "monolithic" tester vendors are adopting VXI to create modular systems that increase the flexibility and expandability of the monolithic option.

The hybrid-system approach incorporates the best compromise between monolithic and rack-and-stack alternatives. Ideally, hardware and software integration resembles the monoliths, as do programming and data-analysis features. Because instrument modules are available from numerous manufacturers, test engineers can select capabilities that best match their needs. As with rack-and-stack choices, users can adopt new instrument products that improve system performance with a minimum of effort.

Disadvantages include the technique's relative immaturity compared to IEEE-488. Instrument choices are, at present, still more limited. Also, this option is still slower than many monolithic products, and users must generally create system-level and other high-level software.

The VXI standard represents a *compromise* among costs, features, and ease of implementation. The same can be said of *de facto* standards for MXI and PXI. They cannot accommodate absolutely every function that every instrument vendor can conceive of. Lead lengths in the cage and other architectural limitations mean that noise and timing can be a concern in critical situations. Board manufacturers requiring low noise and very precise timing may not be able to adopt this solution. In addition, system developers still have to design test fixtures, and users must construct them. Chapter 6 will explore VXI in more detail.

2.3.15 *Other Options*

Manual testing consists of training technicians to analyze board performance using an array of individual instruments and some kind of written procedure. The approach is most appropriate in small companies and other low-volume applications. It can also provide a good way to analyze very early production runs for a new product, where information from manual tests helps programmers develop automatic tests. Startup costs are relatively low—the cost of people, training, and instruments. The method is quite flexible if technicians move easily from one board type to another. Of course, the technique is too slow for many applications. It requires highly skilled and well-trained people, and results may be inconsistent from technician to technician or from day to day.

Inspection techniques are also gaining popularity to complement or supplement traditional test. Chapter 3 will examine this option.

The most interesting alternative board-test strategy is not to test at all, as with the Japanese television manufacturer referred to in Chapter 1. Test engineering is perhaps the only occupation whose ultimate goal is its own demise. Ensuring vendor material quality and monitoring the production process at every step will, at some point, produce 100 percent good boards and systems. Test professionals can rest assured, however, that we will not likely achieve that goal anytime soon.

2.4 Summary

Finding a fault at any production stage can cost 10 times what it costs to find that same fault at the preceding stage. Monitoring processes to prevent problems from occurring at all represents the least expensive option. Failing that approach, manufacturers generally try to remove as many failures as possible at board level, before assembling systems.

Many test techniques are available to accomplish this task. Bed-of-nails methods, such as shorts-and-opens, manufacturing-defects analysis (MDA), and in-circuit testing, operate on individual components, board traces, or functional clusters to ensure correct assembly. Most test failures have one and only one cause, minimizing diagnostic efforts.

Functional-test techniques, which control boards through a few nodes at the edge connector or a test connector, examine overall board behavior and verify it against designer intentions. Because of the logic depth between a failing node and the access point, fault diagnosis can be lengthy and may involve guided probing or other additional steps.

Emulation testers perform a functional-type test on microprocessor-based logic. Hardware pods instead of software models mimic complex circuitry, allowing testers to examine logic states and assess board performance.

Ultimately, the goal is to have processes that are sufficiently in control to eliminate test completely. Fortunately, it will be some time before that goal comes within reach.

CHAPTER 3

Inspection as Test

Living with ever-increasing component complexity and board density, along with decreased nodal access, test engineers must face the reduced efficiency of traditional test strategies. Yet the need to ship good products has become more critical than ever. Customers expect that their electronic products will work the first time without difficulty, and that they will continue to work with a minimum of fuss.

To cope, manufacturers are turning to inspection as complement or supplement to traditional test. Inspection, when it works, offers numerous advantages over test. It requires neither bed-of-nails nor edge-connector fixtures. Good-board criteria against which you compare the board under test may come from a known-good board (or a number of them to establish appropriate tolerances) or from a simulation. Also, most inspection is noninvasive. That is, it does not exercise or otherwise disturb the circuit.

On the other hand, inspection can determine that the board *looks* correct, but it cannot verify that the board *works*. Only a true test can establish functionality. Test and inspection represent a tradeoff, as Figure 3-1 illustrates. Taking maximum advantage of inspection can simplify requirements for subsequent test, which leads to a recommendation that I hope will become an industry *mantra*: "Inspect everything you can, test only what you must."

The inspection equivalent of a test program consists primarily of a representation of a good board's physical-layout specifications and a collection of rules and heuristics to decide whether the board under scrutiny conforms sufficiently. Perhaps the greatest *caveat* that accompanies most inspection techniques is that unless the heuristics allow for sufficient variation in judging what constitutes a good board, the step will produce excessive numbers of false failures.

Test, on the other hand, relies on input signals and output measurements, which, at least to some degree, require exercising the circuit to determine its quality. An improperly designed test on a board containing a catastrophic fault can aggravate existing problems. Powering up a board containing a short, for example, accomplishes little beyond frying working circuitry and verifying the performance of the facility's smoke detectors.

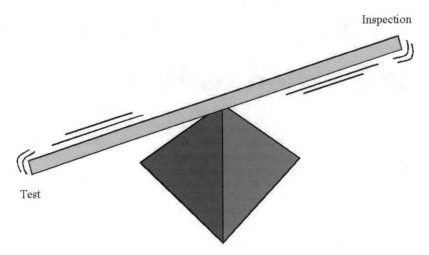

Figure 3-1 Test and inspection represent a tradeoff. Extra effort expended at the inspection stage should reduce the test burden downstream.

Creating inspection algorithms generally takes much less time than developing comparable test programs does. Also, most manufacturers place inspection before test in the process. As a result, finding and repairing faults during inspection generally costs considerably less than uncovering and correcting those same problems during test. In addition, inspection can find faults that defy electrical test. Insufficient solder, mousebites on traces, and other problems do not show up during test because they do not directly affect board performance—yet. Sometimes a solder joint may appear to make adequate contact even if it contains no solder at all. These problems increase the likelihood that a previously functioning board will fail in the field, after it has been shaken during shipping, for example. Repairing boards before they leave the factory improves product reliability. Therefore, if you can economically justify both inspection and test steps, shift as much as possible of the fault-finding burden to inspection.

At the same time, inspection cannot uncover many problems that show up routinely during test. Inspection's ability to detect an incorrect part is extremely limited. Test, however, can find exactly those problems. In-circuit and other bed-of-nails techniques measure single components or clusters to determine if they are correct and function properly. Functional, emulation, or system test examines the behavior of the board (or system) as a whole.

3.1 Striking a Balance

Neither test nor inspection can find all faults for all manufacturing lines. Certainly coverage overlaps, which can lead to the mistaken impression that one technique or the other will suffice. But the areas of overlap are not sufficient. Each approach has strengths and weaknesses that engineers must consider when deciding on the *best* strategy.

Unfortunately for test engineers, inspection often falls into the "not invented here" category, because in many organizations inspection comes under the purview of the manufacturing rather than the test department. Any attempt to replace portions of the test strategy with inspection steps is often perceived as a threat to job security. This attitude emphasizes the necessity of encouraging the "we're all in this together" philosophy. Buckminster Fuller, engineer and Renaissance thinker, contended that the solution to the problem of trade imbalances around the world was to simply draw a line around the planet and call it one market. This eliminates the problem of "them vs. us" because "them *is* us," so there is no imbalance. The same philosophy applies to manufacturing and test responsibilities. All quality steps, including process monitoring and feedback, inspection, test, and repair belong to the same basic task—getting only good products out the door. If "them is us," then including inspection in what would otherwise be "merely" a test strategy creates no imbalance in responsibility—only a shift in timing. And because inspection usually precedes test, finding and dealing with problems during inspection costs less than doing so later.

Lack of access to board nodes often drastically reduces the effectiveness of bed-of-nails test techniques. Shifting the burden to functional and system test usually means increasing the time required for test-program development, debugging, and implementation. Once again, pressures to shorten time-to-market strain the ability to adopt this approach.

Ironically, many common factors both encourage and discourage including inspection as part of an overall "test" strategy, as Figure 3-2 shows. Smaller parts and denser board placement make bed-of-nails access more difficult, but they also complicate the lives of human visual inspectors and require higher-resolution inspection equipment. The need for higher resolution, in turn, increases camera-positioning time and image-processing time, which slows the inspection step and increases the likelihood that lighting or other conditions during inspection will falsely flag good boards as bad. More complex components mean analyzing more solder joints placed closer together. Increasing use of BGAs, flip-chips, and other hidden-node device designs precludes human and automated optical inspection, because those techniques cannot see nodes out of the line of sight. The increased use of complex, low-volume boards in today's products favors inspection because of its shorter program-development time as compared with conventional test. These same boards require fast ramp-up to full production, making it unlikely that engineers will have time to fine-tune the inspection equipment to more easily differentiate between marginally good and marginally bad boards.

An effective inspection step can verify both a product's quality and reliability. As mentioned earlier, *quality* denotes that a product performs to specification. *Reliability* indicates the degree to which the product will continue to perform to specification—and without failure—during actual use. Test, in contrast to inspection, concentrates primarily on quality issues.

Smaller parts
More complex parts
Denser boards
Ever-shrinking device-pin spacings
Increasing use of BGAs, flip-chips, and other technologies with hidden nodes
Increased use of expensive, low-volume boards for many products

Figure 3-2 These factors both encourage and discourage inspection as part of a "test" strategy.

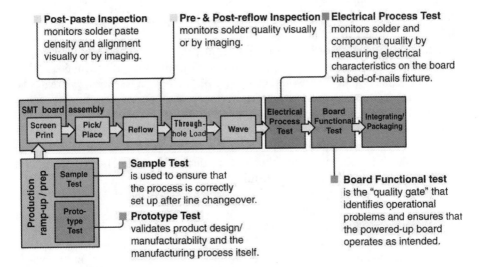

Figure 3-3 The various stages of the manufacturing process, and the test or inspection requirements at each stage. (Courtesy Teradyne.)

Historically, "inspection" evoked images of armies of human beings examining tiny board features looking for problems. In fact, human visual inspection (HVI) remains a huge part of the industry's inspection arsenal. A few years ago, Stan Runyon at *Electronic Engineering Times* speculated that 40,000 human beings still performed this function, despite its declining effectiveness in the face of advances in board technology.

In fact, inspection covers much more ground than the human beings armed with magnifying glasses or microscopes looking for anomalies on the board surface. Figure 3-3 shows the various stages of the manufacturing process, and the test or inspection requirements at each stage. Figure 3-4 shows the types of test and inspection that can serve those needs.

As you can see, inspection can take place at any or all of three process locations. Post-paste inspection examines the board after the paste-printer has deposited solder, before assembly operations have added components. Post-placement inspection occurs after pick-and-place machines, chip shooters, and

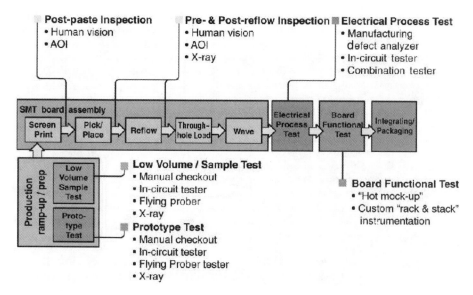

Figure 3-4 The types of test and inspection that can serve the needs depicted in Figure 3-3. (Courtesy Teradyne.)

human assemblers have finished their tasks. Post-reflow (it could also be called "post-wave-solder," but most soldering these days avoids that error-prone step) takes place after boards have emerged from reflow ovens, when the final solder joints are available for qualitative and quantitative analysis.

3.2 Post-Paste Inspection

Examining the board immediately after paste deposition enjoys numerous advantages. Inspection at this stage looks for a number of fault types. A clogged aperture in the printer stencil, for example, can prevent solder from reaching some pads, or the solder deposited may be insufficient to create an acceptable joint later on. This step can also find residual solder where it doesn't belong on the board. Measuring the area of solder on the pad permits a reasonable estimate of the quantity of solder deposited, a reasonable predictor of final solder-joint integrity. Off-pad solder, like that on the board in Figure 3-5, can also cause device-connection problems later on.

One of the primary benefits of post-paste inspection is the ease and low cost of board repair. Fixing the board at this stage requires merely washing the solder from the board and returning it to the "back of the line" in the manufacturing process, ready for another go. This step eliminates many problems before they escape to downstream test processes. Some manufacturers estimate that up to 80 percent of manufacturing faults on surface-mount boards result from problems at the solder-paste step.

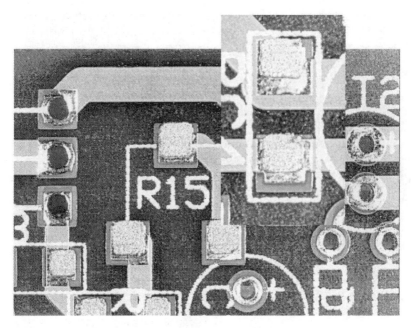

Figure 3-5 Inspecting immediately after solder-paste printing will find problems such as off-pad solder. (Photo courtesy Teradyne.)

Post-paste inspection can take several forms. Some companies still use human inspectors, but the number and size of solder pads and the difficulty of conclusively determining solder-pad registration limits their effectiveness.

The least expensive equipment-supported technique takes advantage of the capabilities of the paste printer itself. Paste printers include their own cameras to detect fiducials on the boards, as well as to facilitate process setup and calibration. These cameras can double as inspectors. They can look down at the board to detect solder anomalies and look up to find clogged apertures and other irregularities in the stencils. This technique is inexpensive and effective within limits (and subject to the ever-present tradeoff between cycle time and inspection resolution). However, it provides only two-dimensional images, limiting the accuracy of solder-volume calculations and, therefore, joint-quality predictions. On the other hand, some manufacturers contend that the area of solder alone correlates well with volume, even without height measurements.

The printer camera cannot look for fiducials and examine solder pads at the same time. The printer deposits the solder, then the camera inspects the results. In high-volume operations, the time required for this two-cycle process generally proves prohibitive. On the other hand, manufacturing engineers in low-volume environments need not concern themselves with the time constraint.

To cope with throughput limitations, some companies use the camera to examine only certain board areas that they consider either typical or critical. Also, some engineers use the camera during paste-print setup to ensure correct stencil loading and positioning, paste-pressure adjustment, and proper paste dispensing. They inspect the first few boards and make any necessary process modifications, then turn off the camera to avoid slowing the process during routine production.

In addition to solder-paste area and position, three-dimensional inline techniques profile solder height to precisely calculate its volume. Proponents contend that irregularities in the shape of the solder on the board make solder-volume calculations from only pad-area coverage inaccurate. Some 3-D systems inspect the entire surface of every board through the process. Slower and less-expensive systems examine critical board areas or production samples.

The more common approach to three-dimensional post-paste inspection scans a laser across each solder pad, starting with a line across the pad edge and moving perpendicularly until the laser has covered the entire pad. The equipment then measures solder height by a technique called *reflective interference*, from which it calculates the volume. Some such systems create a complete volume map of solder paste. To attain adequate speeds for full production, other systems may calculate volume by measuring height from line-scans at one or a few locations.

Another approach projects a diffraction fringe pattern on the board, then calculates height and volume by examining the pattern with a stroboscopic white light. This variation avoids the need to scan the board, thereby increasing inspection speeds. Supporters of this variation contend that it provides as much and as accurate information about the solder joint as the laser scan in less time.

Programming for three-dimensional inspection inline begins with sites that will accommodate fine-pitch components, BGAs, flip-chips, and other chip-scale packages. If time permits, the step can look for clogged stencil apertures and similar problems.

Manufacturers can use this type of inspection as a "real-time alarm"—stopping the line if measurements exceed established tolerances—or merely record the information and move on.

The primary limitation in all three-dimensional inspection techniques is that obtaining a useful reflection for analysis depends on the angle of incidence between the light source and the board under test. Irregularities or reflection variations in the board surface, as well as the reflectivity of the solder itself may affect results. The inspection step must also locate the board surface to determine how much of the measured height is solder. Examining both sides of a double-sided board requires two separate inspections.

3.3 Post-Placement/Post-Reflow

Once components have been assembled onto the board, inspection looks for a number of additional manufacturing conditions: the presence or absence of components, component height, some incorrect components (if the incorrect part is significantly different in size or appearance), as well as the accuracy of pick-and-place machines and chip shooters.

After the reflow oven, inspection performs a final check of solder-joint geometry, solder-joint integrity, and any component movement while the board is in the oven. Realizing that few manufacturers inspect both pre- and post-reflow, this step also looks for component existence and position, component identification (subject to the same *caveat* as above), trace existence, and trace defects such as "mousebites." Figures 3-6 through 3-11 show some common failure types.

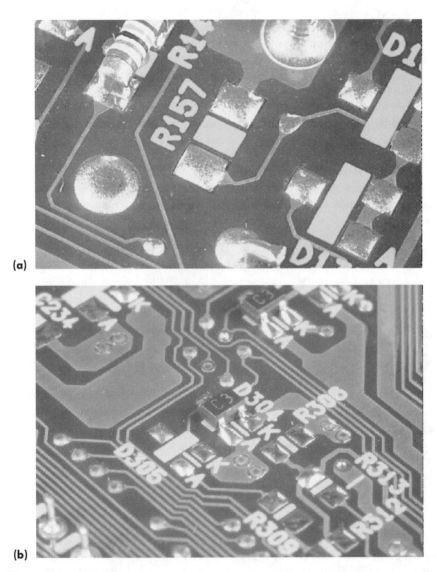

(a)

(b)

Figure 3-6 These two photographs show boards containing missing-component faults. (Photos courtesy Teradyne.)

Figure 3-7 A solder short. (Photo courtesy Teradyne.)

Figure 3-8 Tombstoning. (Photo courtesy Teradyne.)

Figure 3-9 A lifted component leg. (Photo courtesy Teradyne.)

Figure 3-10 An off-registration or off-pad component. (Photo courtesy Teradyne.)

Figure 3-11 Untrimmed leads of a through-hole component. (Photo courtesy Teradyne.)

Some manufacturers use laser or white-light techniques to examine loaded boards. This approach constructs a three-dimensional profile of the board, detecting missing and some off-pad components. The technique is generally too slow for this process stage, however, and one of the others will likely produce more accurate results. Nevertheless, for a cursory check of certain critical board areas, it may be sufficient.

3.3.1 Manual Inspection

As mentioned earlier, despite the preponderance of incredible shrinking electronics, the most common inspection technique remains *manual inspection*, also known as *human visual inspection* (*HVI*) and *manual visual inspection* (*MVI*). Its popularity persists despite a consistent body of evidence that it is less effective than it used to be, always assuming that it was ever effective at all.

Manual inspection consists of a cadre of people examining boards either with the naked eye or aided by magnifying glasses or microscopes. Manufacturers like the technique because it is relatively simple and inexpensive to deploy. Microscopes and magnifying glasses require little up-front investment, and people costs are easy to manage and adjust as situations change. Its flexibility stems from the fact that human beings adapt much more easily to new situations than machines do. Also, manual inspectors are much less bothered by changes in lighting or other environmental conditions. There is no need for programming, and, within the limits of the inspectors' capability, it can be quite accurate.

On the other hand, operating costs are quite high. Labor costs represent a considerable expense, and adjusting the workforce as manufacturing throughput changes is awkward at best. Success of HVI depends a great deal on the experience and diligence of the inspectors. Difficult boards or subtle problems can slow the process and reduce the technique's accuracy and effectiveness. Also, whereas machines make a "yes/no" decision based on established specifications and heuristics, people's judgments are often more subjective. As a result, the consistency of manual inspection leaves much to be desired. According to one study from AT&T, now more than a decade old, two inspectors examining the same boards under the same conditions agreed only 28 percent of the time. With three inspectors, agreement dropped to 12 percent, and with four inspectors, to 6 percent. Even the *same* inspector examining a board twice came up with an identical diagnosis only 44 percent of the time. With today's smaller feature sizes, the situation would likely be worse. In many (if not most) cases, one of the automated techniques would work considerably better.

Human inspection also suffers from inconsistency based on the time of day or the day of the week—the previously mentioned "Monday/Friday syndrome," named after an admonition by Ralph Nader in the 1970s never to buy a car manufactured on Monday morning or Friday afternoon. Manual inspectors often miss failures, while flagging and unnecessarily touching up good joints.

The last drawback to this technique applies to all the visual methods, as well as to laser and white-light approaches when manufacturers use them on loaded boards. They require line-of-sight access to the features they are inspecting. Since one of the reasons for turning to inspection instead of conventional bed-of-nails test is lack of access, the implications of this limitation are significant.

3.3.2 *Automated Optical Inspection (AOI)*

Automated optical inspection consists of a camera or other image input and analysis software to make the pass/fail decision. Implementations include the following range of applications:

- A spot check of critical board areas
- A cursory check for component existence and orientation
- Comprehensive analysis of the entire board surface

AOI systems use several techniques to identify failures. *Template matching* compares the image obtained from a theoretical "golden" image (assuming one is available either from a good board or a CAD simulation). Template matching is somewhat unforgiving of deviations from perceived good-board specifications and of ECOs and other board modifications. The latter remain very common during the early stages of production ramp-up. *Pattern matching* stores examples of both good and bad boards, comparing the board under test to these standards. *Statistical pattern matching* works similarly, except that the pattern represents a compendium of a number of boards, so minor deviations will less likely cause false failures. In fact, its proponents contend that statistical pattern matching can produce orders-of-magnitude fewer false calls than its simpler siblings do.

Humans perform better than machines on recognizing image patterns. Nevertheless, even when features are large enough to be detected by human inspectors, machines succeed better on the monotonous task of inspecting identical products. Machines are faster, and inspection times do not vary from one board to the next. It takes fewer workers to run automated equipment than to inspect boards manually, and automation requires less-skilled (and therefore less-expensive) workers. Therefore, adopting AOI generally lowers labor costs. Also, automated systems can resolve finer features than human beings can, although manufacturers have to trade off resolution against throughput. The finer the required resolution, the longer it takes to inspect a board.

AOI enjoys other advantages over manual inspection. By reducing the number of false failures, it reduces costs and improves the productivity of rework operations. Consistency of results allows feeding information back into the process to prevent future failures, improving quality out of manufacturing and consequently lowering test and repair burdens as the product matures.

When compared to conventional test, AOI can more easily reside in an automated production line with a minimum of human intervention. Because it can detect many faults that otherwise rely on electrical tests, manufacturers can sometimes eliminate the process-test step altogether, reducing capital-investment costs by avoiding a test-equipment purchase or freeing an existing tester for other product lines. AOI provides data that test cannot on parts-placement accuracy, which can affect future product quality. Also, positioning AOI after parts placement and before reflow (and therefore before the possibility of electrical test) can avoid extra reflow steps and thereby lower repair and rework costs.

Of course, as a wise soul once said, "No good deed ever goes unpunished." AOI obviously requires a significantly larger capital investment than does manual inspection. Equipment often costs hundreds of thousands of dollars, including conveyors and other related infrastructure enhancements. Therefore, overall startup costs are higher. The equipment requires "programming" in some form, whereas human inspectors can generally work from existing documentation. For the same reason, implementing engineering changes and product enhancements takes longer and incurs more costs in the automated case.

Humans can make pass/fail decisions even when devices are partially hidden. (A characteristic of humans that machines have thus far failed to match is our ability to draw conclusions from incomplete information. The binary nature of machine logic makes such "fuzzy" decision-making complicated at best.) Human inspectors can also more easily allow for color, size, and other cosmetic variations in the board's parts, as well as lighting variations and other less-than-optimal conditions.

AOI looks for very specific features—part placement, part size, perhaps board fiducials of a certain size and position, and patterns of light and dark, such as bar codes. It can also look at label type fonts and sizes, although this level of resolution slows the inspection step. Unfortunately, many boards include components that exhibit large variations in package sizes and styles. Figure 3-12 shows several examples of electrically identical components that appear different to an AOI system. Automated inspection must also deal with changes in background color and reflectivity and differences in silk-screen typefaces and sizes on allegedly identical boards.

Figure 3-12 Many electrically identical components appear different to an AOI system. (Courtesy GenRad.)

AOI suffers disadvantages when compared to conventional test as well. Inspection times are generally longer because of the time required for x-y positioning and image evaluation. AOI cannot find numerous fault types that test can find easily, and many users report a significant increase in false failures. Unlike conventional test, an AOI system requires that the areas under scrutiny be visible by line of sight, and a board containing components on both sides requires two inspection steps. (Instead of inspecting the same board twice, some manufacturers pass the board between two AOI systems, thereby examining both board sides at the same time.) Constructing a reasonably accurate inspection "program" requires a good board or good-board simulation, not always easy during the often fast-changing period of final preparation for production.

Some companies are turning to AOI systems to examine BGAs before placing them on boards. This inspection step confirms the existence and position of the solder balls, as well as their diameter. Insufficient solder in the balls will likely pass an electrical test after reflow, yet may pose a reliability problem for customers. Even tested BGAs can lose solder balls or experience other problems during handling. Other analysis techniques used after board assembly (such as x-ray) can be expensive or impractical for detecting this situation. Adopting this step can reduce the number of scrapped boards and devices, thereby lowering manufacturing costs.

BGA inspection can also look for excess solder-ball oxidation by gray-level analysis. This condition usually manifests on whole lots of BGAs, rather than on individual components, so identifying it permits repair or returning parts to vendors. Although excess oxidation has never proved to represent a real defect, most manufacturers will reject such BGAs, anticipating possible poor bump bonds and possible poor wetting during underfill. Identifying this situation on BGAs before assembly allows correction at much lower cost than scrapping BGAs or whole boards later.

3.3.3 Design for Inspection

The success of an AOI step depends on the ease with which the camera can distinguish the features that it inspects and the degree to which the board under test conforms to the good-board standard. Several simple steps can make inspection more efficient and more successful.

Maintain consistent size specifications for particular components. For example, 603 discrete component specifications list a size of 60 × 30 mils. Actual devices, however, can vary from 50 × 22 mils to 65 × 34 mils, depending on vendor. If you permit such a wide variation, an AOI system cannot determine whether a device on the board is correct. Therefore, select a single component vendor or a group of vendors who provide visually similar products.

Place components on the board with consistent orientation. This step will make inspection-system programming easier, and will facilitate repair and rework operations later. For the same reasons, it is preferable to select parts with the same pin-1 designator (cut corner, colored dot, stripe, dimple).

Some manufacturers advocate specific techniques to facilitate determining the component's exact position on the board. *Placing fiducials on the component*, as in Figure 3-13, or *painting the component site with a contrasting color*, as in Figure 3-14, makes position measurements more precise and reduces the number of false failures. *Fiducials on the component site*, as in Figure 3-15, help the AOI system make a presence/absence decision.

AOI is becoming increasingly common, experiencing a year-on-year growth rate exceeding 20 percent. Current trends in board technology leave manufacturers with few viable alternatives. Higher speeds, better spatial resolution, and more accurate fault detection also combine to increase its effectiveness, and therefore its popularity.

3.3.4 Infrared Inspection—A New Look at an Old Alternative

Many kinds of board faults exhibit a higher inherent resistance than their faultless counterparts do. When powered, however briefly, these areas heat up, becoming detectable by *infrared* inspection. In the same way, infrared techniques can also reveal marginal components, traces, and solder joints that often surface as early field failures.

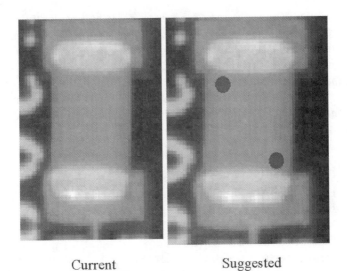

Current   Suggested

Figure 3-13 Fiducials on the component permit more precise determination of its position. (Courtesy GenRad.)

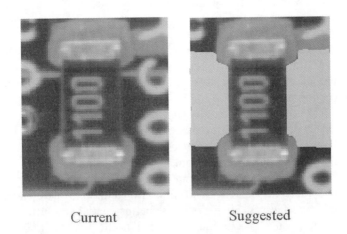

Current   Suggested

Figure 3-14 Painting the component site in a contrasting color allows the AOI system to more easily detect the component edge. (Courtesy GenRad.)

Infrared inspection is not new. For years, manufacturers have used the technique to examine bare boards for hairline shorts, inner-layer shorts, and similar defects by applying power to the board and looking for "hot spots." Spatial resolution hovered in the range of 50μ and reliable defect detection required temperature changes greater than about 1.5°C. In addition, since testing generally took place in the open air, the method had to endure an inconsistent and unpredictable

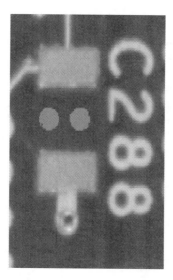

Figure 3-15 The component will cover fiducials on the site, enhancing presence/absence detection capability. (Courtesy GenRad.)

thermal environment combined with the indigenous random noise inherent in infrared camera images. However attractive the theory, these limitations precluded attempts to apply the approach to loaded boards.

3.3.4.1 A New Solution

At least one manufacturer is addressing these concerns, introducing a controlled-environment infrared-inspection station that detects temperature differences as low as 0.025°C on features as small as $5\mu \times 5\mu$. Its proponents claim that in benchmarks and field trials, the system has detected most of the fault types of x-ray techniques, as well as cracked solder joints and other potential reliability problems that no other approach can find directly, at much lower cost than x-ray equipment.

The new solution tests the board in a controlled isothermal chamber, applying power and input patterns and comparing the infrared signature with a standard assembled from a set of known-good boards. Deviations from the standard generally show up as hot or cold spots whose temperature lies outside the three-sigma limits of the normal statistical error function.

One immediate advantage to this technique is that node visibility—either by line of sight for visual inspection or access through a bed-of-nails—becomes irrelevant. The tester hooks up to the board via the edge connector, just as in a traditional functional test. Input patterns can be adopted directly from design simulations to exercise the board and ensure that it works properly. Most companies create such patterns during product development. The manufacturer need not generate *fault* simulations or conventional test programs, saving a significant amount of work and time.

Many fault classes lend themselves to this type of detection. It can find solder voids, misaligned or missing components, insufficient solder, shorts, and broken connections with few false failures. Infrared inspection cannot identify joints with *excess* solder, which most manufacturers regard as process faults and which x-ray techniques will find, because the circuit appears to function normally.

Infrared also detects failures that x-ray misses. A cold solder joint, a faulty ASIC, or an incorrect resistor, for example, could significantly change the board's thermal signature, but would look no different in an x-ray image. In that respect, the infrared technique resembles a functional test more than it does other forms of inspection. The technique's supporters suggest placing it in the process after in-circuit or other bed-of-nails test, possibly *in place of* functional test. This configuration avoids adding a step that would lengthen the manufacturing process, introduce another cycle of board handling, and possibly increase solder-joint breakage and other handling-related failures.

3.3.4.2 Predicting Future Failures

Perhaps the most interesting aspect of the infrared approach is its ability to detect latent defects—defects that do not affect the board's current performance, but which may represent reliability problems after the product reaches customers. Since the components still function and the board as a whole still works, these faults generally defy conventional detection. Cracked solder joints, for example, force the board current through a smaller-than-normal connection, creating a hot spot that this tester will see. Some marginal components also fall into this fault category.

To find such faults, manufacturers traditionally subject their boards to some form of environmental stress screening (ESS), including burn-in, temperature cycling, and vibration, before performing an in-circuit or functional test. The idea is to aggravate the latent faults until they become real faults and therefore visible to subsequent test. Aside from requiring higher costs for equipment, factory floor space, extra people, longer production times, and larger inventories, ESS stresses good and bad components alike. Some authorities suspect that such screening reduces overall board reliability and shortens board life. In addition, vibration—the second most effective screen after temperature cycling—is difficult to control and cannot apply stresses evenly across the board, so results can be inconsistent. (ESS will be discussed in more detail in Chapter 7.)

In contrast, stimulation during an infrared test subjects boards to no more stress than they would experience in normal use. In one field test conducted by an automotive manufacturer, the infrared test found all known failures from a sample group of defective boards, whereas various versions of ESS revealed no more than 42 percent. In addition, the infrared system discovered that 2 percent of the boards contained failures of which the manufacturer was unaware.

3.3.4.3 The Infrared Test Process

To build the good-board model, a test engineer views a computer image of the board, highlights each component and area of interest with the computer's mouse, identifies the component, and enters other pertinent data. From this information and a user-supplied input pattern that mimics the board's actual operation, the system assembles a database.

Next, an operator feeds production boards to the system, positioning each board's tooling holes on matching tooling pins in the isothermal chamber. The tooling pins and I/O connectors represent the only "fixture" that the method requires. Each board generates a set of infrared images when executing the test pattern. A statistical analysis of the resulting signatures produces the standard against which the system will measure each board during testing. The tester will flag any uncovered outliers in the sample as bad boards, and does not include them in the standard.

The production test itself is relatively straightforward. An operator puts a board into the isothermal chamber, positions it on the tooling pins, and connects the edge connector and any other connectors. After reading the ambient infrared image, the tester powers up the board very briefly (to prevent any shorts from frying it) and again examines its thermal signature. Image-analysis software compares this result to the good-board signature and makes a go/no-go decision. Shorts (including power-to-ground shorts, which are nearly undetectable by conventional test) show up as statistically significant temperature differences at particular board locations, as in Figure 3-16. A failing board at this stage would proceed to repair before continuing, as with both in-circuit and functional test.

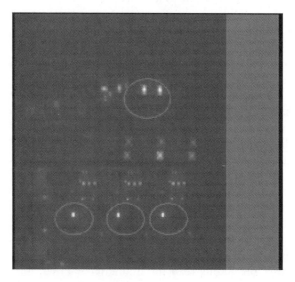

Figure 3-16 Infrared inspection shows "hot spots" or "cold spots" when the board under test differs from the good-board standard. (Courtesy ART Advanced Research Technologies, St. Laurent, Quebec, Canada.)

Once the board is short-free, the tester applies power and an input/output stimulation pattern. Testing a cellular phone board, for example, would verify dialing, "press-to-talk," and other features. Again, any faults should exhibit obvious differences from the good-board thermal signature.

3.3.4.4 No Good Deed . . .

Like other techniques, infrared inspection has its drawbacks. Chief among them is the need for up to 30 production boards from which the system assembles the good-board signature. Many manufacturers never see that many good boards until after production begins, by which time the test must already be in place. Regardless of the cost benefit of anomaly detection, a manufacturer may encounter the "chicken-and-egg" problem—needing a test to generate 30 good boards and needing the 30 good boards to create the test. In high-mix, low-volume, and low-cost situations this requirement could prove prohibitive.

An out-of-tolerance resistor or capacitor will not generally produce a thermal signature sufficiently different from a good one to be detectable. Such failures are relatively rare, however, and an in-circuit test will usually identify them prior to infrared inspection.

An infrared detector cannot see through certain barriers, such as RF shields and heat sinks. Such boards would require testing before attaching these parts, which may be impractical, and will certainly miss any faults induced during attachment.

Infrared inspection can identify the component containing a failure, but its resolution is not always sufficient to identify the exact pin location, especially for solder problems on small-pitch surface-mount boards. In addition, the thermal anomaly might not occur exactly at the fault. However, a repair technician can call up the failing thermal image on a computer monitor to examine it before proceeding. The anomaly's location narrows the search for the actual fault to a few pins. With that information, the technician can pinpoint the actual problem.

Since the technique depends on such small changes in temperature, the board under test must be thermally stable. That is, before power-up, the entire board must be at equilibrium at the room's ambient temperature. A board fresh from wave or reflow solder, for example, or from storage in a room whose ambient temperature is more than ±5°C different from ambient on the test floor, must be allowed to reach equilibrium before the test can proceed. As a result, an infrared detection system may work best in a batch rather than an inline production configuration. A preconditioning chamber where up to 45 minutes of production can reach thermal equilibrium prior to inspection can alleviate this problem. The chamber, however, adds time to the production process and introduces another handling step. Also, current infrared solutions require a human operator to load and unload the boards, precluding their use in unattended high-speed automated production lines.

The application of infrared technology is new to inline loaded-board inspection. Early returns are encouraging, and this alternative deserves consideration. Its availability is still quite limited, however, and the jury is still out on the board types and factory configurations that will benefit most. Still, by offering a new application of an established approach and a way to identify problems that other solutions miss, this technique can provide a viable choice in creating a successful test strategy.

3.3.5 The New Jerusalem?—X-Ray Inspection

Currently the fastest-growing inspection technique, x-ray inspection can detect defects that defy the line-of-sight constraint of optical systems. It can examine hidden components and solder joints on both surfaces of double-sided boards, and can even inspect inner layers of multilayer boards. X-ray is the only available technique that can quantitatively analyze solder joints. Results can indicate problems with the solder and assembly processes, measuring such parameters as solder-joint accuracy, fillet height, solder volume (detecting both insufficient and excess solder), component existence and position, and polarized-capacitor orientation (by the position of the "slug").

Figure 3-17 shows the geometry of a typical solder joint. The x-ray system can measure each of the noted features and compare them with a good-joint standard to permit a pass-fail decision. The gull-wing solder joint in Figure 3-18 will pass both electrical and x-ray inspection. In Figure 3-19, the joint shows an insufficient heel. This joint will pass electrical test because a connection exists. Nevertheless, it represents a potential reliability problem. Finding marginal faults such as this is one of x-ray inspection's strengths.

The x-ray system examines the joint, producing the image shown in Figure 3-20. Combining that information with data that calibrate the darkness of points

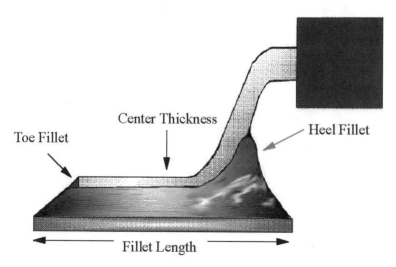

Figure 3-17 The geometry of a typical solder joint. (Courtesy Agilent Technologies.)

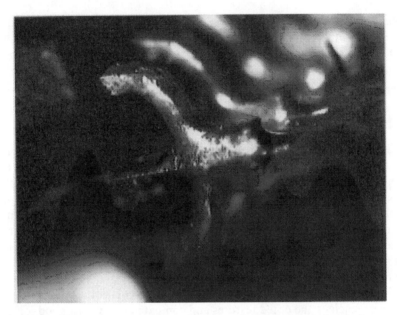

Figure 3-18   This solder joint will pass both x-ray inspection and electrical test. (Courtesy Agilent Technologies.)

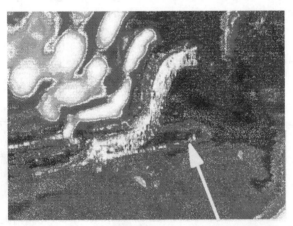

gull-wing with insufficient heel

Figure 3-19   Contrast this joint with one in Figure 3-18. This one will pass electrical test because a connection exists, but will fail the quantitative analysis of x-ray inspection because it represents a reliability problem for the product in the field. (Courtesy Agilent Technologies.)

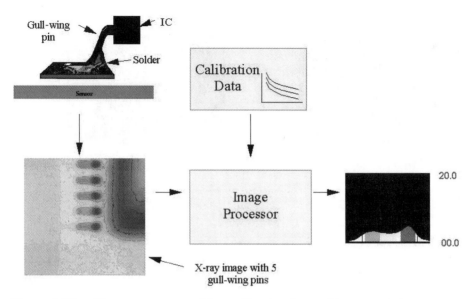

Figure 3-20   The x-ray system combines calibration data and image-analysis techniques to construct a profile of the solder joint. (Courtesy Agilent Technologies.)

in the image with solder thickness, along with some kind of image analysis, produces a quantitative profile of the joint. Figures 3-21 and 3-22 contain actual inspection results. Figure 3-21 contrasts an expected result for a good gull-wing joint with the result from a joint containing insufficient solder. In Figure 3-22, the inspection system detects an open on a J-lead solder joint.

When considering adding x-ray inspection to their test arsenals, prospective users raise two particular issues. The most common question is: *Is the equipment safe?* That is, will harmful radiation escape into the workplace? The x-ray source is carefully shielded from the ambient environment. Governments have established safety standards, which today's equipment vastly exceeds. Study after study investigating x-ray inspection installations have shown no increase in radiation levels.

The second concern involves the migration to lead-free solder. Although solder produces only about 5 percent of the lead escaping into the environment (the majority comes from automobile batteries), environmental protection agencies around the world are pressuring the electronics industry to eliminate it. Lead-free solders present a considerable challenge. Higher melting points, a tendency toward brittleness, and other drawbacks to this approach will require careful consideration and solution. For x-ray inspection, however, lead-free solders present little problem. X-ray relies on the absorption characteristics of several heavy-metal elements (lead, bismuth, silver, tin) to ensure joint integrity. The technique can be adapted to the new solder alloys by recalibrating the inspection system and the image-analysis software.

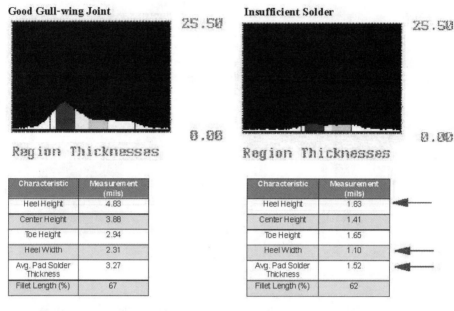

Figure 3-21 Contrasting results from a good gull-wing joint with those from a joint with insufficient solder. (Courtesy Agilent Technologies.)

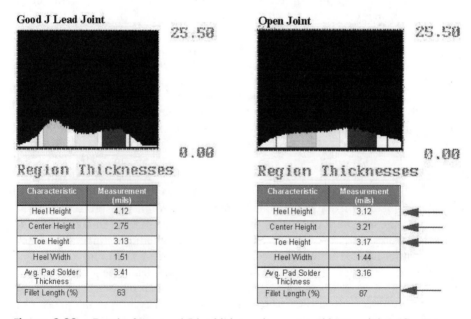

Figure 3-22 Results for a good J-lead joint and a comparable open joint. (Courtesy Agilent Technologies.)

Although x-ray inspection can find many faults that elude other methods, it is not a panacea either. It will not detect a wrong component unless it presents a very different x-ray profile from the correct one. X-rays cannot detect faulty and out-of-tolerance components, and will not notice cracked solder joints that currently appear intact. For x-ray inspection to be effective, you have to combine it with some combination of process control, electrical test, and environmental-stress screening.

3.3.5.1 A Catalog of Techniques

X-ray inspection covers a broad range of capabilities. With *manual* equipment, a human inspector inserts a board into the system, then obtains the relevant image and makes the pass/fail decision.

The image that the inspector sees may contain only the basic x-ray snapshot, but it can also include metrology information, or even complete quantitative analysis. Also, the software may enhance the image in some preset way or provide a level of image processing to make the inspector's decision easier and more consistent.

This approach offers economy, flexibility, and fast implementation. Cost will vary depending on the amount of infrastructure and software support, but begins at less than $50,000. Manual inspection examines boards one at a time, and usually looks only at critical areas, rather than at the entire board. It works better primarily for prototyping or during the ramp-up to full production, for random sampling for process monitoring, and where the nature of the board makes a full inspection unnecessary. Success depends on throughput requirements, and—depending on the level of available software assistance—may vary like other manual techniques, depending on the inspector's experience, the time of day, or the day of the week.

Semiautomated techniques include an x-ray system and sophisticated image-analysis software. This version inspects the board for device placement and solder-joint integrity based on preset gray levels. More expensive than manual alternatives, it requires longer lead times and a software model of the board for comparison. However, it also provides much greater consistency, and generates far fewer false calls.

Most elaborate are the *automated* systems (so-called *automated x-ray inspection*, or *AXI*), where the software makes pass/fail decisions based on established heuristics. Long used for inspecting ball-grid arrays, it has become more popular in the past few years in production because of the difficulty determining board quality using more traditional test techniques. This alternative is faster than manual and semiautomated techniques. It is also considerably more expensive and requires longer startup times. In addition, depending on the throughput requirements of the production line, you may have to compromise between comprehensiveness and cycle time.

Programming x-ray systems can take two forms. Conventional programming involves an image of the board—either a constructed image or one obtained from a board simulation—and metrology tools to establish pass/fail criteria. Some systems learn from a known-good board (subject to the usual *caveats*), and can automatically locate, inspect, and evaluate each solder connection. This approach involves training the system to recognize all of the solder-joint geometries that it must inspect and storing the information in a library with corresponding thresholds and tolerances. The engineer can modify the library criteria to allow for deviations and customizations.

3.3.5.2   X-Ray Imaging

X-ray inspection falls into two broad categories, covering two-dimensional and three-dimensional techniques. Figure 3-23 shows two-dimensional, or *transmission* x-ray, where a stationary x-ray source looks directly through the board, inspecting both board sides simultaneously. Image intensity indicates depth of the feature under scrutiny. The approach works best with single-sided boards.

The mechanics of transmission x-ray equipment are considerably simpler than those of the more complex three-dimensional techniques. It is easier to implement and the equipment is less expensive. Test time is also faster, although proponents of three-dimensional approaches argue that a slower diagnostic time reduces the test-time advantage.

On the downside, transmission x-ray cannot easily distinguish features on double-sided boards because images of the two sides overlap. To compensate, some industry experts recommend staggering components on the top and bottom of the board, as Figure 3-24 illustrates.

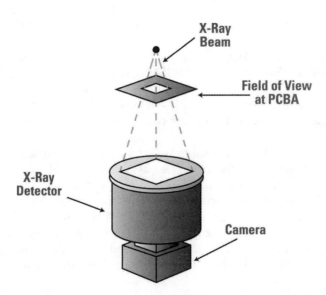

Figure 3-23   Transmission x-ray sees through the board, presenting it as a two-dimensional image. (Courtesy Agilent Technologies.)

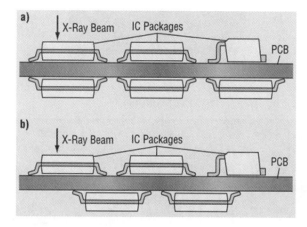

Figure 3-24 Some industry experts recommend staggering components on top and bottom of double-sided boards. (*Test & Measurement World*, June, 2000, p. 16. Used by permission.)

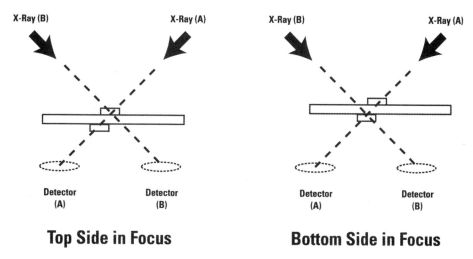

Figure 3-25 Keeping each plane of the board in focus. (Courtesy Agilent Technologies.)

Three-dimensional x-ray techniques, also known as x-ray *laminography*, *tomography*, and *digital tomosynthesis*, permit looking separately at two sides of the board by focusing on one surface while the other surface blurs into the background, as in Figure 3-25. It can even be used to examine inner layers of a multi-layer board. Figure 3-26 shows a two-sided single inline memory module (SIMM), its transmission x-ray image, and its 3-D image.

SIMM Module

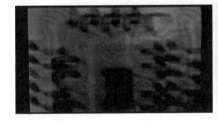 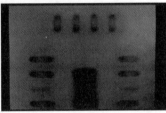

| 2-D X-Ray Image (Transmission) | 3-D X-Ray Image (Laminography) |

Figure 3-26 A two-sided single inline memory module (SIMM), its transmission x-ray image, and its 3-D image. (Courtesy Agilent Technologies.)

Several different mechanisms can achieve the 3-D result. In the approach in Figure 3-27, the x-ray detector moves in a circle around the center of the board section under inspection, while the system mechanically steers the x-ray beam. Alternately, a steerable x-ray beam sends signals to eight stationary detectors. In either case, the system must locate the board surface exactly. The rotating-detector method scans the board with a laser to map the surface before inspection begins. The stationary-detector system performs a dynamic surface-mapping that avoids the test-time overhead of the laser step.

Three-dimensional x-ray can resolve board features individually, examining the two board sides independently—*in one pass*. Therefore, it can provide more precise, higher-resolution analysis than transmission methods can. On the other hand, it also requires a longer setup and longer test time, and is more expensive.

3.3.5.3 Analyzing Ball-Grid Arrays

One advantage of three-dimensional x-ray is the ability to analyze the quality of solder balls and connections on ball-grid arrays. In fact, inspecting BGAs represents the most common justification argument for adopting x-ray inspection. For some manufacturers, that is its only regular application. This inspection step looks for voids, out-of-location solder balls, excess solder, insufficient solder, and shorts. Figure 3-28 demonstrates how by examining slices at the board surface, the center of the ball, and the device surface, 3-D x-ray can create an accurate profile.

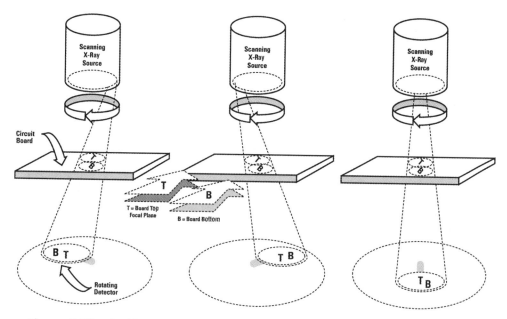

Figure 3-27   In this version, the x-ray detector moves in a circle around the center of the board section under inspection, while the system mechanically steers the x-ray beam. (Courtesy Agilent Technologies.)

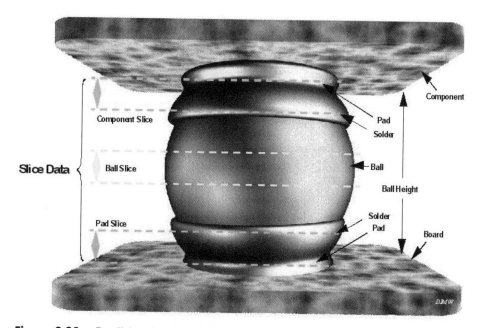

Figure 3-28   By slicing the solder ball on a BGA, an x-ray system can construct an accurate image of the whole ball. (Courtesy Agilent Technologies.)

The voids in Figure 3-29 and the balls containing insufficient solder in Figure 3-30 will very likely pass electrical test. Nevertheless, these weak areas in the joint may succumb to mechanical and thermal stresses from handling and normal operation, and therefore represent reliability problems in the field.

Using multiple images, 3-D x-ray can also verify solder joints and barrel fill in plated-through holes (PTHs), as in Figure 3-31.

Because of the precision of its measurements, x-ray lends itself to process monitoring and feedback, even when conducted only on board samples. In Figure 3-32, UCL and LCL represent the upper and lower control limits, and UQL and LQL the upper and lower *quality* limits. The solder measurement lying outside the quality limit indicates a solder bridge—a short and, therefore, a fault that conventional test should also detect. One of the joints contains insufficient solder, but in this case it lies outside the control limits but within quality limits. This defect will pass electrical test, but x-ray inspection will generate a flag on the process that can initiate an investigation and correction to prevent future occurrences, reducing the number of future failures and thereby increasing manufacturing yields.

Deciding to include inspection in your test strategy represents the beginning—not the end—of the necessary planning. You still have to evaluate what to inspect (critical areas, samples, or all of every board), where in the process to inspect, and which technique or mix of techniques will likely furnish the best results.

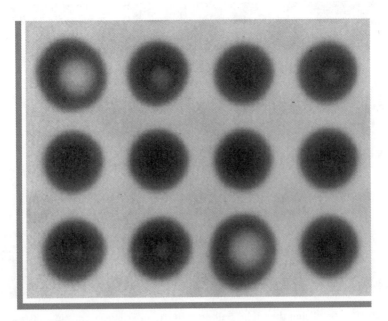

Figure 3-29   Voids in a BGA. (Courtesy Agilent Technologies.)

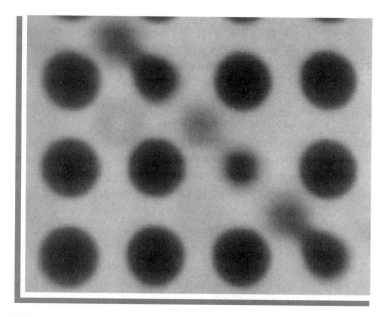

Figure 3-30 Insufficient solder in a BGA. (Courtesy Agilent Technologies.)

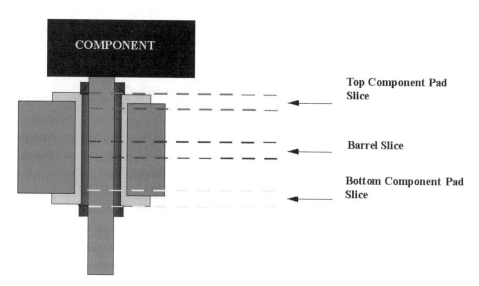

Figure 3-31 Using the 3-D technique allows examining the structure of plated through-holes. (Courtesy Agilent Technologies.)

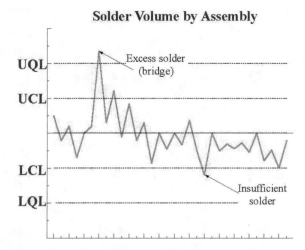

Figure 3-32   Looking at both upper and lower control limits (UCL and LCL) as well as upper and lower quality limits (UQL and LQL) allows the feeding of information back into the process to prevent future defects.

3.4   Summary

The task of finding manufacturing defects in today's boards has become more difficult than ever before. In-circuit and other bed-of-nails techniques suffer from lack of access. For cellular-phone boards and other communications systems, the bed-of-nails itself creates problems because of its RF characteristics. Product evolution and the pace of change combined with skyrocketing complexity make fixture construction and program generation even more painful than in the past. Those same trends drastically reduce the likelihood that we will reach the "perfect process" within the foreseeable future.

Adding inspection to the "test" strategy permits verification when you cannot perform conventional test. It examines the board's structure, ensuring that you have built it correctly. Test determines on some level whether the board works.

Inspection can occur after paste printing, after parts placement, and after reflow. Visual inspection, automated-optical inspection (AOI), and laser and white-light methods work best after paste. After parts placement, most companies select manual inspection or AOI. After reflow, AOI and x-ray inspection—including automated x-ray inspection (AXI)—produce the best results. Choosing among inspection methods, as with test, depends on the nature of the boards and the manufacturing process.

CHAPTER 4

Guidelines for a Cost-Effective "Test" Operation

Many of the methods thus far described, either singly or in combination, can address any particular manufacturer's drive for a high-quality product. As every manager knows, however, a successful "test" strategy (which includes inspection and other nontest quality activities) must be efficient and cost-effective, as well as technically appropriate. In addition, establishing a test operation involves more than "merely" determining test methods. From a project's inception to its conclusion, decisions include evaluating facilities and personnel, planning schedules, and other operational details.

DeSena (1991) applied the program evaluation and review technique (PERT) to better define individual tasks in the planning process. He created a PERT chart that displays critical paths and associated timing. Figure 4-1 shows the first steps of an adaptation of his analysis. Despite the intervening years and the vast changes in both electronic products and test techniques, the principle remains as valid today as when DeSena originally proposed it.

4.1 Define Test Requirements

As discussed in Chapter 1, the first item on a test manager's agenda is to understand the product or products and to define test needs. Neglecting this step, choosing instead to construct a test strategy that encompasses all conceivable test requirements, often results in an excessively expensive solution that lacks important features or capabilities.

In conformance with concurrent-engineering principles, defining test requirements should begin during initial product design. Planning includes evaluating the organization, all product offerings, and management goals, policies, and constraints.

For example, consider product-related issues. Personal-computer manufacturing generally involves high volumes and state-of-the-art technology. Aggressive selling-price competition makes keeping test costs down critical. Personal-digital assistants (PDAs) must cram almost as much technology onto a much smaller motherboard. Appliances such as washing machines and microwave ovens, on the

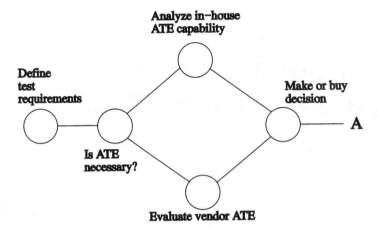

Figure 4-1 DeSena (1991) applied program evaluation and review techniques to define individual tasks in the planning process better. This PERT chart displays critical paths and associated timing for the first steps of an adaptation of his analysis. (DeSena, Art. 1991. "Guidelines for a Cost-Effective ATE Operation," *Test Industry Reporter*, Mineola, New York.)

other hand, may generate equally high volumes, but their electronics technology is usually several generations old, permitting less sophisticated (and generally less expensive) test methods. Cellular phones squeeze all of their functionality into a high-volume, low-profit product that you can hold in your hand. Their counterparts in telephone central offices and cell-switching stations enjoy the luxury of larger boards, lower volumes, and higher profit margins. At the same time, the failure rate for central-office equipment must stay orders-of-magnitude below the rate for the phones themselves.

Oscilloscopes, logic analyzers, and similar measuring instruments are low-volume, high-priced, very-high-precision offerings. For these products, test accuracy is more important than throughput and cost. Testing avionics and military and defense electronics also must emphasize reliability as the primary criterion. In contrast, throwaway products, such as inexpensive calculators, do not even require failure analysis. Go/no-go testing is sufficient in those cases, minimizing test time and costs, but increasing scrap.

Company size also affects test-strategy decisions. A large company can allocate test capabilities among several products or projects more easily than a small company with a more limited product base can. One rule prevails—you cannot buy less than one of anything. The number of board types that a strategy covers dictates the cost of program development. Similarly, the run rate for each board type determines the available program-development budget.

Consider a PC maker producing 100,000 boards per month with a value of $300 each. A budget allocation of 1 percent of the value of one year's production run for programming costs at all test levels permits spending $3.6 million. That allowance includes programmer salaries, office space, amortized costs for testers, CAE equipment, and computer equipment, machine time at each tester station for installation and debug, and other costs. If all boards are identical, then the operation requires only one set of test programs, and the budget is quite generous. If, on the other hand, production includes 10 board types, the allowance for each type is only $360,000 and must include additional programmers, as well as a heavier equipment burden. Still, the budget will probably be adequate in most cases.

In contrast, a small computer manufacturer's boards may be equally complex, but production may total only 5000 boards per month. Because the company's size does not permit taking advantage of certain economies of scale, each board carries a value of $350. Allocating even 2 percent of a year's production for test programming amounts to $420,000. Dividing that sum among 10 board types leaves only $42,000 for each. It is easy to see why, regardless of capital investment in tester hardware, small companies look for ways to minimize programming costs, passing some faults on to later test levels, such as system test, or eliminating some levels altogether.

Today, of course, common practice farms out manufacturing and test operations to a contract manufacturer (CM) (sometimes called an *electronics manufacturing systems* [*EMS*] provider). Yet contract manufacturers face their own challenges. Most large CMs handle boards covering a wide range of sizes, technologies, and functions. It would not be unusual over the course of a single week to watch six different brand-name computers roll off the same assembly line. Efficient costing, therefore, requires enormous flexibility—the ability to deal expeditiously with the next customer's product—whatever it may be. Some CMs may accept only a subset of possible products, ones that match their expertise.

Business conditions may dictate strategic decisions, mandating that any proposed test strategy involve no new equipment purchases, for example, or that new purchases not exceed a few hundred thousand dollars. Management philosophy has an impact, as well. Some managers may minimize test costs by accepting up to, say, 2.5 percent field returns, whereas others may demand fewer than 0.1 percent returns. Any strategic decision that calls for spending money may require a formal return-on-investment (ROI) calculation as part of its justification. (Chapter 10 explores economic issues in more detail.)

An engineer responsible for proposing a test strategy should begin by constructing a matrix of test requirements for each unit under test (UUT), then combine all matrices into a comprehensive list of test specifications. (These are *test* specifications, not test*er* specifications. A strategy to meet them may call for one or more testers, or no tester at all.) If the products are similar, such as different types of PCs, the best overall solution is also often the best solution for each individual product. If the products are sufficiently different, however, as with a contract manufacturer who produces personal-computer boards, disk-drive controller boards, and communications boards, the best and most cost-effective strategy may require strategic compromises for individual board types. A large and varied product base may justify several different strategies in the same company or facility.

Typical products

Telecommunications equipment
Computers and peripherals
White goods (appliances)
Consumer electronics
Military and defense electronics
Capital equipment
Measuring instruments
Automotive
Avionics
Other electromechanical systems

Company mission

Original equipment manufacturer (OEM)
Contract manufacturer (CM)

Company size

$4 million
$40 million
$400 million

Number of board types per year

4
40
400

Volume per board type (per year)

50
500
5000
50,000
500,000

Other parameters

Board sizes
Number of board layers
Node access
Embedded components?
Trace spacing
Analog technology
Number and type of analog components
Level of digital complexity
Number and type of digital components
Hybrid components?
Application-specific integrated circuits (ASICs)?
Multichip modules (MCMs)?
Ball-grid arrays (BGAs)?
Flip-chips and other chip-scale packaging?
Number and type of "jellybean" parts?
Components on both sides?
Surface-mounted components?
Component density
Through-hole nodes

Other strategic factors

Tested bare boards?
Tested components? Where?
Logic simulations?
Fault simulations?
Other CAD and CAE data
Personnel skills
— Operators
— Programmers
— Troubleshooting and repair
— Maintenance

Economic factors

Evaluation costs
Acquisition costs
Overall budget appropriations
Product selling prices and volumes
Likely product revision schedule
— Engineering change orders (ECOs)
— Updates
— Enhancements

Figure 4-2   Some considerations for strategic analysis. The list provides only examples of the important issues. It cannot pretend to anticipate the needs of every company or test facility.

One aspect of the "no-test" option comes into play here. Many boards contain logic pieces that rarely fail. In defining test requirements, test managers may opt to forego testing those segments at board level, especially if testing them is difficult, and concentrate on areas that will more likely contain faults. This tactic improves a test strategy's efficiency and cost-effectiveness, and reduces the production holdup.

Figure 4-2 outlines some considerations for this strategic-analysis step. The list provides only examples of the important issues. It cannot pretend to anticipate the needs of every company or test facility. Note that company size in the figure does not jump to orders of magnitude above $400 million. Large companies do not operate as monoliths. A test engineer in a $4 billion corporation undoubtedly constructs specific test strategies for only one division or facility at a time. For planning purposes, those strategies will likely fall into one of the smaller-company categories.

4.2   Is Automatic Test or Inspection Equipment Necessary?

Corporate managers dislike automatic test equipment (ATE), and its inspection siblings, as expensive solutions to the problem of ensuring product quality. Beyond actual capital expenditures, they require the commitment of facilities, people, and time. Test managers must ask whether another approach can accomplish their goals equally well. Certain portions of a test strategy, such as shorts testing, may necessitate ATE, whereas others do not. Visual inspection, process monitoring, manual testing, self-tests, and rack-and-stack instrument arrays can perform at least part of the job in many situations. Or, as mentioned, entire boards or sections of boards may not fail very often and may, therefore, not require testing at all.

As with all computer-bound tasks, an automated system's value increases with repetition. It is most cost-effective at high board volumes and low mixes. The appropriateness of automated test also depends on the skill levels of technicians and other personnel. Manual and instrument techniques generally demand that people on the production floor be more highly skilled than do ATE alternatives.

One interesting aspect of assessing the need for big equipment is looking at what competitors are doing. Obtaining that information does not require covert surveillance. It is available by observation at trade shows, in technical papers and advertising material from trade publications and conferences, and from conversations with current and potential customers and equipment vendors.

If competitors employ similar test strategies, then those strategies are probably reasonable. If, on the other hand, strategies are significantly different, each company must decide why. Corporate philosophies, product features, or the nature of other tested products in the facility may justify the disparity. Significant strategic differences, however, demand further scrutiny.

This analysis does not necessarily mean adopting competitors' approaches. For example, the vice president at one large systems manufacturer discovered that his major competitor was about to be sold. The buyer was an entrepreneur with a track record of adopting a particular manufacturing-and-test approach. The vice president demanded that his organization modify its procedures to anticipate the competitive change. Unfortunately, he tried to implement his edict by fiat, without considering its impact on his own products and on day-to-day manufacturing operations. The result was a major disruption, causing morale, lost productivity, and reliability problems.

4.3   Evaluate Test and Inspection Options

Once you decide whether ATE or some other alternative or combination of alternatives represents the best solution, the next step examines in-house capabilities and vendor offerings. Examining in-house capability encompasses two issues. First, is there equipment already in place that can test the new products? Second, if not, can you (and do you want to) build what you need?

Incorporating new products into test strategies that employ existing equipment has advantages. It is often the lowest priced approach. Capital expenditures include primarily updates and enhancements. Because programmers, technicians, operators, and managers already understand the equipment, learning curves are short. Also, existing system availability both permits and encourages early test-program development.

Disadvantages include capacity limitations. This problem is particularly acute in small companies and similar situations, where a new product or product line represents a substantial fraction of the facility's manufacturing output. Current capacity may be sufficient, but once sales take off, will test become a bottleneck? Will expansion require adding a work shift, with all of the hiring and other managerial headaches that that entails?

Will the bottleneck necessitate purchasing additional capital equipment later? If that happens, new acquisitions will have to be the same type as older machines. This strategy postpones, rather than eliminates, capital expenditures and severely limits test-method flexibility.

Perhaps the biggest drawback to deciding *in advance* to accommodate a new product on existing equipment is that the approach may not provide the best possible test, even if it does represent the best test for other products. There is a tendency not to evaluate other more innovative test solutions, such as adding an AOI system. The best alternative could involve remaining with the installed equipment, but it could also require a radical departure from that strategy.

If in-place equipment lacks the capability to test the new boards, are field upgrades possible, such as by adding test points, new features, or other options?

Capacity limitations clearly mandate additional equipment, which may be designed in-house. On the other hand, if a facility lacks the skills, the time, or the available staff and other resources to build custom test solutions, then the only alternatives are purchasing a monolithic solution from a single vendor, assembling a rack-and-stack or hybrid system with measurement features from a group of vendors, or hiring a contractor to serve as systems integrator for a multivendor approach.

Should new systems be identical to older ones, similar but updated, or completely different? If a test manager chooses something completely different, can diverting some older products to the new machines benefit the overall operation?

This last question is not as simple as it first appears. Transferring products to an unfamiliar test environment may be painful because of the need to retrain people and create new test programs. If the product line has life left, however, if it will likely expand either in breadth or sales volume, and if new solutions provide better yields or higher throughput, the change may still be worthwhile. Such real-location also opens up time on the old tester to perform some test steps on the new product, thereby minimizing that product's capital costs.

Reallocation has significant implications if you add inspection to a facility that has never employed it before. Some older products may be ideal candidates for this type of verification. You can then reduce or eliminate the burden on one or more subsequent test steps.

Nevertheless, convincing a manufacturer to "change horses" in this way often resembles swimming upstream against a torrent. A couple of years ago, a customer presented a large contractor with the task of manufacturing a particularly complex board. The board measured about 30 inches square, with a manufacturing cost of about $10,000. Production totaled about 400 boards per year. The board presented several problems aside from its size. Barely a quarter of the nodes permitted bed-of-nails access. The board contained BGAs and other components that covered nodes needed during test. First-pass yields hovered around 75 percent.

Because of high board value and low production volumes, a consultant sug-gested that the CM consider x-ray inspection. It would alleviate the problem of hidden nodes and improve the cache of good boards. The CM, however, dismissed the idea, contending that he only applied x-ray inspection to examining BGAs, not boards in production. Had introducing x-ray inspection required a huge capital investment, his reticence would have been understandable. But there, on his factory floor, sat an elaborate high-end x-ray inspection system—*idle*. Clearly, the time had come to examine conventional test-strategy wisdom to come up with effective alternatives.

Evaluating vendor equipment offerings includes matching vendor test methods with test requirements. A vendor who emphasizes bed-of-nails techniques might not represent the best choice to test dense, heavily populated, surface-mount boards. A vendor offering emulation-based functional test cannot adequately address boards not destined for microprocessor-based systems. Of course, vendors must also provide speeds, pincounts, tolerances, and other specifications that meet or exceed test requirements, as well as expandability.

Software issues include test-language flexibility and appropriateness for the board's technology. Human interfaces must be easy to use, powerful enough to take advantage of all machine features, and versatile enough to accommodate external instruments and other add-on features. Reliance on Windows-based tools, for example, does not necessarily equate to ease of use. Much depends upon menu organization, icon arrangement and clarity of meaning, and so on. Transferring files from simulators and automatic program generators to test environments should require a minimum of tweaking and debugging.

Failure-analysis tools and methods should match the manufacturing opera-tion's requirements. Data logging and analysis should be comprehensive, and reports should produce the information necessary to evaluate product and process. Compatibility with Excel or some other standard package makes the data more generally applicable. A data-analysis program that generates 6 pounds of paper but does not provide a specific critical item is worthless.

Beyond product offerings, it is important to examine a vendor's company. Buying test equipment establishes a partnership between vendor and customer that stretches long past warranty expiration. Is the vendor reliable? Does the company have a good reputation, or is there a track record of broken promises, missed delivery dates, advertised but nonexistent features, and other inconveniences? Can you get a list of satisfied customers and testimonials from actual people who will discuss their experiences?

Visits to vendors and other customers can help separate the real world from marketing hype. It is important not to become disillusioned if vendors' strategies for testing their own equipment do not conform to what they recommend. Test equipment presents peculiar problems in low-volume, high-accuracy production with which few customers must contend.

One important aspect of the vendor-evaluation step is that brands of test equipment are not interchangeable. With PCs, a product made by Ipswitch, Ltd., may work identically to a comparable offering from Widgets, Inc. With test equipment, however, no two machines have identical specifications in all particulars, and any tester choice represents some level of compromise. Two machines may do the same job, but they will rarely do it equally well or in the same way. Even when specifications match, each vendor achieves that result by using a unique mix of hardware and software. Therefore, some details of tester behavior will still be different.

Experience with a particular vendor includes equipment already in place, evaluations that were part of a preceding project (even if another vendor was selected), and the experience of employees when they worked for other companies or departments. Is the equipment reliable, or do downtime and maintenance represent major headaches? How does the vendor respond to questions and problems? How promptly does someone return telephone calls? Are applications people knowledgeable and cooperative?

How easy is the equipment to use? What training is available to get new customers started testing boards, and for more experienced types to better take advantage of advanced features and capabilities?

Buying new equipment from a familiar vendor who has already supplied equipment for other projects minimizes training needs and other hand-holding and, therefore, generally reduces the inevitable learning curve that accompanies new-product introduction. In addition, because all test equipment has both limitations and personality quirks, staying with a vendor may keep a company from making unfortunate assumptions about a machine's features and capabilities.

On the other hand, choosing a new vendor may permit test options that the existing vendor's equipment cannot or does not provide. Sticking with a familiar vendor in this case would inherently limit test-strategy choices.

Experience with a particular vendor may be negative and, therefore, not conducive to continuing the relationship. Sometimes, manufacturers remain with certain vendors because of inertia ("We have always done it this way!"), despite the fact that other vendors offer better equipment, better service, or simply a better partnership.

As an example, when planning a new strategy for an existing facility a few years ago, upon noticing that most in-place test equipment came from a single vendor, a consultant asked his client if the practice should be continued with the new acquisitions. The client responded, "Absolutely not! We keep replacing existing equipment with comparable systems because we are familiar with them, but the systems are not performing adequately, and the vendor takes us for granted. For the new product, we will consider any other vendor."

Certainly, all final decisions must satisfy both technical and economic criteria. In some cases, when business times are difficult and budgets are very tight, immediate cash flow issues must predominate. On the other hand, a more effective strategy means a higher-quality, more reliable product. Initial cash outlays may be higher, but, as with design for testability and concurrent engineering, the aim is to minimize *overall* costs.

4.4 The Make-or-Buy Decision

Assuming that a test manager needs additional equipment and that either in-house development or vendor offerings could provide it, he or she must determine which option to take. A decision either way must include consideration of whether new solutions will be monolithic, rack-and-stack, VXI-type, or some combination.

First, a new-test*er* specification outlines existing capabilities and anticipated needs. Tester features should accommodate the new product and any planned enhancement or expansion of that product or product line during the machine's service life. In addition, the specification should include features that existing products would require if strategy changes divert any of them to the new tester.

Once you have completed that analysis, generate cost estimates for each alternative. For in-house-built equipment, important cost considerations include engineering and management time, software development, and production lost to disruption on the factory floor during construction. Even rack-and-stack and VXI solutions require system-level hardware design and software development.

Many make-or-buy considerations do not directly involve money. Designers of in-house test systems have access to product designers and can more closely match tester features to manufacturing-process personality quirks. Vendor offerings represent a more general range of capabilities that may or may not meet a particular application's needs.

Working with vendors to find the best possible test solution generally means completely educating people there about the new product and revealing proprietary aspects of that product's design or technology. Gaps in this communication can compromise test effectiveness. Even companies that dislike such complete disclosure must conform if vendors are to propose the best solutions.

For example, IBM's reticence to discuss intimate details of unannounced products outside departments directly involved is well known. Yet, more than 18 months prior to release of the company's original PC in 1981, several tester vendors had schematics, design information, and board samples to permit recommending appropriate test equipment and developing timely and effective test programs to support the product's launch. The only alternative was to resort to in-house tester design.

Obtaining test equipment from a vendor represents an obligation by that vendor to provide adequate support. Support may involve answering telephone calls or may require on-site presence of an applications engineer or field-service specialist. Waiting for a vendor support person to arrive can delay problem resolution for a few days or longer.

Theoretically, in-house-built equipment permits more immediate help, because tester designers and other experts are near at hand. Unfortunately, the reality is not so simple. Principals may have moved on to other projects, other departments, or even other companies. Because the test equipment is no longer their primary responsibility, support requests may receive less than top priority.

Software development represents another potential bottleneck for in-house systems. Despite all of our sophisticated tools, estimating software-development time and costs remains a black art. Most companies drastically underestimate the effort involved. In-house-developed software also demands bug fixes, updates, and enhancements, straining what, in most companies, are scarce software resources. Standardization of designs around common computer engines and operating systems, such as Windows or UNIX, helps alleviate some of this burden but does not eliminate it. Vendors include system software with their products, along with necessary support.

The make-or-buy decision maker, whether for monolithic, rack-and-stack, or VXI-type testers, must also choose between single-vendor and multiple-vendor solutions. Buying everything from one vendor, if that vendor's products are adequate for the application, offers the advantage of one type of training and one source for problem resolution. Adherents to this approach agree with Mark Twain's adage, "Put all your eggs in one basket—but watch that basket!"

Selecting more than one vendor can mean obtaining instruments or instrument cards from different companies or monolithic in-circuit testers or MDAs from one company and functional testers or inspection systems from another. This alternative permits obtaining the best test solution at every decision point. Users, however, may have to contend with different human interfaces, test-development methods, measurement techniques, and specifications. Test capabilities must not excessively overlap (a waste of money), and there must be no unnecessary gaps in test techniques or fault coverage.

Following a "buy" decision, selecting vendors and equipment uses information on technical and economic criteria as well as history and experience obtained during vendor evaluation. Chosen vendors for each strategy element must provide the solution, support, and training to bring customers the best possible performance.

4.5 Getting Ready

Either a "make" or "buy" decision triggers events that culminate in the installation of testers and related equipment. Figure 4-3 shows the next steps in DeSena's PERT chart model.

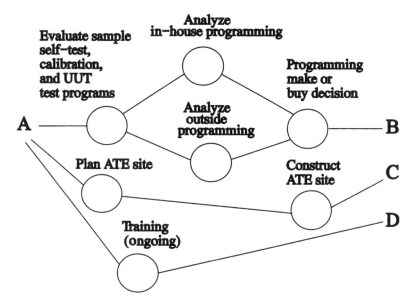

Figure 4-3 The next steps in DeSena's (1991) PERT-chart model. Either a "make" or "buy" decision triggers events that culminate in the installation of testers and related equipment. (DeSena, Art. 1991. "Guidelines for a Cost-Effective ATE Operation," *Test Industry Reporter*, Mineola, New York.)

All test equipment, regardless of source, must offer some kind of self-test to ensure that it works correctly at any particular time. Self-tests can range from simple go/no-go power-on self-tests (POSTs) to more elaborate versions that can cover 80 percent of possible faults. For vendor equipment, test engineers should examine vendor-supplied self-tests carefully, thoroughly understanding what they test for and how completely they verify system calibration. With in-house-developed solutions, test people who will operate and manage the equipment should participate in self-test specification and development.

Equipment calibration should be relatively fast and easy to follow. Cryptic instructions and long procedures will discourage routine use. Vendors and in-house system designers should recommend preventive-maintenance calibration schedules, perhaps providing a quick weekly procedure and a more elaborate version for use when the shorter one fails or when a tester's results are suspect. For example, a weekly calibration might take 15 to 30 minutes, whereas the longer procedure might take several hours.

For a new in-house-designed tester or for an unfamiliar commercial machine, sample test programs are particularly important. Test engineers must fully understand the languages, generation procedures, and tester strengths and weaknesses. For example, for several years one vendor marketed a bed-of-nails tester as an MDA, despite the fact that it came with complete analog measurement and guarding capability. Therefore, although it served only as an MDA on digital circuitry, for a heavily analog board it was almost as efficient as a full (and much more expensive) in-circuit tester. An effective test strategy might send boards directly from such a tester to a functional or hot-mockup test. (The tester was eventually scrapped because of poor sales. Perhaps recognizing its capabilities and marketing them to appropriate customers could have saved it.)

Sample programs for real boards can also help determine the correspondence between tester specifications and actual machine performance. If sample programs are unavailable prior to system delivery (or system startup for in-house development), arranging for "turnkey" delivery—which includes fixtures, emulation pods, and other interfaces, as well as test programs—will permit productive testing to start as quickly as possible.

4.6 Programming—Another Make-or-Buy Decision

Someone has to write test programs, either manually or using automatic tools. Deciding whether to "make" or "buy" test programs requires carefully examining in-house programming capabilities and the cost and availability of outside programming services.

Certainly, truly unique, in-house-designed, monolithic-architecture hardware will likely require in-house program generation. On the other hand, if engineers have created a system around the VXI bus or any of its variants, or from conventional IEEE-488 instruments, outside programming help may be available. Products such as LabWindows from National Instruments in Austin, Texas; VEE from Agilent Technologies in Loveland, Colorado; and ATEasy from Geotest in Irvine, California, can serve as standard software shells and programming aids for systems containing a wide variety of bus-based instrument capabilities from a large number of vendors. These tools permit the creation of instrument drivers and other program primitives that a third party can incorporate into test programs. Primitives represent nonrecurring engineering costs to test departments. Once the contract house has the tools, however, development costs for each program will not significantly exceed the cost for the contractor to develop comparable programs for commercial testers.

These same tools reduce the pain of in-house programming. For example, the IEEE-488 native language has always been one of the great "joys" of the test industry. It consists of single-character ASCII codes, whose meaning may be less than obvious, that initiate setups and trigger measurements. Programmers who do not use a particular instrument or capability regularly cannot construct proper tests without the instrument's documentation. Fortunately, few of today's engineers have to face these Herculean tasks. Vendor or third-party shells provide high-level, user-friendly environments that permit specifying tests either graphically using "point-and-click" or in a more English-like way. Shells then generate correct ASCII codes to stuff down the bus. The chief advantage of this approach is that—to the programmer—measurement commands for all instruments of a given type are identical. Programmers merely choose the instrument by brand name or model number and issue the shell's measurement request.

Any shell's usefulness depends heavily on the level of effort necessary to create instrument drivers. Software vendors offer driver libraries, but they cannot include every instrument and measurement technique that manufacturers unearth to solve their particular problems. Each software product in this category offers a different combination of advantages and disadvantages. Test engineers and managers must carefully weigh how well each product addresses their applications.

Vendor-supplied testers present several program-generation options. In addition to providing turnkey programs with initial system purchases, vendors may furnish programming services separately or as part of a comprehensive support package. Many small contract houses offer test programs for common tester architectures. For in-circuit testing, purchasing programs and fixtures from one vendor—either the tester manufacturer or a third party—is generally less expensive than obtaining them from separate vendors or buying one and creating the other in-house.

In-house program development requires intimate knowledge of test methods and programming languages, as well as a comprehensive familiarity with the product. If such expertise already exists within a facility, and if there are sufficient human and machine resources, then this alternative represents the best combination of cost, product confidentiality, and other factors.

If test people do not already possess the appropriate expertise, vendors offer training courses, either at their facilities or on-site. Vendors often bundle tuition for some or all such courses with the tester's purchase price.

Beyond initial hardware installation, some tester vendors do not encourage customers to obtain test programs directly from them. Vendors are primarily in business to sell hardware, and the most effective allocation of their programming resources is in support of that activity. In high-mix situations, vendors cannot usually supply enough programs to operate testers at full capacity. For a customer requiring 50 programs, the vendor may supply only three or four. In this case, cultivating an internal programming group may represent the best alternative. Small third-party contract houses can also provide cost-effective solutions.

Choosing a contract house for this purpose requires no less diligence than selecting a tester vendor or contract manufacturer. Programmers must demonstrate thorough understanding both of the product under test and the specific target tester. In addition, outsourcing reduces a manufacturer's control over how a test program actually exercises the board. Test-program accuracy and comprehensiveness may be more difficult to monitor and maintain than with an in-house approach. An in-house engineer who develops a test program is strictly accountable for both the program and ultimate product quality. Carelessness or inaccuracy can cost the programmer his or her job.

For a third party, an unacceptable test program can result in the loss of a customer and they may lose others if word gets out. The programmer, however, only rarely bears direct responsibility. More often that person moves to another project more amenable to his or her particular talents, and the contractor markets the newly available time to other companies.

Manufacturers considering farming out programming projects must specifically address timeliness and currency—that is, how quickly a contractor can generate acceptable programs, and how well those programs incorporate the frequent changes in design, layout, and other factors affecting board test that accompany new product introduction. Certainly, communication is generally more difficult and slower between a manufacturer and a contractor than between members of the same engineering staff. Still, communication paths work well enough. In addition, dealing with the communication gap encourages manufacturers to manage their programming tasks more carefully, such as by creating more explicit instructions and specifications on product function, desired quality levels, and appropriate tests.

Manufacturers must diligently monitor third-party programmers' activities. At least some of the necessary test expertise must reside in-house to maximize the likelihood that a contractor will develop appropriate tests that correctly distinguish good boards from bad ones. People resources spent for this type of monitoring represent additional costs of contracting that managers must include in the make-or-buy decision. Advance work to reduce a new contractor's learning curve can significantly improve productivity and reduce the pain of producing efficient programs.

Attempting to perform all programming in-house can result in an enormous peak-load burden. Programming requirements often run in spurts. Consider the cyclic nature of the automobile industry, for example. Most car makers introduce new models in the autumn of each year. Therefore, test-programming loads tend to peak sometime the previous winter or spring. The pattern is well known and predictable. Maintaining sufficient staff to cope with peak periods requires that programmers sit idle for much of the year or at least work at less than full tilt. The only other in-house option is to maintain a smaller staff, avoiding the bottleneck during peak periods by demanding that managers plan more lead time for each program and schedule development accordingly.

Companies often take a compromise position between in-house and contract programming. The dividing line can depend on timing or board-under-test technology. For a cyclic environment such as the automobile industry, companies may do all of their own test programming except during peak periods.

Outsourcing also works for companies and industries that experience peaks at more irregular intervals. Managing such situations demands strategic flexibility and carefully anticipating workload variations.

Some companies do all or almost all of one type of programming in-house, farming out the rest for technical or political reasons. A manufacturer may be sufficiently expert to handle the programming for complex digital products containing numerous ASICs, for which (we hope) test programs generally come from device vendors anyway, but have more difficulty with high-speed or RF boards. A company that farms out most of its programming might prefer to develop tests for an innovative new product in-house to assert better control over sensitive information. In addition, the design and layout of such products often change rapidly as the release date nears. Keeping development in-house can drastically reduce the time needed for test programmers to react to those changes.

Large companies sometimes set up centralized internal programming groups that act like outside contractors. Contract manufacturers themselves can adopt this same approach. Those groups build fixtures and develop programs for all of the company's operations nationwide or even worldwide. A responsible programmer oversees a project's development stages at the main facility, then delivers appropriate materials to the manufacturing location and supervises installation and initial production runs before returning to home base to begin another project.

Spreading workloads across a corporation in this way allows better management of peaks and troughs than standard in-house solutions do. Programmers generally understand both products and resident test equipment better than outsiders and have more experience with the nuances of program development than engineers within a single manufacturing facility. Engineering managers and project leaders can control schedules and other programming decisions more tightly than with more conventional third-party solutions, and learning curves can be significantly less steep. This approach works particularly well with custom test equipment, rack-and-stack, and VXI-based systems.

On the other hand, corporate programming groups can suffer from the same communications gap between designers and test developers as contractors do. In addition, there may still be periods of overload or insufficient work to keep everyone busy.

4.7   The Test Site

As Figure 4-3 shows, preparation and construction of the test site parallel tester construction for either in-house or purchased equipment. Site planning and preparation include allocating sufficient floor space, as well as installing raised flooring, cooling, power, and air-filtration equipment when necessary.

If people are to ferry good and bad boards from station to station during actual production, operational layout requires time-and-motion analysis to maximize throughput and minimize delays. Similarly, if conveyors and other automatic handlers are to transfer material, floor-space arrangements must accommodate tools, people, and equipment. Automated methods generally consume more floor space than do manual alternatives.

At this stage, operations managers must decide whether to group process steps by function—all assembly in one place and all test in another, for example— or by product. In the latter case, assembly and test for each product would reside in a coherently organized independent space. Process paths would intersect only at shipping. This arrangement can encourage adopting several different test strategies, designing each one to serve its production line best. Low-volume, high-mix applications and applications involving automated board handling generally choose functional groupings because they minimize equipment duplication. High-volume and manual-handling applications can benefit from product-based solutions.

How closely are assembly and test steps to related operations on the factory floor? Such operations include burn-in, board repair, test-program development, and test-equipment calibration and maintenance. How much storage space is available for test fixtures, tester spares, and external instruments? Will adjacent space accommodate product or product-line expansion? Optimizing time and motion and planning for the future when introducing a new product into an existing facility is significantly more complicated than when planning an entire factory from scratch.

Furnished utilities must be clean. Line voltages, for example, should be as spike-free as possible. An episode a few summers ago in an industrial park in Toronto, Canada, illustrates this last point. A board manufacturer complained that analog-measurement results from his in-circuit tester were, at best, inconsistent. The tester vendor's troubleshooter, observing the pattern of failures, suggested monitoring the line voltage for excessive variation. The manufacturer replied, "Oh, we know the power is dirty." The entire industrial complex depended on a single bank of air conditioners. Every time the compressors kicked in, they caused a momentary line-voltage drop of over 10 percent. Appropriately placed line conditioners and surge suppressors could have isolated testers from the anomaly. Careful planning prior to test-system delivery could have anticipated the problem and prevented it altogether.

An adequate communications network should already be in place before equipment installation (equipment *completion* for in-house-built equipment). This network facilitates equipment and program-version control, routing of good and bad boards, and system assembly with a minimum of incomplete products and other work-in-process inventory complications. Networks also handle data analysis and provide statistics that monitor and ensure process performance.

Test-site construction should not interfere with ongoing operations. It should not, for example, reduce adjacent-production-line efficiency. This constraint is less critical if the facility has excess capacity that can be "spent" during site setup. If adjacent lines operate only one shift, for example, site construction can proceed on second shift, weekends, or other downtimes.

Completion of in-house-built equipment should occur on-site whenever possible. Similarly, although engineers can check out components of VXI-based and rack-and-stack testers practically anywhere, on-site assembly avoids damage and other problems that transporting systems can cause. For commercial testers, the site should be ready before system delivery.

4.8 Training

All projects require adequately training involved personnel. Training is not an isolated activity that occurs only once. It is an ongoing commitment that should begin no later than the initial make-or-buy decision and continue throughout a project's life. Trainees include not only tester operators and programmers but also repair technicians and equipment-maintenance people. Supervisors and managers must understand how a new strategy fits in with older approaches and how to make the new one properly pay off.

When assessing training needs, determine whether existing staff levels are adequate. If not, consider redistributing employee skills and responsibilities as part of the recruitment process, just as earlier steps might have involved redistributing equipment. Pulling existing staff onto new projects and assigning new people to older projects where schedules and other demands are less critical may represent the best approach.

If adding staff becomes necessary, the planning process must allow sufficient lead time to find the right people. Many managers underestimate both the length and the difficulty of this task. Most professional "head-hunters" will tell you that their supply of available jobs often far exceeds the number of well-qualified applicants.

Use recruitment to complement the existing staff's aggregate skills. That is, look for someone who has skills that other employees lack. A disk-drive manufacturer whose test strategies have always included in-circuit test but not functional test might hire an experienced functional-test engineer to ensure that functional-test options are fully explored. Similarly, if a manufacturer has always dealt with Vendor A, hiring someone with Vendor B experience could produce a more effective *next* test strategy. Note that new staff members require longer learning curves to get up to speed on any project because, in addition to project particulars, new people must learn the company's management style and personality, as well as what is expected of them.

4.9   Putting It All in Place

Figure 4-4 shows the last steps in the PERT chart model. This section covers actually placing the tester and support equipment in service.

Test-program generation, whether in-house, vendor, or third-party supplied, should begin well before system delivery. This is particularly important if the new tester (or significant tester elements for rack-and-stack and VXI-based systems) is

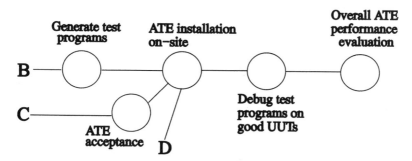

Figure 4-4   The last steps in DeSena's (1991) PERT-chart model. This section covers actually placing the tester and support equipment in service. (DeSena, Art. 1991. "Guidelines for a Cost-Effective ATE Operation," *Test Industry Reporter*, Mineola, New York.)

substantially different from the factory's installed base. Vendors often provide considerable help beyond formal training courses for new customers learning to program unfamiliar equipment, even when those customers do not purchase test programs directly from them. Such help falls under the category of keeping customers happy and supporting future sales.

Program-generation steps include constructing beds-of-nails, edge connectors, emulation pods, and other interfaces, as well as creating failure-analysis software, such as inspection image-analysis tools, fault dictionaries, and GFI logic. For inspection, it may require carefully building a number of "golden" boards.

Before accepting a piece of purchased equipment for delivery, perform an acceptance evaluation at the vendor's factory, establishing that the machine meets all agreed-to criteria. With in-house-designed or in-house-assembled equipment, this initial acceptance can occur in place on the manufacturing floor.

At this point, testers may contain features—especially software features—that were not part of the original specification. Such features may be included instead of or in addition to the specified list. Vendors and system designers will contend that all changes are beneficial. Only you can ultimately make that decision. Any deviations that you agree to, as well as promises to correct problems or add capabilities and features at a later date, must be *in writing* to avoid misunderstandings. There is no substitute for written checklists, instructions, specifications, and agreements. Verbal communication and human memories are rarely as accurate or complete.

Whether or not system delivery includes fixtures and programs, part of the acceptance should include testing real boards if at all possible. Of course, in many cases, appropriate materials are unavailable. For example, to create proper test programs successfully, one contractor required a Mylar drawing, a "blue," a parts list, an x-y node-location list, a schematic, and five good boards. Unfortunately, many times the Mylar does not agree with the blue, which does not agree with the parts list, which does not agree with the x-y list, which does not agree with the schematic, which does not agree with the five boards, which do not agree with one another. As a result, programming costs are often higher and schedules are longer than customers would like.

Commercial equipment undergoes an extra acceptance step once it arrives on the customer's manufacturing floor. On-site installation and acceptance require conformance to the same checklist of specifications and conditions as the earlier acceptance at the vendor site. One important aspect of this step creates interfaces with conveyors, other product lines, and similar aspects of the overall factory system. For in-house developed equipment, vendor-type acceptance and on-site acceptance generally occur simultaneously.

This step should involve tester operators, board-repair and equipment-maintenance technicians, programmers, managers, and anyone else who will interact directly with the systems. Training under these conditions is fairly informal, but this step represents a good opportunity to ensure that all parties understand at least the rudiments of effective, efficient, and *safe* tester operation. Again, exceptions to the acceptance, covering undelivered items and other irregularities, must be in writing.

On-site acceptance should include debugging test programs on known-good boards. (Unfortunately, the difficulty of *finding* known-good boards, especially on new products and product lines, cannot be overstated.) This step ensures that programmers understand the tester's personality and can work with it. Debugging activities also serve to train operators and other support people. In addition, with debugged test programs already in place, productive testing can begin immediately, and the new acquisition can start to earn its keep. Few elements of a manufacturing strategy are as frustrating to engineers and managers as the proverbial "white elephant"—the machine that sits on the factory floor like a large paperweight, patiently awaiting test-program completion and related tasks.

The final step is an overall equipment performance evaluation, verifying that the machine does what you expect and what you need, which are not necessarily the same thing. During early months after installation, users should carefully monitor tester activities, for example, by tracking throughput, product yields, and uptime, and documenting whether downtime results from tester malfunctions or personnel inexperience.

For in-house-designed systems or unfamiliar architectures, a resident or on-call applications engineer can provide incremental training and can solve most simple startup problems easily, with only minimum impact on productivity and schedules. This person can be a vendor, development-team staff member, or a thoroughly trained member of the production team. Preparing this person in advance for project startup can significantly reduce learning curves and associated costs.

Unfortunately, the best-laid plans. . . . Controlling the controllable does not necessarily mean on-time delivery from vendors or on-time test-program completion. Incentives and disincentives can better the odds. Penalty or bonus clauses in purchase contracts make timely performance in the vendor's best interest. For in-house tester and program development, site planning, construction, and other employee tasks, incentives help raise morale and increase productivity. These incentives can include monetary rewards, such as bonuses, but nonmonetary alternatives are often just as effective. Some small companies, for example, throw Friday-afternoon beer-and-pizza parties to celebrate successful project completion.

At all stages of setting up a cost-effective test operation, *written* documentation represents an indispensable tool. Although no one likes to write it, and good documentation rarely exists in the real world, there is no substitute. Designers and manufacturing engineers should prepare complete documentation on boards to be tested. Vendors or in-house development teams must supply information on the testers themselves. Once documentation is available, managers must ensure that people read it and follow it.

4.10 Managing Transition

It is important to note that anytime a test strategy changes, to either a new vendor, a new class of test systems, a new test technique, or a new philosophy (such as testing processes instead of products or adding an inspection step), there will be a painful transition. No amount of preparation can completely eliminate this pain. Employees are generally most comfortable maintaining the *status quo*. Any change will (temporarily) reduce worker productivity until everyone becomes acclimated to the new environment.

The best way to deal with problems of transition is to recognize that they are unavoidable, handle each situation according to its particular merits, and approach each person involved with an understanding of his or her specific agenda. In a healthy work atmosphere, where everyone is convinced that they "are all in this together," solutions become apparent, and difficulties dissipate over time. Patience is the key.

For example, consider the actual case of a computer and printer manufacturer who was changing his operation from an old-style in-circuit tester to a more modern version from a different vendor. The new equipment was faster, its signals were cleaner, and the line lengths from tester driver to board node were much shorter, increasing digital-test accuracy.

The programmers and engineers were dismayed by differences in programming techniques, languages, fault coverages, hardware configurations, and other personality traits. The manufacturer defused the problems by first convincing his people that the new machine would indeed detect more failures. He then sent them to the new vendor's training courses, where they became both more familiar and more comfortable with the new solutions.

The same problem can occur when changing vendors within a tester generation. In this case, both testers may provide equivalent fault coverage and test results, but programming and specific test approaches are quite different. One large computer manufacturer had to replace an entire installed base of one tester type when its vendor stopped making it. The chosen successor (from a different vendor) was regarded as at least as good a machine, yet a small revolt ensued because employees regarded its programming and operating environments as cryptic and hostile. Was the replacement machine actually more difficult to use than the installed variety? The answer does not really matter. The change was unavoidable, and the two testers performed approximately the same job. A change in the other direction would have been no less painful.

Moving experienced people to a completely new test system is very much more difficult than teaching inexperienced people completely from scratch. In many ways, the problem is analogous to teaching typing on a new keyboard layout. Many alternatives to the conventional "QWERTY" key arrangement exist. One version, known as the Dvorak keyboard, is reportedly easier to use and would save an average person typing 8 hours a day *miles* of finger travel. All current typing-speed recordholders use this alternative. Research has shown that a nontypist can learn the Dvorak keyboard in about 18 hours. For a skilled typist, the number is more than 40 hours because of the need to unlearn old habits. So, the "QWERTY" keyboard persists. In the tester case, the most efficient way to bring people up to speed is to treat them to some extent as brand-new users and train them accordingly.

It is important not to allow the challenge of changing vendors to prevent a company from doing so. The number of variables involved in strategic decision making is large enough without introducing a constraint that could be counterproductive.

One manufacturer has suggested that the most effective reaction to a change is to transfer key people completely to the new environment. This approach creates a dedicated cadre immersed in the new technology and reduces transition efforts. He noted that the only alternative would be to hire all new people, and resource limitations prohibit that solution.

4.11   Other Issues

All of the preceding discussion assumes considerable flexibility in establishing a test strategy or test facility. Analyzing or modifying an existing strategy requires many more compromises between the desirable and the possible. For example, few managers will sanction purchasing new equipment if capability and capacity are already available. Therefore, make-or-buy decisions and vendor evaluations would be irrelevant in that case.

Nevertheless, adding a product to an existing line still requires logistical decisions on product flow, equipment and personnel utilization, and efficient scheduling. A facility with high volumes and low mixes can make these decisions on a board-by-board basis. Managers in high-mix operations have to create groups of boards that share a test strategy and product flow, then generate plans and schedules around those groups.

At the same time, even a proverbial shoehorn cannot always force a particular board into an existing strategy. At times a strategic change is necessary, even if spare capacity exists. In these cases, justifying additional capital acquisition may depend on demonstrating the cost impact on current products, as well as the likelihood that future product developments can adopt the new solutions.

4.12   Summary

Regardless of selected test strategies and individual tactics, creating a cost-effective test operation involves numerous steps and decision points both before and after test-equipment delivery or in-house assembly. People must clearly understand test requirements, as well as the capabilities and experience of test-program and equipment vendors and manufacturers' own employees.

One debate centers around the necessity of testing products at all. Other approaches, such as visual inspection and process monitoring, can often serve the same purpose at lower cost. Some boards or parts of boards rarely fail, and therefore, may not need testing until after product assembly. Instrument-based manual testing and rack-and-stack instrument arrays can also replace expensive ATE.

Equipment and test programs can come from outside vendors or in-house developers. Available tools can provide user-friendly programming environments even for rack-and-stack or VXI-based solutions. These tools include a single high-level language for access to a range of similar instruments and measurement capabilities from a wide variety of vendors, as well as drag-and-drop menus, graphical test generation, and means to create drivers for new or uncommon test options.

Preparing a new ATE site or adding a product or product line to an existing facility requires carefully examining staffing levels, as well as the training and experience of current staff. Both training and recruiting can add to the operation's aggregate knowledge base and experience level.

If test tactics are compatible, manufacturers can apply excess capacity to new products, although this choice limits test flexibility. On the other hand, some new products contain technology that installed equipment and strategies cannot accommodate. In those cases, acquiring additional equipment is generally necessary, even if the capacity would otherwise be available.

Changing strategies or introducing unfamiliar strategies into an existing facility can involve a painful transition. Moving people completely over to the new project and immersing them in the equipment and technology can ease that pain.

CHAPTER **5**

Reducing Test-Generation Pain with Boundary Scan

Perhaps the most significant impediment to building a successful board-test strategy is the need for efficient, effective, and timely test programs. Unfortunately, constant increases in board complexity run at cross-purposes to meeting that need. Bed-of-nails techniques can alleviate some of the pain by reducing the logic depth that a particular test element must exercise to confirm a correct response to a particular input pattern, but today's boards often lack the necessary node access.

In addition, many boards serving military and other high-reliability applications require conformal coatings to prevent contamination during normal service. In those cases, beds-of-nails and guided-fault-isolation probes must pierce the coating, so manufacturers must recoat boards before shipment. At least one large-system manufacturer has been seeking a strippable coating process, permitting coating removal for bed-of-nails testing of boards that fail functional test. A technique that permits examining internal board states without physical contact would offer numerous advantages.

As with many test challenges, one class of solutions involves rethinking circuit *design*. That is, designers may organize device and board architectures so that internal signals can propagate unchanged to the board's edge connector or other convenient access point.

5.1 Latch-Scanning Arrangements

In his discussion of the evolution of testability design, Tsui (1987) describes access to component latches as a first step—setting them to specific values before testing, then observing them after test completion. The most critical constraint in this technique is the need to limit the number of I/O points dedicated to test, thereby minimizing device and board real-estate penalties and system performance degradation. To accomplish this, most approaches serialize data transfer before shifting data bits in and out of individual devices. Tsui characterizes all such solutions as latch-scanning arrangements (LSAs).

The serialization concept originated during system-level testing of IBM series 360 mainframe computers more than four decades ago. A test program shifts patterns into internal system registers ("shift registers'") before initiating a test (scan in), then shifts register contents to an observation point after test (scan out).

IBM introduced a formal version of its scan-path technique, known as *level-sensitive scan design (LSSD)*, in the late 1970s. Level-sensitive-design circuit responses are independent of timing and propagation delays. Moreover, if an output results from more than one input signal, it is independent of their order. Design considerations include internal logic storage through polarity-hold latches that can operate as shift registers (hence the name *shift-register latches* [*SRLs*]). When the output of one latch feeds the input of another, their clocks must be independent and nonoverlapping.

Practitioners of these methods enjoy a number of benefits. Test-generation complexity reduces almost to the level of a purely combinational circuit. Tsui notes that including more than one shift register shortens the test-sequence logic path further. In addition, system performance during test is independent of circuit parameters such as rise time, fall time, and minimum path delay. The only requirement is that the longest delay not exceed some specified value, such as the period between dynamic-memory refresh cycles.

Circuits designed in this way permit the examination of the states of all internal storage elements, in real time if necessary. This capability is a tremendous aid during both circuit design verification and test-program debugging. Modularity and the test circuit's relative insensitivity to timing variations reduce the pain of building working circuits from simulations and minimize the number of engineering changes. In addition, because scan-path testing isolates board components at least as effectively as a bed-of-nails does, engineers can adapt chip and module test programs to run at board level, system level, and in the field.

The technique's biggest disadvantage is the serialization process itself. Because every shifted bit occupies an entire clock cycle, scan testing can be extremely slow. A 32-bit bus, for example, requires 64 clock cycles just for each test vector's I/O, not even counting test time.

Tsui describes several scan techniques besides LSSD. *Scan/set logic* on an IC provides an output data path separate from the signal path during normal device operation. Incorporating this approach into edge-connector-based functional board testing allows the diagnosis of many complex failures down to the chip level, reduces test-generation efforts (and costs), and facilitates guided-fault isolation using standard probes or clips.

With a *random-access scan*, every internal storage element is always available for control and observation. The method utilizes existing outputs in a selectable random serial sequence in the time domain. Separate serial input and output pins are unnecessary.

Shift test-control logic is somewhat similar to scan-set, except that it works primarily at the device level. The periphery of each chip includes a shift-register chain that optionally connects a latch for each signal's I/O pad. Additional test-only circuitry adds seven extra pads that require on-silicon probe contact to permit testing chips while they still reside on wafers, sorting them for speed and other quality parameters.

Each of these techniques takes advantage of the fact that most devices already contain latch-type sequential elements. Tsui considers LSAs the foundation of all device-based testability designs.

At the board and system levels, LSAs can furnish sufficient circuit-segment visibility to permit functional testing with in-circuit-like fault isolation. Data buffering minimizes the test impact of circuit timing and synchronization. Between test steps, the board or system halts in a logic state defined by the latch contents.

In practice, systems containing more than one fault will defeat an LSA-only testability design. Multiple faults can propagate incorrect signals throughout a system, causing effects unlike anything that a good circuit or a circuit with any specific single fault will produce. Isolating actual failing devices in this case may be difficult or impossible.

Most of these techniques are proprietary to one manufacturer (or at most, a few). Small companies generally lack the resources to develop their own scan-test approaches and implement them at all design-and-test levels. The plethora of available options and the relative dearth of useful tools to help designers apply them can make adoption prohibitively expensive. Even the most popular of these early methods, LSSD, had few adherents aside from its chief proponent, IBM.

The existence of multiple scanning schemes also makes designing end products more difficult. Components come from a variety of vendors and may contain different types of scan circuitry. Many boards contain ASICs and other complex devices. A *standard* scanning approach permits board and system designers to work together with all vendors to get the best functional performance from their products without sacrificing the ability to generate reliable, comprehensive, cost-effective test programs. The best option would combine design-for-testability at component, board, and system levels, so that higher-level testing can take advantage of component test structures, regardless of device manufacturer.

5.2 Enter Boundary Scan

An initiative to cope with these issues within Philips in Europe led to the creation of the so-called *Joint Test-Action Group (JTAG)* in 1985. The group consisted of representatives from makers and users of components and boards in Europe and North America, who recognized that only a cooperative effort could address the mounting testability problems in a coordinated way. Its mandate was to propose design structures that semiconductor makers would incorporate into device designs to aid in testing boards and systems, then to encourage their proliferation throughout the electronics manufacturing industry.

The proposed approach had to complement and supplement existing test methods, rather than supplant them. It had to work with in-circuit, functional, built-in, and other test techniques. Most important, it had to allow maximum use of component-level test patterns at board and system levels with a minimum of modification.

The JTAG's original goal was to produce a standard for digital, analog, and mixed-signal circuit designs. The group published its first attempt in 1986, calling

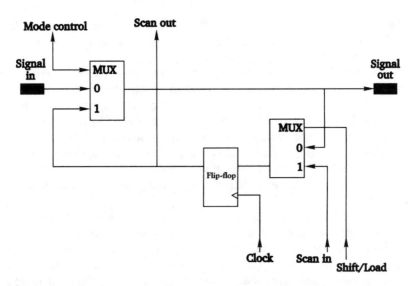

Figure 5-1   A sample boundary-scan cell implementation for an IC input. (IEEE Std 1149.1–1990, *IEEE Standard Test Methods and Boundary-Scan Architecture*, Institute of Electrical and Electronics Engineers, Inc., Piscataway, New Jersey.)

it P1149. In 1990, after two revisions, the IEEE adopted the portion of the proposal relating to IC-based implementation as IEEE Standard 1149.1. Its stated purpose was to test interconnections between ICs installed on boards, modules, hybrids, and other substrates. Manufacturers adopting the standard could also test the IC itself and observe its behavior during normal circuit operation.

Other portions of the original proposal included the extended serial digital subset (1149.2), which defines particular scan-test implementations such as LSSD and scan-path, and real-time digital subset (1149.3). Real-time analog subset standard IEEE 1149.4 addresses mixed-signal boards. IEEE 1149.5 serves primarily for testing multiple-board systems and backplanes. All of these variations share the same basic principles as 1149.1. Since the earlier standard remains by far the most common, and the purpose here is to introduce the subject and not to provide a substitute for comprehensive sources and courses, this discussion will focus on IEEE 1149.1.

The crux of the IEEE 1149.1 proposal is a standard testability bus that implements the *boundary scan* technique. Designers of conforming devices must include a shift-register latch within a boundary-scan cell adjacent to each I/O pin, permitting serialization of data into and out of the device and allowing a tester or other engine to control and observe device behavior using scan-test principles.

Figure 5-1 shows a sample boundary-scan cell implementation for an IC input. Mode-control signals applied to the multiplexers determine whether to load data from the normal device input ("signal in") into the scan register or from the register through the scan-cell output into the device logic. Scan cells for all device

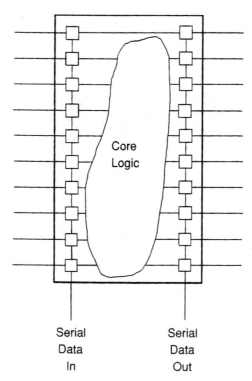

Figure 5-2   Scan cells for all device pins connect to form a chain around the core logic. Hence, the name "boundary scan."

pins connect to form a chain around the core logic, as Figure 5-2 shows. Hence, the name "boundary scan."

Connecting boundary-scannable components in series on a board, as in Figure 5-3, produces a single path through the design for test purposes. Some designs may feature several independent boundary-scan paths, whereas others may include boundary-scan devices as well as sections that do not conform to the standard. Notice that in the figure the board's TMS and TCK signals are connected in parallel to each conforming device.

If all of a board's components include boundary-scan circuitry, the resulting configuration allows testing the devices themselves and the interconnections between them. Proper boundary-scan design can even permit a limited slow-speed test of the entire circuit. Loading a boundary-scan device's inputs and outputs in parallel and serially shifting out the results allows sampling device data without interfering with its operation. The 1149.1 standard suggests that such a sampling test allows debugging designs and isolating unusual faults more easily.

Boundary scan offers numerous advantages. Because a standard exists, both the benefits and burdens are predictable. Like all scan approaches, it provides access to internal logic nodes on devices and boards. Fault diagnostics are node-to-node,

Boundary Scan Board

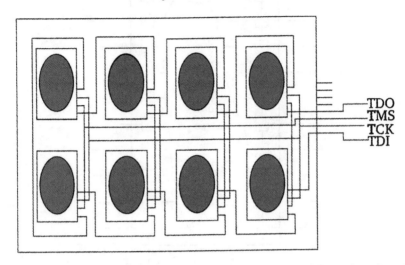

Figure 5-3 Connecting boundary-scannable components in series on a board produces a single path through the design for test purposes. Signals: TDI, test data in; TDO, test data out; TCK, test clock; TMS, test mode select.

rather than nail-to-nail. Most important, it reduces circuit-simulation and test-generation efforts and can minimize a product's time to market.

Disadvantages include possible performance penalties caused by extra circuitry immediately prior to device output buffers. In Figure 5-4, the multiplexer between the system pin and the interior system logic could add two gate delays. Input loading of the boundary-scan register causes additional delays. These delays may degrade device activity to the point where performance falls below specification targets. Careful design, however, such as combining boundary-scan cells with device input buffers, can reduce the adverse impact.

Extra circuitry and the need for additional signal lines for testing can also increase board-level design and construction costs. Of course, all boundary-scan accommodation requires allocating precious real estate that designers surrender reluctantly, if at all.

In addition, possible faults with the scan logic itself can reduce product yields and increase test, inventory, and other costs. Test programs must verify boundary-scan performance before employing it to examine other logic.

For the standard to work during board test requires that device manufacturers adopt it first. Unfortunately, because of development-resource constraints, costs, and possible performance problems, few device manufacturers have introduced boundary-scan versions of their standard products. Also, with individual logic devices on most boards yielding to large-scale parts, these "jellybeans" no

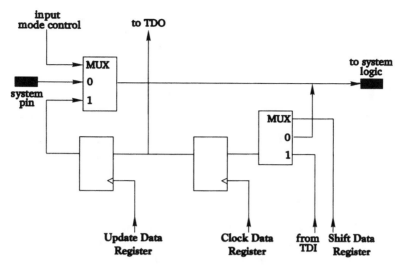

Figure 5-4 In this sample circuit, the multiplexer between the system pin and the interior logic could add two gate delays to device timing. Input loading of the boundary-scan register causes additional delays. (IEEE Std 1149.1–1990, *IEEE Standard Test Methods and Boundary-Scan Architecture*, Institute of Electrical and Electronics Engineers, Inc., Piscataway, New Jersey.)

longer represent the optimum choice for board design. Their popularity will continue to wane, making major commitments from device manufacturers to upgrade these products unlikely, at best. Therefore, few standard parts (other than massively complex devices of the microprocessor class and above) will ever offer the feature. They could not provide sufficient return on the design investment.

IEEE 1149.1 contains no analog capability. Although this advance is addressed in other current and proposed parts of the standard, few manufacturers at any product level have embraced these alternatives. Therefore, there generally is no way to completely test most boards through boundary-scan alone.

One place where boundary scan is becoming increasingly common is on ASICs and other custom parts. ASIC vendors will often incorporate boundary-scan circuitry into any of their parts on request from customers for a premium on each device produced. Because many board designs contain little of significance aside from microprocessors, ASICs, flip-chips, and so on, the technique can prove quite useful despite the dearth of conforming standard parts.

In many respects, boundary scan for device and board testing is much like high-definition television (HDTV) in consumer markets. No one can deny that it is better than what we already have, but until people recognize that it is *a lot* better, and unless people are willing to pay for it in money, time, training, and other factors, few are going to buy it. As long as not enough people buy it, it will remain more expensive and less well understood than necessary.

Achieving 100 percent boundary scan with a design containing primarily non-ASIC devices will not occur in the foreseeable future. Once again, test professionals can rest assured that their jobs are secure.

5.3 Hardware Requirements

To conform to the boundary-scan standard IEEE 1149.1, a device design must contain the following:

A test-access port (TAP)
A TAP controller
A scannable instruction register
Scannable test-data registers

In addition, the standard mandates that instruction and test-data registers be parallel shift-register-based paths, connected, as in Figure 5-2, to a common serial-data input and a serial-data output, and ultimately to the associated TAP pins. The TAP controller selects between instruction and test-data-register paths. Figure 5-5 shows a typical implementation.

According to the standard, the TAP *must* contain the following four signals, each available through a dedicated device pin:

Test-data in (TDI): Test instructions shift into the device through this pin.

Test-data out (TDO): This pin provides data from the boundary-scan register or other registers.

Test clock (TCK): This input controls test-logic timing independent of clocks that normal system operations employ. As a result, test data and "ordinary" data can coexist in the device. The TDI shifts values into the appropriate register on the rising edge of TCK. Selected register contents shift out onto TDO during the TCK's falling edge.

Test-mode select (TMS): This input, which also clocks through on the rising edge of TCK, determines the state of the TAP controller.

An optional, active-low test-reset pin (TRST*) permits asynchronous TAP-controller initialization without affecting other device or system logic. Asserting this pin inactivates the boundary-scan register and places the device in normal operating mode. Theoretically, in this mode, the circuit operates as though the test circuitry were not there.

The TMS and TCK inputs program the TAP controller as a 16-state machine, generating clock and control signals for the instruction and data registers. Only three events can trigger a change of controller state: a test-clock rising edge, assertion of a logic 0 onto TRST* (if it exists), and system power-up.

The *instruction register* receives an instruction through the TDI, decodes it, and selects the appropriate data register depending on the state of the TAP controller. An instruction register contains at least two cells, each of which includes a shift-register flip-flop and a parallel output latch. Instruction-code width must match register width. Instructions passing through this register reside on the

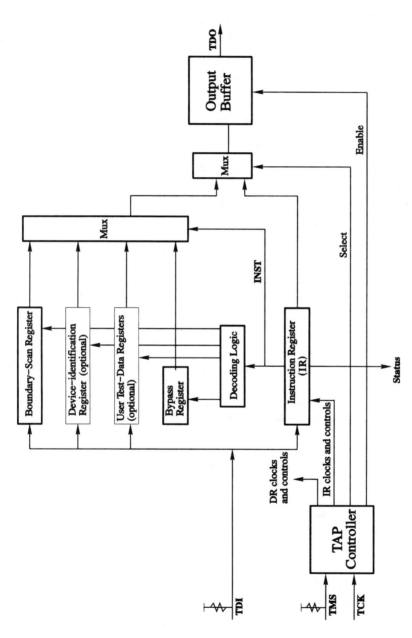

Figure 5-5 Typical IEEE 1149.1 block diagram. (IEEE Std 1149.1–1990, *IEEE Standard Test Methods and Boundary-Scan Architecture*, Institute of Electrical and Electronics Engineers, Inc., Piscataway, New Jersey.)

flip-flops. Output latches hold the current instruction. The standard defines a number of mandatory and optional instructions.

Data registers include the *boundary-scan register* and the *BYPASS register*, which are mandatory. The boundary-scan register includes one cell for each I/O pin on the device, permitting observation and control of device activity. The BYPASS register, consisting of a single cell, reduces the scan chain to one cell during test of other devices on a board or module.

Optionally, a device may also contain a 32-bit *device-identification register*, which can store the device type, manufacturer, lot number, date code, and other information. This register permits an "Is it there, and is it the correct device?" type of device-by-device test. Programmable devices such as EPROMs can be checked for the correct program and revision level. One or more optional *user test-data registers* can execute manufacturer-specified self-tests and perform other functions.

The TAP controller, instruction register, and associated circuitry must remain independent of system function. That is, they should operate in device or board test mode only. Normal system operation may share test-data registers unless specific tests require independence.

Boundary-scan proponents estimate that a TAP controller, 3-bit instruction register, BYPASS register, and 40-bit boundary-scan register will occupy only about 1 mm^2 of device surface, or about 3 percent of a 36-mm^2 device. Integrating data-bit cells with the circuit's input or output buffer or placing the TAP and boundary-scan registers in the "dead" area around a circuit's periphery can reduce even this level of overhead.

IEEE 1149.1 defines the following TAP-controller states:

Test-Logic Reset. Holding TMS high for five consecutive TCK rising edges executes this mode, which causes the device to ignore the test circuitry altogether and allows the system to function normally. In this state, the instruction register holds the BYPASS instruction or the IDCODE instruction if the device-identification register exists. Applying a logic-low to the TRST* input, if there is one, will also place the TAP controller in this state.

Run-Test/Idle. If the instruction register contains certain bit patterns, this state selects the test logic and executes the appropriate instruction. Otherwise, all active test-data registers retain their previous state. As an example, the standard suggests that a RUNBIST instruction or user-designated alternative can trigger a device self-test when the controller enters this state. While the controller is in this state, the instruction cannot change.

The remaining states are divided into two groups. One controls the instruction register, and the other accesses the data registers. *SELECT-IR-SCAN* decides whether to operate on the instruction register or return to test-logic reset. *SELECT-DR-SCAN* determines whether to operate on one or more data registers or to SELECT-IR-SCAN.

CAPTURE-IR and *CAPTURE-DR* states initiate parallel loading of the instruction register or the data registers, respectively. CAPTURE-IR assigns the two least-significant bits to the value 01. *SHIFT-IR* and *SHIFT-DR* send serialized information from the appropriate register out through TDO, simultaneously shifting the same number of data bits in from TDI. *UPDATE-IR* and *UPDATE-DR* take instructions or data that a previous state has shifted in and latch them onto the appropriate register's parallel output.

PAUSE-IR and PAUSE-DR temporarily halt shifting of data into the register from TDI or out to TDO. These states might allow reloading ATE memory, for example, before continuing with a boundary-scan test.

Temporary states *EXIT1-IR* and *EXIT1-DR* combine with the state of TMS. A HIGH forces the controller into the corresponding UPDATE state, which terminates scanning. A LOW puts the controller into the PAUSE state. Similarly, *EXIT2-IR* and *EXIT2-DR* terminate scanning and enter an UPDATE state in response to a TMS HIGH. LOW returns the controller to a SHIFT state.

5.4 Modes and Instructions

All boundary-scannable devices can operate in either *normal* mode or *test* mode. As the name implies, normal operation means that, with respect to other devices and the system, the device behaves as though the test circuitry did not exist. Certain independent test instructions, however, can execute. The *SAMPLE* function of an instruction called *SAMPLE/PRELOAD* taps the data that are present on the device at a particular time, then shifts them out through TDO for verification or examination. The *PRELOAD* function determines an output-register data-bit pattern in anticipation of a subsequent scan operation. *BYPASS* allows examining data from the device inputs unchanged, even if the device is operating. Optional *IDCODE* and *USERCODE* instructions shift information out from the device-identification register.

In *test mode*, boundary-scan logic can execute a number of instructions and functions in addition to those available during normal mode, and manufacturers can opt to emulate or extend them. The mandatory *external test* (*EXTEST*) allows examining interconnections between boundary-scan cells and some other access point, such as an edge-connector pin, fixture nail, or another boundary-scan device, thus verifying that the boundary-scan register works and that the device connects properly to the board. Loading a logic 0 into all instruction-register cells initiates an EXTEST.

EXTEST also permits checking simple interconnections between two boundary-scan devices, simulating a conventional shorts-and-opens test. Data shifts into the first device through TDI and the BYPASS register, loading up the cells that drive device output pins. Similarly, the next device's input pins can load the associated boundary-scan cells, which then shift information out through TDO for examination. One advantage to this approach is the ability to know immediately which of a group of input pins is not properly connected to the board—without having to analyze an output pattern that device logic has modified. This alternative to the opens-testing techniques discussed in Chapter 2 enjoys the advantage of not requiring bed-of-nails access, an increasingly necessary consideration. Like all boundary-scans, however, its serial nature introduces a test-time penalty.

Because EXTEST connects all internal device logic to the boundary-scan register, testing can proceed without regard to the scannable device's actual function. The instruction steps the TAP Controller through the CAPTURE, SHIFT, and UPDATE functions. CAPTURE deposits input signals onto the boundary-scan cells, SHIFT serially shifts the values into the output cells or out through TDO, and UPDATE sends output-cell values through device output pins.

Internal test (*INTEST*) is an optional instruction that permits examining the behavior of a boundary-scan device that resides on a board without regard to board topology or surrounding circuitry. Signals SHIFT into input cells through TDI, and UPDATE applies them to the device core logic. CAPTURE places responses on the output cells for SHIFT out through TDO. Because it delivers test data to the component and examines test results, INTEST can also allow a limited, static test of all board or system logic. This time, SHIFT and UPDATE apply the input pattern to the first device in the test path, and CAPTURE and SHIFT permit examining the overall output through TDO of the last device in the path.

INTEST offers advantages and disadvantages in comparison with in-circuit alternatives. The test can include any non-speed-dependent steps from device-level development. In-circuit implementations must consider how the device is wired to the board, such as by inputs tied to V_{CC} or ground. INTEST allows ignoring those constraints. Only TDI and TDO require nails or other access, reducing fixturing costs compared to a full bed-of-nails, along with the performance penalties that beds-of-nails can cause. Also, design engineers will more likely agree to add a few test points to the board to permit INTEST than add an entire set to test the device in-circuit.

Like all scanning techniques, however, INTEST shifts vectors serially into the device under test, so the test process can be many times longer than an in-circuit-test's parallel methods. Some manufacturers have expressed concern that such long test times may cause device damage from excess overdriving. Keeping a device operating without "forgetting" data or logic states may also be difficult.

Optional *RUNBIST* executes a device-resident self-test. Like INTEST, the self-test proceeds regardless of surrounding board circuitry. In this case, a binary response to self-test execution (go/no-go) alleviates the test-time penalties caused by input and output signal serialization. The 1149.1 standard permits simultaneous self-test executions from multiple devices. A device manufacturer may specify a sequence of self-tests for execution in response to the RUNBIST instruction, as well as other self-tests that do not execute in boundary-scan modes.

The standard also permits user-defined commands. Two of these have been incorporated as additional optional instructions into what is referred to as 1149.1a (Andrews [1993]).

Consider a device that is not actively involved in the current round of tests but that must provide a particular logic pattern to allow those tests to proceed. The test program shifts the appropriate pattern into the boundary-scan register, then transfers it to the UPDATE register. The *CLAMP* instruction holds it there and forces its value onto system logic, providing the BYPASS register as the path between TDI and TDO. CLAMP allows testing to proceed without repeatedly loading the same test vectors onto the uninvolved device, speeding operations considerably.

The other instruction, *HIGHZ*, places all of an IC's system outputs into a high-impedance state, keeping them in that state until the test program loads another instruction or resets the test logic. This instruction provides a convenient way to isolate a component from a boundary-scan test of another device or group of devices, or for isolating parts of a board's boundary-scan logic from a bed-of-nails test.

Some proponents of the technique advocate using boundary-scan circuitry to program PALs, FLASH memories, and similar parts on the assembled board. Current practice involves pulling blank devices from stock, programming them, then returning them to stock, with each program or program revision requiring a separate stock bin. The programmed devices would be pulled from inventory during board assembly.

Alternately, you could build blank devices onto boards, *then* burn the programs. This version requires only one inventory item—for the blank parts—drastically reducing the logistics of inventory tracking. It allows programming changes to the last possible moment, and reduces the likelihood that a device will contain the wrong program or revision. Of course, the devices must include the boundary-scan circuitry, and the throughput requirements of the manufacturing operation must be able to accommodate the additional test time.

5.5 Implementing Boundary Scan

IEEE Standard 1149.1 establishes a target. Successful implementation requires cooperation between board and device manufacturers. As designers have become increasingly comfortable with the technology, its advantages, and its constraints, it has become more common. Adding it to ASICs and other custom logic has become routine in some cases, while other manufacturers ignore it altogether. The automobile industry, for example, goes to extraordinary lengths to avoid raising per board manufacturing costs. For them, boundary-scan would represent an extremely unlikely ally.

For the foreseeable future, systems will include both boundary-scan and non-boundary-scan parts. Logic segmentation to take advantage of whatever boundary scan resides on a board will simplify test-program generation and execution compared to a board containing no boundary scan, but it will not offer as much improvement as would a complete boundary-scan approach.

If all analog and other non-boundary-scan devices are accessible through a bed-of-nails, test generation for those devices reduces to standard in-circuit tests or, at worst, to cluster approaches. The only remaining issues are test generation for boundary-scan devices and interconnects.

Within the boundary-scan portion of the board, a shorts-and-opens test consists of applying test vectors to boundary-scan nodes at device outputs at one end, receiving them at device inputs at the other end, and shifting them out through TDO for analysis. According to Jarwala and Yau (1989), a *parallel test vector*

(PTV) is one set of values applied to a group of independent boundary-scan networks simultaneously. The captured values constitute a *parallel response vector (PRV)*. Detecting all possible faults requires a number of PTVs. The pattern applied over time at a single node is called the *sequential test vector (STV)* for that node. Therefore, the *sequential response vector (SRV)* is the set of output vectors from one node over time.

The researchers describe four fault classes for boundary-scan interconnect networks:

1. *Single-net faults* include stuck-at-1 and stuck-at-0, which affect an entire network, and opens, which may affect only a single node.
2. *OR-type multiple-network shorts* result from network drivers shorted so that a 1 on any driver produces a 1 on the entire network.
3. *In AND-type multiple-network shorts*, a 0 on any network driver forces all other network drivers to that level as well.
4. *In strong-driver multiple-network shorts*, the network's value exactly follows the value of the dominant member.

STVs and SRVs represent identifiers for each end of the network under test. Thorough network testing requires choosing a comprehensive set of unique identifiers for each node. All-1 and all-0 cases provide no new fault information and are therefore unnecessary.

On a fault-free board, no two SRVs will be the same. Shorted nets produce identical responses. Ideally, analyzing those responses will pinpoint the corresponding failure. Unfortunately, networks producing the same SRV are not necessarily shorted together, because of two syndromes known as *aliasing* and *confounding*.

For example, Figure 5-6 shows a set of PTVs and the associated STVs for diagnosing a circuit containing 10 networks. If network n_3 is shorted to network n_4 with an OR-type short, both will produce SRVs of 0111. Because network n_7 produces this result in a good circuit, this test cannot determine whether n_7 is shorted to n_3 and n_4. This phenomenon is called *aliasing*.

Confounding occurs when two independent faults exhibit the same output signature. An OR-short between n_4 and n_{10} and an OR-short between n_6 and n_8 both produce an output of 1110. These vectors cannot determine whether the two faults exist independently or all four nodes are shorted together. The vector set in Figure 5-6 can identify shorted networks only when test results exhibit neither syndrome.

Other test sequences can correctly diagnose faults in a wider variety of circumstances. Jarwala and Yau offer the set in Figure 5-7, called a *true/complement* test sequence, to handle aliasing. Every bit in the second vector group complements the corresponding bit in the first group. Shorted networks n_3 and n_4 still give the same true response as n_7, but their complement vectors produce responses of 1111 and 1000, respectively.

Even this set, however, does not help in the confounding case. Both shorted-network pairs respond to the complemented vectors with 1111.

Nets	Parallel Test Vectors				Sequential Test Vectors
	v_1^T	v_2^T	v_3^T	v_4^T	
n_1	0	0	0	1	v_1
n_2	0	0	1	0	v_2
n_3	0	0	1	1	v_3
n_4	0	1	0	0	v_4
n_5	0	1	0	1	v_5
n_6	0	1	1	0	v_6
n_7	0	1	1	1	v_7
n_8	1	0	0	0	v_8
n_9	1	0	0	1	v_9
n_{10}	1	0	1	0	v_{10}

Figure 5-6 A set of parallel test vectors (PTVs) and the associated serial test vectors (STVs) for diagnosing a circuit containing 10 networks. This set is susceptible to *aliasing* and *confounding*. (Jarwala, Najmi, and Chi Yau. 1989. "A New Framework for Analyzing Test Generation and Diagnosis Algorithms for Wiring Interconnects," International Test Conference, Institute of Electrical and Electronics Engineers, Inc., Piscataway, New Jersey.)

Nets	True Vectors				Complement Vectors			
n_1	0	0	0	1	1	1	1	0
n_2	0	0	1	0	1	1	0	1
n_3	0	0	1	1	1	1	0	0
n_4	0	1	0	0	1	0	1	1
n_5	0	1	0	1	1	0	1	0
n_6	0	1	1	0	1	0	0	1
n_7	0	1	1	1	1	0	0	0
n_8	1	0	0	0	0	1	1	1
n_9	1	0	0	1	0	1	1	0
n_{10}	1	0	1	0	0	1	0	1

Figure 5-7 Applying this vector set, called a *true/complement* test sequence, to the 10-network circuit in Figure 5-6 will handle aliasing but not confounding. (Jarwala, Najmi, and Chi Yau. 1989. "A New Framework for Analyzing Test Generation and Diagnosis Algorithms for Wiring Interconnects," International Test Conference, Institute of Electrical and Electronics Engineers, Inc., Piscataway, New Jersey.)

The researchers note that selecting identifiers such that only one node has a short's dominating value at any time would eliminate much ambiguity. Unfortunately, such identifiers are generally too long for routine use. To reduce test times, a set of precalculated identifiers finds as many faults as possible, then additional, more complex vectors diagnose those situations that the precalculated patterns cannot.

5.6   Partial-Boundary-Scan Testing

Testing interconnects between boundary-scan and non-boundary-scan components requires more innovative approaches. According to Robinson and Deshayes (1990), the biggest impediment to creating such tests is the fact that boundary-scan logic must be under power to function properly. Therefore, conventional logic responds to changes on boundary-scan nodes, such as nodes 1 and 3 in Figure 5-8. In addition, free-running clocks, initialization problems, and other circuit elements can complicate test generation.

A reliable test-and-diagnostic program must assume the worst conventional-logic configuration—that is, that boundary-scan circuitry cannot initialize or control conventional nodes. Shorts between conventional and boundary-scan devices will produce unpredictable and unrepeatable results. The researchers also note several other factors that complicate the partial-scan board-test problem. Boundary-scan components may reside in a single chain or in multiple chains. Multiple chains may share boundary-scan pins at the board edge, or each chain may offer independent access.

Some nodes respond to a combination of boundary-scan and non-boundary-scan activity. Nodes such as 3 in Figure 5-8 have a boundary-scan output but only conventional input, whereas for nodes such as 6, the situation is

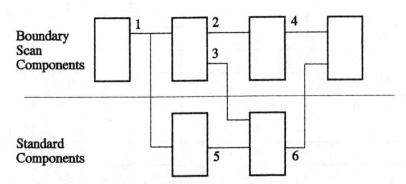

Figure 5-8   Most boards contain some components that conform to IEEE 1149.1 and others that do not. Because boundary-scan nodes must be under power to function properly, conventional nodes, such as 1 and 3, react to logic changes on boundary-scan circuitry. (Robinson, Gordon, and John Deshayes. 1990. "Interconnect Testing of Boards with Partial Boundary Scan," International Test Conference, Institute of Electrical and Electronics Engineers, Inc., Piscataway, New Jersey.)

reversed. Any non-boundary-scan node that does not offer tester access through a bed-of-nails will likely confuse test results.

Although Jarwala and Yau's methods work on boundary-scannable circuitry, they cannot detect shorts between boundary-scan nodes and conventional nodes. In addition to complicating fault diagnosis, such mixed environments require more manual test-pattern generation, increasing time and cost burdens for that step in a test strategy.

All shorted networks exhibit a dominant and a nondominant bit value. Robinson and Deshayes note that a boundary-scan node at the nondominant level shorted to a conventional node will reside at the wrong logic level while the conventional node is at the dominant level. Boundary-scan techniques cannot adequately control or initialize conventional logic, so many test results are unrepeatable.

The researchers propose the following goals for an interconnect test on a board containing both boundary-scan and non-boundary-scan logic:

It should encourage automatic program generation using available information.

The test should successfully identify shorts between nodes with either boundary-scan or bed-of-nails access.

It should identify the specific open driver on nodes attached to several drivers. Shorts to inaccessible nodes should only rarely give incorrect results.

The test should not unduly stress or damage the board.

To accomplish these goals, Robinson and Deshayes suggest a four-step test approach.

5.6.1 Conventional Shorts Test

A *conventional shorts test* identifies problems with all nodes that permit physical tester access. This test comes first because it is safe and accurate and detects faults that cause damage to the board if they are not corrected before power-up. In addition, when the tester can access all of a device's boundary-scan pins, this simple test allows the correction of many problems that would otherwise confuse later steps.

5.6.2 Boundary-Scan Integrity Test

A *boundary-scan integrity test* verifies that the boundary-scan circuitry itself on each device is functioning properly and that no short or open exists in the scan path, allowing its use in subsequent testing. For example, as mentioned, the standard requires 01 as the two least-significant bits of a CAPTURE-IR instruction. If the scan path contains a broken connection, those bits would produce 11, whereas a short to ground would generate 00.

One simple test loads the BYPASS instruction into each boundary-scan device, then observes the captured instruction-register value at as many points as possible. This test must also verify that the scan path can be disabled when appropriate. Shifting a pattern containing all necessary transitions, such as 11001, through all BYPASS registers ensures that no scan-path connections are either broken or shorted.

Other possible tests include using SAMPLE to examine the boundary-scan-register path. If boundary-scan devices incorporate IDCODE and USERCODE, these instructions can check that the assembly process has loaded the correct components onto the board. This test also allows monitoring the performance of supposedly identical parts from different lots, manufacturers, or date codes. Cooperation between vendors and customers to resolve problems identified at this step can improve overall product yields and reduce system manufacturing costs.

IDCODE also permits one test program to work with several versions of programmable devices. The test reads the code, loads the appropriate test program for the actual installed device, and continues.

5.6.3 Interactions Tests

Interactions tests look for shorts between boundary-scan nodes and conventional nodes with bed-of-nails or edge-connector access, as well as opens between tester nails and boundary-scan input pins. The EXTEST instruction latches boundary-scan nodes to the state that permits easier in-circuit backdrive (a logic HIGH for TTL). The tester then looks for node movement when it forces non-boundary-scan nodes to their opposite states.

Applying this technique to a single conventional node places a HIGH on that node and scans out boundary-scan-input states, then injects a LOW onto the test node and scans again. A short between the test node and a boundary-scan node will show up as a failure. An open connection will cause both scanning operations to produce exactly the same output pattern.

Some shorts other than those between test and boundary-scan nodes can cause this operation to fail. Only rarely, however, will such a faulty node follow the test node at both the latched-HIGH and latched-LOW states. Changing test-node states several times and declaring a short only when the suspect boundary-scan node exactly follows these transitions further improves the likelihood of a correct diagnosis.

Engineers can reduce test execution times by applying unique sequential patterns to several nodes at once. Each pattern should contain several HIGH-to-LOW and LOW-to-HIGH transitions. In Figure 5-9, output *J* is shorted to node *D*, which is the input to IC U3. Therefore, node *D* will follow the pattern on node *E*. Because of the intervening buffers, this test will indicate that nodes *D* and *E* are shorted, rather than finding the actual short between nodes *D* and *J*. Subsequent tests that stimulate nodes *H* and *J* will further clarify these results, ultimately reporting that node *D* is shorted to one or more of nodes *E*, *H*, and *J*.

In some cases, failure diagnoses from test results are less specific. If sensing a driven node produces a pattern different from the test pattern, the node itself is faulty. Perhaps there is a short between the test node and several other nodes, so

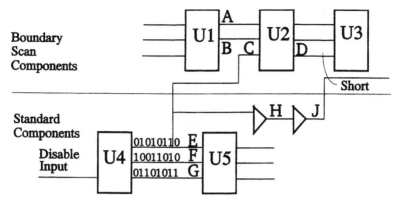

Figure 5-9   Output *J* is shorted to node *D*, which is the input to IC U3. Therefore, node *D* will follow the pattern on node *E*. (Robinson, Gordon, and John Deshayes. 1990. "Interconnect Testing of Boards with Partial Boundary Scan," International Test Conference, Institute of Electrical and Electronics Engineers, Inc., Piscataway, New Jersey.)

that the in-circuit tester's backdrive capability cannot overcome the total current capacity of the shorted network. If the boundary-scan-node response is different from any of the applied test vectors, that node is definitely *not* shorted to one of the test nodes. However, a response vector that contains unexpected values indicates some kind of fault and must therefore be included in the failure report.

5.6.4   Interconnect Test

The final step is an *interconnect* test on boundary-scan-accessible portions of the board logic, including pure boundary-scan nodes and such other nodes as edge-connector pins. The approach uses standard in-circuit techniques to disable non-boundary-scannable board logic, then proceeds as though the board contained only boundary-scan parts. Robinson and Deshayes first subject the nodes under test to a set of precalculated patterns that identify node candidates where a short may be present. Based on these results, a set of adaptively generated patterns walks dominant logic values through all suspect groups. Test results showing different response vectors at different points along the same node indicate either an open circuit or a node with an ambiguous voltage.

Some boards may contain nodes that are neither boundary-scannable nor accessible through a bed-of-nails fixture. A short between such a node and a boundary-scan node can cause indeterminate results. Analysis of these situations requires considerable care. A fault dictionary can help, but it must be kept small by including only those situations with inaccessible nodes. The interactions test will already have identified shorts between tester-accessible nodes and boundary-scan nodes, so the fault dictionary can safely exclude them. Otherwise, the dictionary grows too large for practical fault diagnosis.

Each test step provides evidence of the following:

A node has a problem. Subsequent steps may clarify this conclusion but will never refute it.

A set of nodes may be shorted together. Additional evidence may establish that some of the nodes are not shorted, as in the aliasing and confounding examples. The test may or may not establish why a node problem exists, it may remove a node from suspicion (as in aliasing), or it may break a fault into subset faults.

A node may have one or more pins open.

A node driver may never be active, such as with an open on a driver pin.

A driver may always be active.

Analysis of the evidence proceeds only after completion of all test steps. Combining information from all four stages permits accurately diagnosing interconnect failures. Because this technique is so methodical, program generation is significantly simpler than with other approaches, lowering overall test costs for designs containing at least some boundary-scan circuitry. These cost reductions encourage boundary-scan use.

5.7 Other Alternatives

Arment and Coombe (1989) offer several other circuit configurations to deal with the problem of partial boundary-scan and no universal bed-of-nails access. The bidirectional buffer between scannable devices and the nonscannable microprocessor in Figure 5-10 allows the use of EXTEST on Bus 1. Reading data propagated directly through the system or SAMPLE-ing data through the buffer's TAP will verify Bus 2.

Figure 5-11 eliminates the extra propagation delay that the buffer in Figure 5-10 contributes to the system. This version reverses the microprocessor and buffer positions. The buffer tristates and halts the processor and then performs an EXTEST on the interconnects. These researchers suggest the SAMPLE mode to verify the behavior of the microprocessor's output register.

When a circuit contains conventional logic surrounded by boundary-scan devices, as in Figure 5-12, the EXTEST mode of the border devices performs either an internal or a sampling test on the nonscannable logic. In this case, the output from one boundary-scan device serves as input to the logic under test, and the input to the next boundary-scan device provides access to the test output. Shifting alternating 1s and 0s into the input drive register provides a system clock.

Critical to the success of board test with partial boundary scan is tester speed because of the technique's serial character. The number of tester I/O pins must accommodate the boundary-scan logic (at least four pins) and bed-of-nails or edge-connector access points for non-boundary-scan logic. Nevertheless, this approach reduces tester and fixturing costs by avoiding the high pin counts that more conventional test solutions demand. Some testers are emerging that specifically address boundary-scan applications.

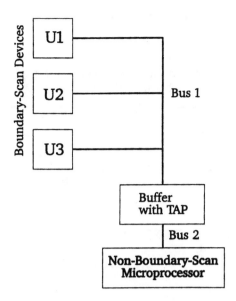

Figure 5-10   The bidirectional buffer between scannable devices and the nonscannable microprocessor allows using EXTEST on Bus 1. Verifying Bus 2 involves reading data propagated directly through the system or SAMPLE-ing data through the buffer's TAP. (Arment, Elmer L., and William D. Coombe. 1989. "Application of JTAG for Digital and Analog SMT," ATE & Instrumentation Conference, Miller-Freeman Trade-Show Division, Dallas, Texas.)

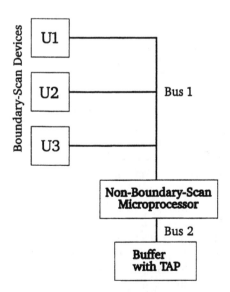

Figure 5-11   This version of the circuit in Figure 5-10 eliminates the buffer's extra propagation delay by switching its position with that of the microprocessor. (Arment, Elmer L., and William D. Coombe. 1989. "Application of JTAG for Digital and Analog SMT," ATE & Instrumentation Conference, Miller-Freeman Trade-Show Division, Dallas, Texas.)

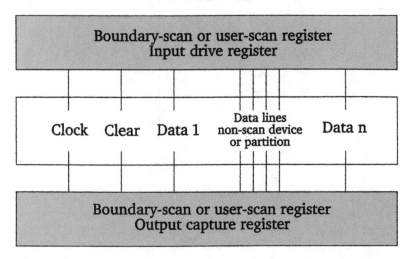

Figure 5-12 When a circuit contains conventional logic surrounded by boundary-scan devices, the EXTEST mode of the border devices can perform either an internal or sampling test on the nonscannable logic. Shifting alternating 1s and 0s into the input drive register provides a system clock. (Arment, Elmer L., and William D. Coombe. 1989. "Application of JTAG for Digital and Analog SMT," ATE & Instrumentation Conference, Miller-Freeman Trade-Show Division, Dallas, Texas.)

Boundary-scan techniques are not panaceas. They do, however, begin to reduce the test-program generation burden and simplify the test process.

5.8 Summary

Proponents of design-for-testability have invented scan techniques that permit direct tester access to board logic that is not available through an edge connector, test connector, or bed-of-nails. Several schemes have been proposed. A standard version of one of them, known as IEEE 1149.1, provides a means for device and board manufacturers to take maximum advantage of scan features.

IEEE 1149.1 requires that every device contain a Test-Access Port (TAP) with four dedicated I/O pins: Test-Data In (TDI), Test-Data Out (TDO), Test Clock (TCK), and Test-Mode Select (TMS). Conforming devices also need a TAP controller, a scannable instruction register, and scannable test-data registers. Although testing through scan circuitry is slow because serializing data for input or output takes significant time, the availability of this technique simplifies the tasks of test-program generation and fault diagnosis.

For the foreseeable future, most boards will contain both scannable and nonscannable components. Researchers have proposed numerous strategies for testing such boards.

CHAPTER 6

The VMEbus eXtension for Instrumentation

One challenge of building a board-test strategy is the rapidity with which board and test technologies change. Test equipment that cannot adapt easily to new developments quickly becomes obsolete. More than 20 years ago, a few farsighted engineers advocated *modular* test systems tailored to fit particular applications. Both vendors and users could modify machine capabilities in the field to accommodate evolving test needs. Aside from rack-and-stack systems designed around the IEEE-488 bus, few test solutions adopted the modular approach.

In addition, commercial-tester modifications must come from the vendor. If a vendor cannot supply all necessary enhancements, customers find third-party partners or revert to the IEEE-488 external-instrument alternative.

An ideal solution would utilize a standard chassis with individual instrument functions from a variety of vendors on plug-in cards. This approach avoids many of the constraints and redundant features of stand-alone-instrument architectures while providing a scheme that allows instruments to communicate easily with one another as well as with a computer host.

In the early 1980s, companies such as Tektronix and Wavetek advocated one version of this approach, *computer-based instrumentation (CBI)*, which placed instrument functions on conventional PC expansion boards. Unfortunately, the PC-bus interface's slow data-transfer rate limited this proposal's success.

The VMEbus offers a faster standard interface for boards in a computer-driven card cage, but board-to-board spacing does not accommodate electromagnetic-interference (EMI) shielding or high-profile components. In addition, signal definitions leave much to users' discretion, so individual VME implementations are often incompatible with one another.

In 1987, five companies—Hewlett-Packard (now Agilent Technologies), Racal-Dana, Tektronix, Colorado Data Systems (now part of Tektronix), and Wavetek—announced a VME extension for Instrumentation, dubbed VXI, to address these issues. Its stated purpose is "to define a technically sound modular-instrument standard based on the VMEbus that is open to all manufacturers and is compatible with present industry standards." The specification defines an architecture for attaching conforming board-based instruments from any vendor to a standard chassis, creating something close in performance to a monolithic system. In 1993, the IEEE adopted Revision 1.3 of the VXI specification as standard IEEE-1155.

Although still in its infancy compared to the 30-year-old IEEE-488, VXI enjoys significant technical advantages and has generated considerable interest in the test-and-measurement community. Unlike the original IEEE-488, which was invented by a single vendor (Hewlett-Packard) and then adopted by the industry at large because it addressed test needs better than its *ad hoc* predecessors, VXI is a *compromise* among vendor electronics companies. The five original sponsoring companies were otherwise fierce competitors. Because it is a compromise, VXI may not handle a particular application as well as a proprietary solution would, but it can serve in most situations requiring modularity, flexibility, and better signal performance than IEEE-488. The IEEE did revise the IEEE 488 standard in 1999. Users will find the new version very similar, although it does address numerous shortcomings in the old standard, such as I/O speeds and the number of instruments permitted on the bus.

VXI enjoys two distinct lives in the test-and-measurement world. First, electronics manufacturers attempting to design and build their own test equipment prefer its feature, control, and communication choices over older alternatives. Second, many tester *vendors* are creating new products around VXI to accommodate their customers' ever-changing needs. VXI also offers some possible solutions for analog functional testing, where IEEE-488 often provides the only viable alternative.

VXI-based testers address a wide variety of measurement functions and applications. Standardizing around a single architecture minimizes development costs. Several vendors already offer VXI-based models optimized for specific manufacturing situations. With instrument choices constantly expanding, this approach will become increasingly common.

The fact that VXI is still in its infancy is important. Practitioners should not expect product offerings to be as extensive or as mature as their IEEE-488 counterparts.

This book is not intended as a comprehensive VXI reference text. The following discussion introduces IEEE-1155 to people who will encounter the standard in designing or purchasing test systems for particular strategies. Readers can refer to the "Works Cited and Additional Readings" section for more detail. In addition, magazines, technical seminars, conferences, and trade shows will disseminate the most current information. VXI has spawned offspring architectures, such as MXI and PXI, that specifically address additional needs of the standard's users. These versions derive from the VXI standard, but are not standards themselves.

6.1 VME Background

The VME standard defines a hierarchical open system consisting of one or more subsystems, each of which includes one or more modules, which in turn

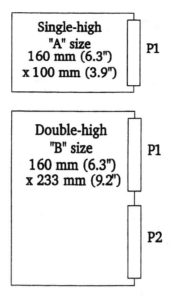

Figure 6-1 VME boards come in two sizes, referred to as *single-high* and *double-high*. The VXI specification renames them "A" and "B." (IEEE Std 1014–1987, *IEEE Standard for Versatile Backplane Bus: VMEbus*, Institute of Electrical and Electronics Engineers, Inc., Piscataway, New Jersey.)

contain one or more devices. The approach aims primarily at computer rather than real-time applications. The architecture is electronically and logically similar to that of 68000-family microprocessors, but it does not restrict designers to those devices. The bus supports many microprocessor architectures, including Pentium-class systems. Also, many simple VMEbus boards contain no microprocessor at all.

VME boards come in two sizes, referred to as *single-high* and *double-high*, as Figure 6-1 shows. The VXI specification renames them "A" and "B." Both are about 6 inches deep. The A-size board is about 4 inches high, and the B-size board is about 9 inches high. Each contains a 96-pin connector known as P1, with the pins arranged in three rows of 32 pins on 100-mil centers. The B-size board also includes a P2 connector of similar design.

P1 offers all handshaking, bus arbitration, and interrupt servicing for a minimum VMEbus system. It incorporates 16- and 24-bit addressing, as well as 8- and 16-bit data paths. Figure 6-2 shows the pin assignments for a VME P1 connector, which VXI has adopted intact. Row *b* of the P2 connector expands VME address and data lines to 32 bits each, and rows *a* and *c* are completely user-defined, as Figure 6-3 shows. User-defined pins permit boards to address an internal disk drive, for example, or to communicate with one another.

The VMEbus specification allows a maximum of 21 boards spaced 0.8 inch apart in a 19-inch chassis, sometimes called a frame. In 19-inch rack installations, the practical limit is 20 boards. Although boards are mounted vertically on most

Pin	Row a	Row b	Row c
1	D00	BBSY*	D08
2	D01	BCLR*	D09
3	D02	ACFAIL*	D10
4	D03	BG0IN*	D11
5	D04	BG0OUT*	D12
6	D05	BG1IN*	D13
7	D06	BG1OUT*	D14
8	D07	BG2IN*	D15
9	GND	BG2OUT*	GND
10	SYSCLK	GEIN*	SYSFAIL*
11	GND	BG3OUT*	BERR*
12	DS1*	BR0*	SYSRESET*
13	DS0*	BR1*	LWORD*
14	WRITE*	BR2*	AM5
15	GND	BR3*	A23
16	DTACK*	AM0	A22
17	GND	AM1	A21
18	AS*	AM2	A20
19	GND	AM3	A19
20	IACK*	GND	A18
21	IACKIN*	SERCLK(1)	A17
22	IACKOUT*	SERDAT*(1)	A16
23	AM4	GND	A15
24	A07	IRQ7*	A14
25	A06	IRQ6*	A13
26	A05	IRQ5*	A12
27	A04	IRQ4*	A11
28	A03	IRQ3*	A10
29	A02	IRQ2*	A09
30	A01	IRQ1*	A08
31	-12V	+5VSTDBY	+12V
32	+5V	+5V	+5V

Figure 6-2   Pin assignments for a VME P1 connector, which VXI has adopted intact. (IEEE Std 1014–1987, *IEEE Standard for Versatile Backplane Bus: VMEbus*, Institute of Electrical and Electronics Engineers, Inc., Piscataway, New Jersey.)

frames, the standard does not define board orientation. Many VMEbus systems accept horizontally mounted boards.

The standard makes no special provision for multiple-chassis systems or for chassis-to-chassis communication. Electrically buffering the bus permits assembling such systems, although this technique inherently reduces signal bandwidth. Standard data-communication links that mask the VMEbus architecture also

Pin	Row a	Row b	Row c
1	User Defined	+5V	User Defined
2	User Defined	GND	User Defined
3	User Defined	RESERVED	User Defined
4	User Defined	A24	User Defined
5	User Defined	A25	User Defined
6	User Defined	A26	User Defined
7	User Defined	A27	User Defined
8	User Defined	A28	User Defined
9	User Defined	A29	User Defined
10	User Defined	A30	User Defined
11	User Defined	A31	User Defined
12	User Defined	GND	User Defined
13	User Defined	+5V	User Defined
14	User Defined	D16	User Defined
15	User Defined	D17	User Defined
16	User Defined	D18	User Defined
17	User Defined	D19	User Defined
18	User Defined	D20	User Defined
19	User Defined	D21	User Defined
20	User Defined	D22	User Defined
21	User Defined	D23	User Defined
22	User Defined	GND	User Defined
23	User Defined	D24	User Defined
24	User Defined	D25	User Defined
25	User Defined	D26	User Defined
26	User Defined	D27	User Defined
27	User Defined	D28	User Defined
28	User Defined	D29	User Defined
29	User Defined	D30	User Defined
30	User Defined	D31	User Defined
31	User Defined	GND	User Defined
32	User Defined	+5V	User Defined

Figure 6-3   Pin assignments for the VME P2 connector. (IEEE Std 1014–1987, *IEEE Standard for Versatile Backplane Bus: VMEbus*, Institute of Electrical and Electronics Engineers, Inc., Piscataway, New Jersey.)

permit building systems with more than one frame, but this approach limits system flexibility.

For instrumentation applications, P2's user-defined pins represent both VME's greatest strength and its greatest weakness. Chassis and board manufacturers can assign these pins to a particular activity on a proprietary basis. Because pin assignments are nonstandard, however, two vendors' instrument boards will not generally play together on the same bus. Such incompatibility makes designing test *systems* around the bus much more difficult. In addition, relatively few test products support the VMEbus at all.

VMEbus also offers no guidelines for conducted or radiated electromagnetic compatibility (EMC) or electromagnetic interference (EMI). EMI has become a particular concern in Europe, where strict regulations have emerged to protect the wireless infrastructure from interference from noisy products. Power consumption, heat dissipation, and cooling are left to system integrators. Because these issues are important to test and measurement, VXI extensions address them more rigorously.

6.2   VXI Extensions

The VXIbus specification begins by increasing the mandatory spacing between boards in a frame from 0.8 inch to the Eurocard standard of 1.2 inches, thereby accommodating high-profile components, heat sinks, and EMI shielding. As a result, a standard 19-inch chassis can contain a maximum of 13 boards (also called "modules").

The board in slot 0 must perform certain system-controller functions in addition to any instrument capabilities. System-controller resources include power monitors, a bus timer, timing generators, and IEEE-488 or RS-232 communications ports.

Slot 0 also functions as a systemwide *resource manager* that *may* perform the following:

Configure the A24 or A32 address maps, and assign memory blocks to devices as needed.

Determine commander/servant hierarchies to prevent more than one processor from trying to control the same device.

Control an operating system to allocate shared memory blocks or other hardware resources such as trigger buses.

Allocate the VMEbus interrupt-request (IRQ) lines.

Perform system self-test and diagnostics.

Initialize all system commanders on power-up or on command.

Begin normal system operation.

The standard also adopts two larger 13.4-inch-deep board sizes from the Eurocard standard, dubbed "C" and "D," as Figure 6-4 illustrates. A D-size board can include an additional 96-pin connector, called P3. A VXI-based system may contain up to 255 devices (or "instruments") in one or more subsystem frames.

As for the connectors themselves, VXI retains P1 and the defined center row of P2. Instead of leaving P2 rows *a* and *c* to user discretion, however, VXI assigns them as well, adding a 10-MHz ECL clock, ECL and analog power-supply voltages, two ECL and eight TTL trigger lines, an analog summing bus, and a module-identification line, as Figure 6-5 shows. P2 also includes 24 local-bus pins (12 lines

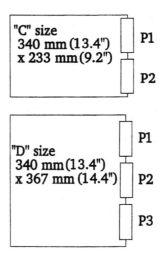

Figure 6-4   The VXI standard adopts two larger 13.4-inch-deep module sizes from the existing Eurocard standard, dubbed "C" and "D." (IEEE Std 1155–1992, *IEEE Standard VMEbus Extensions for Instrumentation: VXIbus*, Institute of Electrical and Electronics Engineers, Inc., Piscataway, New Jersey.)

in and 12 lines out) that can pass signals from any board to the adjacent one on either side. A- and B-size VXI cards remain compatible with their VME counterparts, as long as the VME boards do not assign P2 rows *a* and *c* in a manner that conflicts with their VXI existence.

P3 provides numerous high-performance features, adding 100-MHz clock and sync signals, additional power pins, six more ECL trigger lines, and 48 more local-bus lines (24 in and 24 out). In addition, P3 includes access to a STAR bus that permits direct communication with the system-controller functions in slot 0. Figure 6-6 shows the P3 pin assignments.

Note that both P2 and P3 contain pins identified as "reserved." These are not user-definable. The specification allocates them for future expansion of the standard.

Because of their special functions, slot-0 connectors P2 and P3 have different pin assignments from their counterparts in slots 1 through 12. Note, for example, that in the P2 slot-0 pin list in Figure 6-7, rows *b* and *c* are identical to Figure 6-5. Row *a*, however, replaces the local bus pins with a set of 12 MODID pins, one for each other slot in the frame. On Pin 30a, where other P2s contain their own MODID, the slot-0 module also contains a MODID, primarily for situations where the slot-0 card includes instrument functions.

As Figure 6-8 shows, the slot-0 version of P3 replaces both local buses with the center of the STAR bus. With STARX+, STARX–, STARY+, and STARY–, every module supporting the STAR bus can have four signal lines that connect to slot 0. Support for the bus, however, either by the slot-0 module or other modules, is *not* mandatory.

Pin	Row a	Row b	Row c
1	ECLTRG0	+5V	CLK10+
2	-2V	GND	CLK10-
3	ECLTRG1	RSV1	GND
4	GND	A24	-5.2V
5	LBUSA00	A25	LBUSC00
6	LBUSA01	A26	LBUSC01
7	-5.2V	A27	GND
8	LBUSA02	A28	LBUSC02
9	LBUSA03	A29	LBUSC03
10	GND	A30	GND
11	LBUSA04	A31	LBUSC04
12	LBUSA05	GND	LBUSC05
13	-5.2V	+5V	-2V
14	LBUSA06	D16	LBUSC06
15	LBUSA07	D17	LBUSC07
16	GND	D18	GND
17	LBUSA08	D19	LBUSC08
18	LBUSA09	D20	LBUSC09
19	-5.2V	D21	-5.2V
20	LBUSA10	D22	LBUSC10
21	LBUSA11	D23	LBUSC11
22	GND	GND	GND
23	TTLTRG0*	D24	TTLTRG1*
24	TTLTRG2*	D25	TTLTRG3*
25	+5V	D26	GND
26	TTLTRG4*	D27	TTLTRG5*
27	TTLTRG6*	D28	TTLTRG7*
28	GND	D29	GND
29	RSV2	D30	RSV3
30	MODID	D31	GND
31	GND	GND	+24V
32	SUMBUS	+5V	-24V

Figure 6-5 VXI version of P2. (IEEE Std 1155–1992, *IEEE Standard VMEbus Extensions for Instrumentation: VXIbus*, Institute of Electrical and Electronics Engineers, Inc., Piscataway, New Jersey.)

The VXI specification grants individual modules nonconflicting address space. System configuration information resides in the upper 16 KB of the 64-KB address-space limit of a 16-bit architecture. Each device is allotted 64 bits in this space, which is sufficient for many simple devices. Devices requiring more memory place their requests in a defined register. During power-up, the system reads these requests and allocates the necessary memory in the 16 MB of space permitted by 24-bit addressing or the 4 GB of 32-bit addressing space.

Pin	Row a	Row b	Row c
1	ECLTRG2	-24V	+12V
2	GND	-12V	-12V
3	ECLTRG3	GND	RSV4
4	-2V	RSV5	+5V
5	ECLTRG4	-5.2V	RSV6
6	GND	RSV7	GND
7	ECLTRG5	+5V	-5.2V
8	-2V	GND	GND
9	LBUSA12	-5V	LBUSC12
10	LBUSA13	LBUSC15	LBUSC13
11	LBUSA14	LBUSA15	LBUSC14
12	LBUSA16	GND	LBUSC16
13	LBUSA17	LBUSC19	LBUSC17
14	LBUSA18	LBUSA19	LBUSC18
15	LBUSA20	+5V	LBUSC20
16	LBUSA21	LBUSC23	LBUSC21
17	LBUSA22	LBUSA23	LBUSC22
18	LBUSA24	-2V	LBUSC24
19	LBUSA25	LBUSC27	LBUSC25
20	LBUSA26	LBUSA27	LBUSC26
21	LBUSA28	GND	LBUSC28
22	LBUSA29	LBUSC31	LBUSC29
23	LBUSA30	LBUSA31	LBUSC30
24	LBUSA32	+5V	LBUSC32
25	LBUSA33	LBUSC35	LBUSC33
26	LBUSA34	LBUSA35	LBUSC34
27	GND	GND	GND
28	STARX+	-5.2V	STARY+
29	STARX-	GND	STARY-
30	GND	-5.2V	-5.2V
31	CLK100+	-2V	SYNC100+
32	CLK100-	GND	SYNC100-

Figure 6-6 VXI pin assignments for P3. (IEEE Std 1155–1992, *IEEE Standard VMEbus Extensions for Instrumentation: VXIbus*, Institute of Electrical and Electronics Engineers, Inc., Piscataway, New Jersey.)

The IEEE-1155 specification outlines mechanical and electrical bus-structure requirements, low-level module communication protocols, slot-0 controller functions, maximum EMI susceptibility and radiation levels per module, and required cooling. It leaves other issues up to users. For example, it does not cover functionally controlling individual modules, determining necessary power levels, integrating power supplies into the system, or specific techniques for cooling the system.

Pin	Row a	Row b	Row c
1	ECLTRG0	+5V	CLK10+
2	-2V	GND	CLK10-
3	ECLTRG1	RSV1	GND
4	GND	A24	-5.2V
5	MODID12	A25	LBUSC00
6	MODID11	A26	LBUSC01
7	-5.2V	A27	GND
8	MODID10	A28	LBUSC02
9	MODID09	A29	LBUSC03
10	GND	A30	GND
11	MODID08	A31	LBUSC04
12	MODID07	GND	LBUSC05
13	-5.2V	+5V	-2V
14	MODID06	D16	LBUSC06
15	MODID05	D17	LBUSC07
16	GND	D18	GND
17	MODID04	D19	LBUSC08
18	MODID03	D20	LBUSC09
19	-5.2V	D21	-5.2V
20	MODID02	D22	LBUSC10
21	MODID01	D23	LBUSC11
22	GND	GND	GND
23	TTLTRG0*	D24	TTLTRG1*
24	TTLTRG2*	D25	TTLTRG3*
25	+5V	D26	GND
26	TTLTRG4*	D27	TTLTRG5*
27	TTLTRG6*	D28	TTLTRG7*
28	GND	D29	GND
29	RSV2	D30	RSV3
30	MODID00	D31	GND
31	GND	GND	+24V
32	SUMBUS	+5V	-24V

Figure 6-7   Slot-0 pin assignments for connector P2. (IEEE Std 1155–1992, *IEEE Standard VMEbus Extensions for Instrumentation: VXIbus*, Institute of Electrical and Electronics Engineers, Inc., Piscataway, New Jersey.)

Similarly, module insertion and ejection and chassis mounting are left to individual tastes. Therefore physical construction may differ from one chassis to the next, and some modules may not fit into a particular vendor's chassis.

Each subsystem in a multiframe system contains its own slot-0 controller and up to 12 additional instrument modules. Many VXIbus systems reside in a single frame. In fact, several VXI manufacturers offer frames that hold only five or six modules to reduce the cost of assembling a small system.

Pin	Row a	Row b	Row c
1	ECLTRG2	-24V	+12V
2	GND	-12V	-12V
3	ECLTRG3	GND	RSV4
4	-2V	RSV5	+5V
5	ECLTRG4	-5.2V	RSV6
6	GND	RSV7	GND
7	ECLTRG5	+5V	-5.2V
8	-2V	GND	GND
9	STARY12+	-5V	STARX01+
10	STARY12-	STARY01-	STARX01-
11	STARX12+	STARX12-	STARY01+
12	STARY11+	GND	STARX02+
13	STARY11-	STARY02-	STARX02-
14	STARX11+	STARX11-	STARY02+
15	STARY10+	+5V	STARX03+
16	STARY10-	STARY03-	STARX03-
17	STARX10+	STARX10-	STARY03+
18	STARY09+	-2V	STARX04+
19	STARY09-	STARY04-	STARX04-
20	STARX09+	STARX09-	STARY04+
21	STARY08+	GND	STARX05+
22	STARY08-	STARY05-	STARX05-
23	STARX08+	STARX08-	STARY05+
24	STARY07+	+5V	STARX06+
25	STARY07-	STARY06-	STARX06-
26	STARX07+	STARX07-	STARY06+
27	GND	GND	GND
28	STARX+	-5.2V	STARY+
29	STARX-	GND	STARY-
30	GND	-5.2V	-5.2V
31	CLK100+	-2V	SYNC100+
32	CLK100-	GND	SYNC100-

Figure 6-8   The slot-0 version of P3 replaces both local buses with the center of the STAR bus, connecting each module directly to the controller. (IEEE Std 1155–1992, *IEEE Standard VMEbus Extensions for Instrumentation: VXIbus*, Institute of Electrical and Electronics Engineers, Inc., Piscataway, New Jersey.)

Modules can contain more than one instrument, and an instrument may reside on more than one module. A system may consist of any combination of subsystems. As an example, the specification describes a system that includes one frame with a slot-0 card and 12 additional instrument modules extended to a frame with a slot-0 card and three instrument slots and another frame containing a slot-0, five instrument slots, and four standard VMEbus slots whose P2 is undefined.

6.3   Assembling VXI Systems

Most engineers designing systems around VXI choose a chassis first and then buy modules to plug into it. Unfortunately, not all modules fit, and power consumption and cooling requirements for a collection of modules may exceed chassis capacity. The resulting system may not offer the best feature mix and may not work at all.

For example, obviously a C-size chassis will not accommodate D-size cards, so choosing a C-size chassis prevents taking advantage of any module requiring the P3 connector. To avoid this limitation, some engineers automatically buy a D-size box. D-size frames are expensive, however, and if current applications do not involve D-size cards or P3, the system will cost significantly more than necessary.

Therefore, building a VXI-based system should begin with decisions on instrument capabilities and the specific module requirements needed to perform the testing, an appropriate slot-0 controller module, and a host computer. Then and only then, based on module size, power-consumption, and airflow specifications published in vendor catalogs, can customers select a correct chassis.

VXI equipment vendors provide tools to help customers through the specification process. For example, the Colorado Data Systems division of Tektronix offers the configuration worksheet shown in Figure 6-9. I_{pm} represents peak module current, in amps, as defined in the standard. After settling on an appropriate slot-0 module and accompanying instrument modules, a user locates power-consumption and cooling requirements in the grid in Figure 6-10, entering them in the worksheet and computing totals. Then, the catalog indicates chassis power capacities and offers cooling curves, such as those in Figure 6-11, to permit the selection of a chassis sufficient for the task without overbuying.

VXI flexibility supports a wide range of system configurations. Figure 6-12 shows some typical examples. The system in Figure 6-12a consists of a host computer connected to a slot-0 controller via an IEEE-488 cable. The four other occupied slots contain three instrument modules and extra RAM. In this case, the system's only CPU resides in slot 0, so that it must handle most controlling functions, probably through either the local bus or the STAR bus.

Figure 6-12b shows a similar arrangement, except that individual instruments contain their own CPUs. The instruments come from different manufacturers, but because they conform to the specification, the system does not care. Two additional slots contain display drivers and RAM. Notice that the modules need not occupy consecutive slots.

Figure 6-12c contains no external computer host. The host in this case, called an *embedded computer*, resides within the chassis itself. Several manufacturers offer such computers. These computers are compatible with conventional PCs, and contain a keyboard and display, as well as IEEE-488 and RS-232 interfaces for connection outside the chassis. The version in the figure shows the computer on a double-wide module with a separate disk drive. This configuration comes from the

Slot #	Model #	Power Requirements							Suffic. Cooling
		+5V I_{pm}	+12V I_{pm}	-12V I_{pm}	+24V I_{pm}	-24V I_{pm}	-5.2V I_{pm}	-2V I_{pm}	(Y/N)
0									
1									
2									
3									
4									
5									
6									
7									
8									
9									
10									
11									
12									
Total									
Chassis									

Notes

Figure 6-9 The Colorado Data Systems division of Tektronix offers this configuration worksheet to help customers through the system-specification process. (*Card-Modular Instruments Information & Ordering Guide*, Tektronix, Englewood, Colorado.)

Model #	Power Requirements[1]							Cooling Requirements[5]		Note 9
	+5V Ipm	+12V Ipm	-12V Ipm	+24V Ipm	-24V Ipm	-5.2V Ipm	-2V Ipm	Airflow l/s	Backpressure mm H₂O	Approx. Weight
VX4820	3.2	–	–	.66	.66	.45	.1	2.00	0.75	3.1 lbs
VX5260A	8.40	.70	.90	.30	.10	5.4	3.4	Note 7	Note 7	7.3 lbs
VX5521	Note 7	–	–	–	–	–	–	Note 7	Note 7	Note 7
VX5535	10.0	.25	.06	.003	.003	6.0	.003	4.00	0.05	8.0 lbs
VX5790A	4.40	–	–	.50	.50	4.6	1.8	6.50	0.63	4.5 lbs
73A-151B[2]	2.5	.012	.012	.01	.003	.06	.01	1.50	0.04	2.8 lbs
73A-155[2]	3.0	.012	.012	.01	.003	.20	.07	1.30	0.04	3.0 lbs
73A-156[2]	2.5	.012	.012	.01	.003	.06	.01	1.50	0.04	2.8 lbs
73A-270	3.3	–	–	.12	.10	.035	.026	1.80	0.10	3.1 lbs
73A-308	2.0	–	–	–	–	–	–	0.80	0.10	2.9 lbs
73A-425	2.5	–	–	.10	.10	–	–	1.40	0.17	3.1 lbs
73A-426	2.5	–	–	.10	.10	–	–	1.40	0.18	2.9 lbs
73A-451	1.6[8]	Note 8	Note 8	Note 8	Note 8	Note 8	Note 8	Note 6	Note 6	1.3 lbs
73A-452	Note 8	Note 8	Note 8	Note 8	Note 8	Note 8	Note 8	Note 6	Note 6	1.0 lbs
73A-453	2.3	–	–	.07	.09	–	–	1.10	0.02	2.9 lbs
73A-455	5.0	–	–	.36	.36	.08	.02	3.40	0.18	4.6 lbs
73A-541	3.0	–	–	.3	.16	–	–	1.50	0.05	3.0 lbs
73A-575	1.9	.022	–	.17	.145	–	–	1.78	0.17	3.3 lbs
73A-576	1.9	.022	–	.18	.15	–	–	1.78	0.17	3.3 lbs
73A-850	–	–	–	–	–	–	–	–	–	1.6 lbs
73A-851[3]	.80	–	–	–	–	.09	.09	0.40	0.15	2.2 lbs
73A-853[4]	1.20	–	–	.01	.01	–	–	0.50	0.13	1.8 lbs

[1] All current specified in Amps. Ipm is the peak module current as defined in the VXIbus specification.
[2] Module only requires +5V and -5.2V to properly operate. Additional current requirements are for on-board voltage monitoring only.
[3] The power and cooling requirements listed are for the module only. Additional power and cooling requirements should be added for the VME or VXI module installed.
[4] To maintain a 10°C maximum temperature rise across the module, approximately 0.08 liters/sec of airflow is required for each Watt of power dissipated on the module. Current and cooling requirements are based on the one 73A-853 only. Additional current and airflow required to operate installed CDSbus (53A) card must be added to the values given. Consult factory for CDSbus card data.
[5] Cooling specifications assume 10°C maximum temperature rise across module under maximum loading conditions. For multi-slot modules, airflow is specified per slot.
[6] Cooling requirements depend on user circuitry installed. To maintain a 10°C rise across the module, approximately 0.08 liters/sec of airflow is required for each watt of power dissipated on the module. For a fully populated module with front and back shield installed, a typical pressure drop is 0.04 mm H₂O.
[7] Complete data for this module was not available at time of printing.
[8] Total power consumed will depend upon additional user-installed circuitry.
[9] Actual weight will depend upon specified option(s) ordered.

Figure 6-10   After settling on an appropriate slot-0 module and accompanying instrument modules, users locate their power-consumption and cooling requirements in this grid, entering them into the worksheet. (*Card-Modular Instruments Information & Ordering Guide*, Tektronix, Englewood, Colorado.)

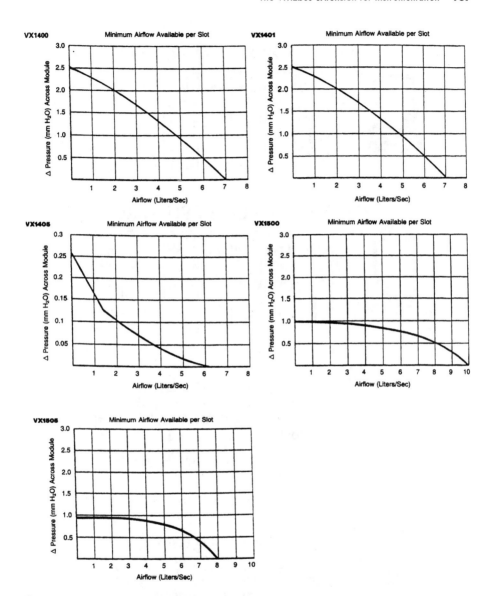

Figure 6-11 These chassis cooling curves permit selecting a chassis sufficient for the test task without overbuying. (*Card-Modular Instruments Information & Ordering Guide*, Tektronix, Englewood, Colorado.)

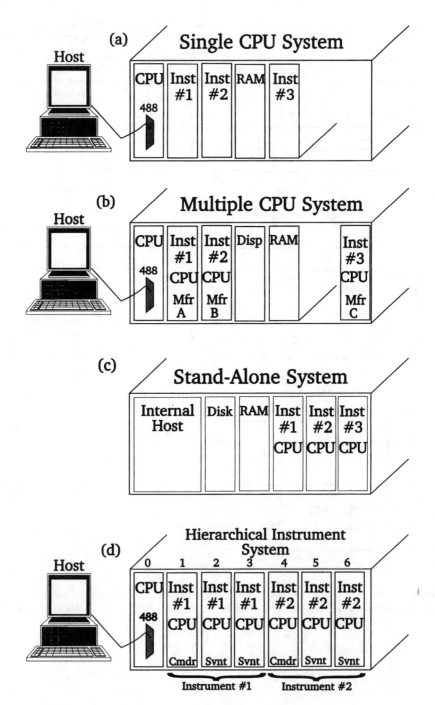

Figure 6-12   Some typical VXI system configurations. (IEEE Std 1155–1992, *IEEE Standard VMEbus Extensions for Instrumentation: VXIbus*, Institute of Electrical and Electronics Engineers, Inc., Piscataway, New Jersey.)

VXI standard. The inexorable progress of the industry's "incredible shrinking electronics," however, allows the embedded computer to be put onto a single module. Similarly, memory today occupies little real estate and no longer requires a module of its own. Nevertheless, the purpose here is merely to demonstrate the different alternatives.

To enhance system functionality, VXI devices communicate through a hierarchical structure involving *commanders* and *servants*. A commander may have multiple servants, but a servant can have one and only one commander. A commander initiates servant functions and exercises complete control over its activities and data. IEEE-1155 permits a virtually unlimited number of commander/servant levels. Any commander may act as a servant at the next higher level.

Figures 6-12a, 6-12b, and 6-12c show systems that include only one hierarchical level. The slot-0 controller functions as the commander, while displays, disks, RAM, and instruments are servants.

In Figure 6-12d, the slot-0 controller acts as commander (Cmdr) for the modules in slots 1 and 4. The module in slot 1, in turn, controls servant (Svnt) modules in slots 2 and 3, and the module in slot 4 controls slots 5 and 6. Therefore, modules in slots 1 and 4 are both commanders and servants. In effect, this system consists of two multimodule instruments and a slot-0 controller, all subject to commands from the PC host over the IEEE-488 link.

The standard mandates certain minimum capabilities from every device regardless of its actual function. P1 permits controlling a set of *configuration registers* that identify a device's class, manufacturer, and model, as well as address and memory needs. A *register-based device* contains only this level of capability. Controlling register-based devices requires either purchased or home-grown software drivers. Register-based devices can be servants only and cannot function as commanders in the VXIbus hierarchy.

Memory devices offer blocks of either temporary or permanent storage in RAM, ROM, or other memory types. Configuration registers contain the addresses of such blocks, generally arranging them by speed, memory type, or application.

The standard also describes *message-based devices* that provide higher levels of communication. In addition to configuration registers, message-based devices contain *communication registers* that other modules can access. With data-communication protocols such as the *word-serial protocol*, modules can communicate easily with one another. Such protocols permit any manufacturer to build a module with complete assurance of its compatibility with other modules in the system.

Commanders *must* be message-based devices. Therefore, servants that act as commanders for a lower hierarchical level must also be message-based. In Figure 6-12b, each of the instruments is a message-based device that can receive instructions from a host or common host interface. Controlling the wide range of possible VXI systems requires the layered communication topology depicted in Figure 6-13.

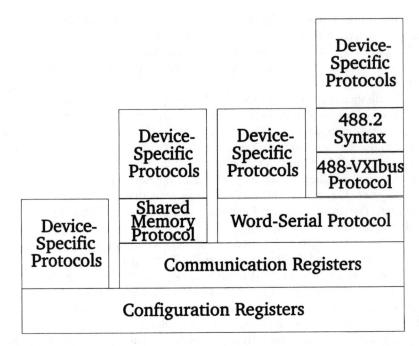

Figure 6-13 Controlling the wide range of possible VXI systems requires this layered communication topology. (IEEE Std 1155–1992, *IEEE Standard VMEbus Extensions for Instrumentation: VXIbus*, Institute of Electrical and Electronics Engineers, Inc., Piscataway, New Jersey.)

One common link between VXI instruments and a host computer is the 488-VXIbus interface device, defined as a function-specific message-based device that includes communication protocols supporting the IEEE-488 standard. All instruments on the VXIbus can share this device, giving each instrument direct access to the host. The VXI specification suggests designing other host or peripheral interfaces, such as RS-232s or local area networks (LANs), in the same way.

6.4 Configuration Techniques

As Figure 6-12 shows, placing modules in consecutive frame slots is unnecessary. Every module contributes a feedback resistance to its slot, so the resource manager can detect whether a slot is occupied and knows when a slot is occupied but its module does not function.

Like IEEE-488, VXI permits users to assign particular devices to specific bus addresses that the resource manager identifies during the power-up sequence. This approach is referred to as *static configuration*.

The newer standard is considerably more flexible, however. An optional technique called *dynamic configuration* permits the resource manager to assign devices to any unoccupied logical addresses during power-up.

If a system includes dynamic configuration, users assign any device capable of this function to logical address 255 (FF_{16}). During power-up, the resource manager identifies device assignments other than 255 considering them to be static-configuration addresses that it cannot change. It then assigns dynamic-configuration devices to other logical addresses. Note that although support for the MODID line on each module is ordinarily optional, dynamic-configuration modules *must* support it.

Dynamic configuration offers numerous benefits. A manufacturer may have several VXI systems dedicated to different product lines or test applications, but those systems may share certain instrument modules. A person removing modules from a frame, adding modules to a frame, or moving modules from one system to another need not remember to place them in any particular slot. Only the module in slot 0 must be statically configurable because it contains the resource manager itself. Also, because dynamic configuration includes MODID support, replacing a malfunctioning module with a good one requires no manual reconfiguration by the engineer, even if the new module comes from a different manufacturer and offers different specifications and other features. On power-up, the resource manager simply reads the MODID information and knows what instruments and other capabilities reside on that module.

The specification makes no provision for the resource manager's method of polling modules or the order in which it assigns logical addresses. Therefore, if a particular device's logical address must be restricted in some way, it should be statically assigned and ineligible for dynamic configuration.

6.5 Software Issues

Although the VXIbus standard presents hardware requirements in considerable detail, it leaves both operating and instrument software entirely to system integrators and users. This decision permits maximum flexibility for applying VXI techniques, but it also means that creating a VXI system extends far beyond the specification of controllers, modules, instruments, and frames.

Fortunately, numerous VXI-module manufacturers and third parties offer software-development tools. Graphical user interfaces and other conveniences help reduce test-program development time and costs. Some of the packages run in conventional Microsoft-Windows environments; others include proprietary software shells. These tools allow test engineers to write programs in high-level languages such as BASIC and C or, using "point-and-click" graphical techniques, automatically provide instructions in instruments' native languages through software drivers. Examples include VEE from Agilent Technologies, LabView, LabWindows and WaveTest from National Instruments, TekTMS from Tektronix, and ATEasy from a small software company called Geotest. Products such as EP Connect from Radisys, makers of an embedded VXIbus PC, facilitate communication between the host and slot 0, and between slot 0 and other modules.

Features include flowchart and interactive techniques for creating test procedures, along with waveform editors and runtime generators. All packages come equipped with built-in drivers for popular test instruments to simplify the programming process. In addition, instrument-driver generators guide users through the complicated task of creating drivers for new, unusual, or proprietary instrument devices.

Several of these products allow creating identical instrument drivers and test programs for RS-232, IEEE-488, and VXIbus-based systems, easing transition to the newer integration standard. This convenience permits developing and debugging program segments on stand-alone instruments, then transferring them to the system only when they are complete. Test departments can therefore develop program pieces in parallel, reducing initial time to market and minimizing the impact on test (and therefore the pain) of implementing engineering changes.

Graphical environments for programming a particular instrument can resemble that instrument's front panel—including display, dials, and switches—providing a link with more familiar benchtop and IEEE-488 rack-and-stack testing. Users can also create their own front panels, specifying the physical layout, type, grouping, and interaction of controls. In graphical approaches, programmers can manipulate images that resemble schematics or flowcharts. Of course, for the more masochistic, ASCII programming strings are still available for initiating tests.

Wolfe (1992) defines an instrument driver as a piece of software that contains at least some of the following elements:

A driver body or core that exactly describes the instrument's functions. An operator interface that allows users to interact directly with the driver from the computer keyboard and display.

A program interface that allows test-program access to the driver.

An I/O interface between the driver and the instrument itself.

A subroutine interface that allows the driver to call other software modules, including the operating system and software libraries for data formatting and analysis.

Wolfe cautions that although the more convenient menu-driven driver generators save development time, they may create "black-box" drivers. Such drivers do not permit examining source code. Programmers may not understand their operation and applicability to a particular test situation and, therefore, cannot change them.

Other packages create driver code as a user manipulates the computer-monitor representation of the instrument front panel, mimicking manual procedures on a benchtop. In these cases, an inexperienced test engineer who does not know how to exercise the instrument efficiently will create an inefficient driver. For example, selecting a measurement mode may invalidate a previously specified range setting. Optimizing and debugging procedures generated in this way may be very difficult. To alleviate this problem, some manipulative generators permit programmers to interact and optimize instrument operations before actually producing any code.

Driver generators that require programming in a conventional language may be more difficult to use than more graphics-oriented alternatives, but Wolfe contends that they provide developers with more direct control over instrument activity and that the resulting drivers are more robust. He advocates that manufacturers planning VXI-based test systems consider access to instrument-driver source code a primary concern.

There are several ongoing efforts to standardize instrument-control languages, thereby simplifying the programming process. Many military applications support CIIL, which has been around for more than 15 years and has spawned a version of ATLAS. Commercial manufacturers prefer the standard computer programming interface known as SCPI (pronounced "skippy"). Again, a standard language cannot incorporate every conceivable instrument function. Nevertheless, a common method to address all spectrum analyzers or all waveform generators would further reduce test-programming efforts. In high-speed, complex, and specialized situations one may still have to resort to instrument-native code.

6.6   Testing Boards

As mentioned, anyone assembling a board-test system based on the VXIbus must first evaluate the product's test needs. Testing a complex digital board, for example, requires power supplies, a digital stimulus/response module to simulate the operating environment, pattern generators, word recognizers, a voltmeter for DC measurements, and a counter/timer to measure clock frequencies. Individual module specifications have to meet or exceed the board-under-test's performance parameters. The stimulus/response module must include a sufficient number of I/O pins and adequate clock speeds, and the counter should measure frequencies to about five digits.

The chassis's power rating must exceed the sum of all module power requirements. Cooling requirements depend on the hottest module. For testing some boards, necessary power supplies might violate maximum chassis cooling specifications. Some experts recommend that if the chassis's cooling does not exceed the hottest module's requirements by at least 50 percent, engineers should prepare a detailed system cooling profile to ensure sufficient cooling for that application.

Large boards containing a lot of power-hungry devices might require a rack-and-stack power supply as large as the chassis itself. One VXI-based test-system manufacturer facing this problem positioned his system power supplies *outside* the cage—an awkward arrangement logistically but the only option that worked.

When choosing a slot-0 controller, users should consider whether it offers IEEE-488, RS-232, and direct VME interfaces to accommodate capabilities that VXI cannot provide or to permit incorporating instruments and other tools that the manufacturing facility already has but that do not conform to the VXI architecture.

After assembling instrument modules into the chassis, users set module addresses statically or allow dynamic configuration to perform the task. In either case, the resource manager occupies address 0.

Software can come from instrument vendors. These vendors know their own products better than anyone else does. Noninstrument vendors provide software that is not biased toward or against particular products but also may not take best advantage of unusual features. Shells, data-analysis and statistical-process-control programs, and test generators can also come from third parties or in-house development teams. Note that in-house development consumes scarce software resources not only during the initial project but also on an ongoing basis for bug fixing, updates, and enhancements.

Some instrument companies, such as Agilent Technologies, offer system-integration services, acting as though they were independent third parties and assembling appropriate hardware and software into complete VXI systems to solve customers' test problems. This activity requires that these companies select instruments *other than their own* when necessary to provide the best overall system performance. Customers availing themselves of this service have the advantage of the vendor's expertise but must be careful of natural biases that may surface.

6.7 The VXIbus Project

Between December 1990 and April 1993, the editors of *Test & Measurement World* conducted a comprehensive experiment in the creation of a VXIbus-based test system. They formulated and evaluated a test problem, examined hardware and software options, assembled the system, ran tests, and reported results. The 12 articles in the series, listed in the "Works Cited and Additional Readings" section under "VXIbus Project," demonstrate the steps involved, the successes, the pitfalls, and the steepness of the learning curve. Even a decade later, they provide an excellent introduction for anyone considering a VXI test solution.

The editors selected a Kenwood R-5000 communications receiver covering the RF spectrum from 100 kHz to 30 MHz as a typical unit under test (UUT). This choice included a mix of analog and digital circuitry and, therefore, a wide range of test requirements. As in the real world, the editors discovered along the way that some of their initial test goals were unrealistic. Therefore, the completed system executed a slightly scaled-down version of the original test plan. For example, although the intent was to test the UUT down to the board level, time constraints and resource availability permitted only a system-level test.

There were numerous other impediments, as well. There was no suitable VXIbus-based RF generator, so the system included an IEEE-488 alternative, an HP 8656B. The original DC-power-module vendor, VereTest, went out of business before project completion, requiring a search for an appropriate VXIbus substitute, in this case a Model S60 from Advanced Power Designs of Irvine, California. Unable to find a 1-kHz notch filter, the editors built one using Racal-Dana's model 7064-prototyping board.

The other VXI modules were a Wavetek 1362 DMM and a Tektronix VX4223 frequency counter. The project required a team effort, including vendors and others with expertise in hardware, software, interconnects, and system integration, as well as a thorough understanding of how to test the UUT. In most real situations, in-house test-system designers also have access to product designers and documentation.

As part of their investigation, the participants examined numerous available hardware and software VXI options, including embedded computers and software packages: WaveTest (then from Wavetek, but now owned by National Instruments), TekTMS from Tektronix, and National Instruments' LabView and Lab-Windows. Their analysis indicates their impressions of each product and the apparent advantages and shortcomings of each. These observations do not constitute any kind of endorsement (which is why I have not detailed them here) but, rather, show how to evaluate candidates for a position in a particular VXI system. For example, the editors divide software packages into two categories, those that provide ease of use (and therefore a relatively flat learning curve) and those that permit engineers with expertise and experience to create more sophisticated test programs.

Based on the experience, the project team offers numerous recommendations. For example, test engineers must carefully evaluate connection methods between the test system and the UUT. This solution employed individual cables, although more complex situations might require other alternatives.

The editors also suggest obtaining vendors' demonstration packages for all software under consideration for a particular application. These demos will help you understand the software's capabilities, as well as its look and feel. Moreover, if you are comfortable programming in a particular language, such as BASIC or C, look for a software environment that supports that option, and be sure to get instrument drivers in that language.

During the development process, the editors maintained frequent contact with vendors' technical support staff. They especially found software documentation difficult to understand and lacking in programming examples.

Many software packages allow users to create their own operator screens. Project participants strongly suggest keeping such screens as simple as possible, perhaps providing only a start/stop button and a pass/fail indicator. Showing other information to test operators, such as actual measurement results, is often unnecessary.

In Windows-like environments, avoid confusing screens. At one point in the project, the display contained two overlapping windows, each of which included a "continue" button. An operator could easily click on the wrong one. In any event, design useful screens rather than spending time and effort making them pretty. When designing screens for more than one test product, keep the screens as similar as possible to keep learning curves down and avoid confusion. Be sure there is a means both to save and print test data.

Most important, the editors caution that you need to appreciate the scope of the development task. When they realized they had drastically underestimated the amount of work involved, they scaled down their test goals, as mentioned. In the real world, that decision is probably not an acceptable option. In final analysis, the editors found the VXIbus well suited for applications requiring both accuracy and flexibility.

6.8 Yin and Yang

The VXIbus standard offers considerable advantages over the original IEEE-488. It handles data-transfer rates of up to 40 MB per second, instead of only 1 MB per second. It permits up to 255 instruments per system, against the older standard's 14. Whereas only one IEEE-488 instrument can be a "talker," there is no such limitation on their VXI counterparts. Second-generation standard IEEE 488(2) alleviates some but not all of these limitations. Like rack-and-stack solutions, VXI systems can incorporate conforming instruments from any vendor. Users can replace an instrument capability, such as a logic analyzer, when a better or more advanced alternative comes along, also from any vendor. With available features such as dynamic configuration, however, integration of new VXI devices generally involves considerably less effort.

VXI systems' fewer and shorter cables than IEEE-488 approaches mean that the architecture will work in more situations. The local bus, STAR bus, and MODID also permit better data transfer between instruments and better communication between an instrument and the computer host. Because individual modules are usually smaller than stand-alone instruments (and a single module can carry many instruments), military and other space-sensitive field operations can carry more spares, reducing downtime. As more and more VXI equivalents of popular IEEE-488 instruments become available, companies will increasingly consider adopting VXIbus solutions.

Other VXI capabilities will tend to further reduce test-system size. A digitizer, for example, can perform the function of several individual instruments. At some point, test-system designers may replace many traditional instruments with software-driven A/D and D/A converters. The D/As will generate all test stimuli, while A/Ds will take all measurements.

Like all compromise standards, however, IEEE-1155 contains inherent limitations. High-end power supplies, for example, may never migrate over to VXI because of cooling and connection considerations. Similarly, test applications demanding very precise timing and very short leads will still require monolithic solutions.

Another drawback to VXI implementations is their cost. Chassis remain relatively expensive, as are individual modules. A VXI system can cost 50 percent more than its rack-and-stack counterpart. Cost represented one primary incentive to develop alternatives, adopting the same principles. PXI, for example, uses PC-expansion–type boards in a considerably smaller chassis.

VXI offers features and benefits to satisfy many situations. A time will undoubtedly come when tester manufacturers are themselves VXI system integrators, incorporating their own test expertise with test capabilities from third parties, tying everything together with a comprehensive software package that provides a coherent user environment. A tester vendor who is expert at digital board testing need not ask customers to settle for inferior analog capability and need not expend precious internal resources to develop capability that someone else does better. Similarly, a company that does analog measurement well can obtain digital-test technology elsewhere. Product differentiation will come from integration and software. Such expertise sharing will benefit both vendors and customers and avoid the ancient pitfall—reinventing the wheel.

6.9 Summary

The IEEE-1155 VXIbus standard provides a framework to build modular test systems with many of the performance characteristics of monolithic alternatives. It extends the popular VME computer bus, incorporating aspects of the existing Eurocard standard, including two 13.4-inch-deep board sizes and board-to-board spacing that allows for high-profile components, heat sinks, and EMI shielding. As a compromise standard originally proposed by a committee of competing instrument vendors, it cannot include every conceivable capability, but it offers a good solution for many test applications.

The board occupying slot 0 in the VXI card cage must perform certain systemwide controller functions, serving as a resource manager on power-up. Special standard VXI features include a local bus that allows any board to communicate with its nearest neighbors and an optional STAR bus that permits any module to communicate directly with slot 0. An optional feature called "dynamic configuration" allows the resource manager to assign logical addresses to individual VXI instruments on the fly, so that users disassembling, reassembling, or otherwise rearranging the cards in a VXI frame need not remember to place specific modules in particular slots. MODID permits the system to identify an instrument's manufacturer and other important information.

VXI-based systems generally are faster, are more flexible, and offer better signal integrity than IEEE-488 alternatives. The standard leaves many aspects of system operation to users, however, including cooling, EMI protection, and most software issues. A VXI system remains more expensive than other test alternatives.

A time will come when tester vendors are in fact VXI system integrators, incorporating the best test features available from any supplier. Product differentiation will come from integration success and from software power and ease of use. Such expertise sharing will benefit both vendors and customers, avoiding that age-old pitfall—reinventing the wheel.

CHAPTER **7**

Environmental-Stress Screening

The ultimate goal of any test strategy is to remove faulty products before they reach customers. But what about products that are apparently good on the factory floor but fail because of stresses that they experience during shipping or during early months in the field? Examples include cold-solder joints and other bad contacts, oxide and silicon defects, and failure from various kinds of surface contamination. If a method could be found that uncovered those problems before the end of the manufacturing cycle, products would be more reliable, customers would be happier, and warranty-repair and other costs would be lower. Inspection will find some of these problems, but not all of them. A cracked solder joint, for example, will likely pass both inspection and electrical test. Aggravating it until it becomes a hard failure allows much easier diagnosis and repair. That, in a nutshell, is the principle behind *environmental-stress screening (ESS)*.

7.1 The "Bathtub Curve"

Aside from death by heat, static electricity, power surges, severe mechanical shock, and other instant killers, almost all electronic-product failures fall into two categories—infant mortality and old-age fatigue. In between those events is generally a long and useful life, during which the product performs reliably. Therefore, a graph of failure rate against age takes on the characteristic shape of Figure 7-1, known as the *bathtub curve*. ESS subjects electronic components, boards, and systems to conditions that encourage infant-mortality failures, so that normal test-and-repair methods can eliminate them and products leaving the factory have already begun their "useful-life" phase.

Davis and Davis (1989) found that the decay rate in the infant-mortality portion of the curve depends on the mean time between failures (MTBF). At any time t, the failure rate $F(t)$ is:

$$F(t) = e^{(-t/\text{MTBF})} \qquad \text{(Eq. 7–1)}$$

These researchers divide the infant-mortality region into *internal failures* that occur before shipment, *external failures within warranty*, and *out-of-warranty fail-*

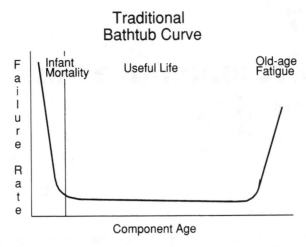

Figure 7-1 A graph of failure rate against age takes on this characteristic shape, known as the *bathtub curve*.

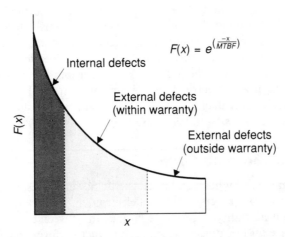

Figure 7-2 Davis and Davis divide the infant mortality region of Figure 7-1 into *internal failures* that occur before shipment, *external failures within warranty*, and *out-of-warranty failures*. (Davis, Don and Brendan Davis. 1989. "The Economics of Stress Screening," Nepcon West, Reed Exhibitions, Stamford, Connecticut.)

ures, as shown in Figure 7-2. More reliable products featuring higher MTBF experience shallower failure slopes. Ironically, fewer such products normally fail before shipment than do their less-reliable counterparts, and relatively more failures occur in the field.

To illustrate this point, consider a product with an MTBF of 1000 hours and a warranty of 1000 hours. Suppose that the manufacturer operates it for 40 hours in the factory before shipment. Theoretically, 96 percent of all failures will occur in the field—59 percent under warranty and 37 percent after the warranty expires. For a product with an MTBF of only 200 hours, internal failures will total 18 percent, with warranty failures at 81 percent and out-of-warranty failures representing only 1 percent. This second product will likely fail four or five times during the warranty period, incurring high warranty-support costs and annoying customers. In addition, the small number of post-warranty failures will generate little compensating service-related income.

By accelerating product failures so that they occur in the factory, an ESS program offers the following benefits:

Reduces in-warranty service costs. Davis and Davis contend that these costs alone can justify ESS.

Improves overall product reliability, thereby improving customer turn-on rate and post-warranty performance, reducing returns, and enhancing vendor reputation and customer goodwill.

Permits better in-factory data collection and failure analysis for process improvement.

Combined with statistical process control, provides better process-problem information than SPC alone.

7.2   What Is Environmental-Stress Screening?

Environmental-stress screening in itself is not a test. Rather, manufacturers generally perform tests during or after each screening step. ESS encompasses a wide variety of specific techniques, depending on the nature of each product and its target applications. Appropriate ESS steps for a personal computer, for example, may depend on whether that computer will be used in a conventional office or a factory environment, as well as whether it is a desktop or laptop machine.

ESS is not a sampling technique. Because it attempts to find latent faults that occur randomly in every product lot, stresses must apply to *all* products coming through the factory. Every uncovered fault must be fixed or the product scrapped.

Although the procedure occasionally reveals design problems that make field failures more likely, design verification is not a primary ESS mission. Instead, it looks for failures that relate to the manufacturing process, including parts quality and board and system assembly techniques.

It can take place at any process stage. One manufacturer stresses bare boards, verifying aspects of their construction—ensuring trace integrity on the multiple layers, for example, and that the layers do not come apart. Loaded boards undergo another round of screening. Boards that fail subsequent testing steps undergo comprehensive failure analysis. Results permit production-process improvement and provide information to vendors that helps improve the quality of vendor-supplied material.

An ESS program must not damage good products or create defects that would not otherwise occur. Actual stress conditions depend on operating-condition specifications. If an electronic system permits a maximum temperature of 75°C, exceeding that temperature during screening is unacceptable. Maximum permissible stresses depend on the least tolerant components, and stress conditions will be different at component, board, and system levels. Although the idea is to accelerate aging, the procedure must not consume more than an insignificant portion of a product's useful life.

7.3 Screening Levels

ESS authorities define a number of levels of screening complexity. Unfortunately, neither the terms nor their definitions are standardized. The following list represents a compromise from a number of sources, with major discrepancies noted.

Static screening generally means subjecting an unpowered board to a constant stress, as in traditional burn-in. Typically, testing occurs before and after the screen. Some sources also call for continuous power during a static screen.

Dynamic screening requires power-on. Sources that require power for static screening define dynamic screening to include power cycling. Dynamic screens can use power levels both above and below nominal values for the board under test, cycling up to 3.2 V for a 3-V power supply, for example. In some sources, dynamic screens also exercise board logic.

Exercised screening powers the board, applies input signals, and loads outputs. *Full functional screening* operates the board as though it were in its natural environment. This screen variety applies power and any necessary signals and loads, possibly (although not necessarily) including cycling.

Any powered stress screen also permits *monitoring*. A monitored screen actually tests board outputs during the screening step.

This variation offers several advantages. Monitoring permits the identification of "soft" errors that cause a device to fail but then continue to function properly, as with retries on a hard-disk drive. The technique permits logging product conditions and the time of a failure, which may be useful for failure analysis or process improvement. Because ESS-precipitated faults generally fall off over time, monitoring also allows manufacturers to determine optimum screen length, ensuring adequate screening results without spending excessive time and money with little or no additional benefit.

7.4 Screening Methods

7.4.1 Burn-in

Without doubt, classical burn-in remains the most common ESS technique. Burn-in is a usually static screen that holds devices, boards, or systems at a constant high temperature for a predetermined time, from a few hours up to a week. As noted previously, burn-in temperature must not exceed the maximum safe operating temperature. This procedure accelerates circuit aging according to the Arrhenius rate equation:

$$A = e^{\left(\frac{E_a}{k}\right)\left(\frac{1}{T_1} - \frac{1}{T_2}\right)}$$

(Eq. 7–2)

where A is the acceleration factor, E_a is an empirically determined failure-mechanism activation energy in electron-volts (eV), T_1 is the normal operating temperature in °K, T_2 is the burn-in temperature in °K, and k is Boltzmann's constant $(8.617 \times 10^{-5} \text{ eV/°K})$. Therefore, burn-in theoretically subjects circuits to thousands of hours of normal use in a relatively short time, revealing true infant-mortality failures. It can also precipitate flaws from chemical reactions and other indirectly temperature-dependent conditions.

Some burn-in procedures, especially for military and other critical applications, raise humidity levels as well as temperatures. Certain military specifications, for example, require a soak at 85°C and 85 percent relative humidity. These methods are generally referred to as *highly accelerated stress tests*, or *HAST*. Often they apply to only a sample of products, trying to determine a good product's tolerance for such conditions. This "beat-it-up-until-it-dies" approach is a design-verification procedure, but, strictly speaking, not ESS.

Test during burn-in represents a classic example of a monitored screen. Suppose a burn-in procedure specifies a screen time of 72 hours. Testing throughout that time may determine that by 20 hours, all or almost all boards that are going to fail have already done so. The manufacturer can then shorten the screen to that level, substantially increasing product throughput, while decreasing floor-space requirements and work-in-process inventories.

In practice, testing does not occur constantly on all boards. A tester cycles through boards one at a time in a predetermined sequence. This approach minimizes capital-cost requirements.

As products have become more reliable, burn-in precipitates fewer failures than it once did. Device manufacturers routinely boast 50 ppm and lower defect levels. Very few if any of these devices will succumb to a burn-in procedure. Nevertheless, board manufacturers continue to employ it. The fact that few boards will fall out during this step does little to diminish the concern that for some mysterious reason, ending the practice will increase field failure rates. Recent evidence indicates that hot-to-cold and cold-to-hot temperature transitions are far more effective at revealing problems than is maintaining a constant temperature.

Low-temperature burn-in is similar to its high-temperature counterpart, except that it occurs substantially *below* ambient. Some installations achieve low temperatures with liquid nitrogen or dry ice; others rely on cold air or mechanical refrigeration. Powering boards during this screen improves its effectiveness by increasing the temperature contrast between the hot devices and the cold environment. Certain solder anomalies are susceptible to this scheme, but it does not usually detect many problems. Its primary application is for products aimed at outer space, high altitudes, and other cold and unprotected environments to ensure circuit integrity under "normal" operating conditions.

7.4.2 *Temperature Cycling*

Much more effective than burn-in, *temperature cycling* subjects boards to multiple transitions between two temperatures at a predetermined rate of change. Expansion and contraction of solder joints and other circuit elements with different thermal coefficients of expansion create considerable stress. Bending-moment responses to temperature changes of materials in multilayer boards can cause separation of board layers or tombstoning of surface-mounted parts, for example, and can aggravate solder cracks. Power to the board should be off during hot-to-cold transitions to avoid reducing the cool-off rate, which would reduce screen effectiveness. Current information indicates that temperature cycling is the most effective single screen, although other methods will find certain failures that cycling will not.

Note that, as Figure 7-3 shows, product temperature varies somewhat more slowly than the chamber's air temperature. A product's actual temperature profile depends on its heat-dissipation capability, temperature gradients between the product and surrounding air, and board-surface characteristics. These factors, in turn, depend on air velocity and direction. Up to a point, the higher the airflow, the better the heat transfer and the more closely product temperatures follow air-temperature changes. Increasing airflow beyond that point merely increases screen equipment and operating costs with little additional benefit.

Evidence indicates that dwell or "soak" time at temperature extremes does not contribute significantly to product stress. Therefore, some manufacturers hold products at extremes only long enough for testing.

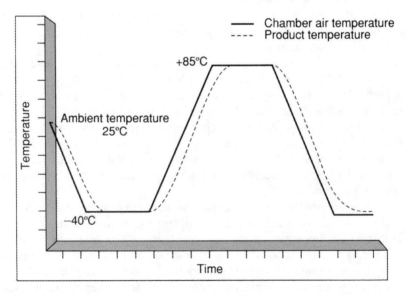

Figure 7-3 During temperature cycling, product temperature varies somewhat more slowly than the chamber's air temperature. (*Environmental Stress-Screening Handbook*, 1988, Thermotron Industries, Holland, Michigan.)

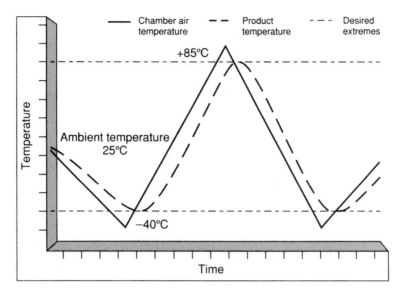

Figure 7-4   Some manufacturers eliminate soak time completely. (*Environmental Stress-Screening Handbook*, 1988, Thermotron Industries, Holland, Michigan.)

Certain manufacturers take this idea even further, eliminating soak time completely. Because under these circumstances the board surface never reaches equilibrium, spiking high and low temperatures in the chamber beyond product specifications is necessary for the board to reach its limits, as shown in Figure 7-4. Note that engineers must carefully monitor total heat added to and removed from the product/chamber environment. Loading the chamber with less than a full complement of boards without modifying other conditions could permit those boards to follow the chamber's temperature transitions more closely than normal, with disastrous results. In addition, powering boards contributes heat to the overall system.

The chamber temperature's time-rate-of-change significantly affects the screen's effectiveness, and can drastically affect its length. Smithson (1990) describes a case study that illustrates this point. Cycling between −40°C and +125°C at 5°C/minute revealed all defects in a particular surface-mount class in 400 cycles totaling 440 hours. Increasing the rate to 10°C/minute reduced the number of cycles to 55 and the corresponding time to 30 hours. At 15°C/minute, the same result required only 17 cycles and 6 hours.

At those rates, implementing temperature cycling involves batch-type processes, contributing to high capital costs and creating large amounts of work-in-process inventory. Such processes reduce manufacturing flexibility and prevent taking full advantage of just-in-time inventory-control procedures.

Increasing temperature-change rates further to 20°C/minute revealed all relevant problems in seven cycles and less than two hours. Four cycles and less than one hour sufficed at 25°C/minute. At 40°C/minute, the entire process required only one or two cycles and as little as 8 minutes.

These conditions permit employing continuous processes, which can reduce operating costs significantly. Manufacturers can even eliminate "boxes" altogether, subjecting products to temperature changes on conveyors during transfer from one test station to the next. Such high change rates, however, may exceed the capacity of mechanical-refrigeration methods, necessitating the use of liquid-nitrogen–based equipment.

7.4.3 Burn-in and Temperature-Cycling Equipment

Evaluating burn-in or temperature-cycling systems before purchase or installation requires examining a large range of alternatives. Modular systems allow simultaneously processing boards of various dimensions, electrical connections, and power requirements. Many modules of different designs can coexist in a single ESS chamber. An operator can change from one board type to another merely by swapping modules.

In power-cycling situations, some manufacturers request soft-start, soft-stop, and zoning capability. Soft-start and soft-stop ramp power up and down rather than allowing line conditions and the laws of physics to control rise times and fall times. These approaches prevent the damage that voltage spikes can cause.

Zoning applies power to only some boards in a burn-in or cycling module at any particular time. In addition, distributing powered boards rather than bunching them together in one location ensures more even heat dissipation and a more consistent temperature profile. Zoning also alleviates sags in wall voltages, particularly important if the chamber uses three-phase power.

A technical flyer from Micro Instrument Company (circa 1989) recommends that chamber power supplies should be rated at least 20 percent above the maximum that the screen requires. Voltage monitoring to permit screen interruptions helps protect devices from spikes. To protect boards during setup, the chamber may have to execute a specific on-and-off sequence of several supplies. Also, fuses at power-supply points can protect supplies from damage caused by board shorts. The flyer suggests that ESS systems sense voltage levels at the board, rather than at the supply, to accommodate line losses. It also recommends locating sense displays so that they are easily visible from outside the chamber.

Employing a computerized chamber-control system ensures accurate screen conditions. Parameters such as temperature, time, number of cycles, ramp speed, ramp duration, and total time in the chamber are available for quick retrieval and analysis. The computer engine can also store test programs for execution during burn-in or temperature cycling, as necessary.

According to Thermotron's *Environmental Stress-Screening Handbook* (1988) and the more recent *Fundamentals of Accelerated Stress Testing* (1998), thermal cycling provides the most effective single screen from the standpoints of total defects found, cost, and screen time per defect. The technique subjects boards to uniform stresses, facilitating control, evaluation, and modification for particular circumstances. Unlike some other approaches, it is unlikely to create problems that did not already exist.

To provide the best thermal-cycling operation, the handbook recommends setting temperature extremes as far apart as possible within the product's tolerance limits, starting with differences of at least 100°C. Equipment should provide a temperature rate of change of at least 5°C/minute. To permit the product's temperature profile to closely mimic the air temperature in the chamber, airflow velocity should exceed 750 fpm (feet/minute) at the product. Of course, modify these numbers and determine the appropriate number of cycles to precipitate latent failures for your particular situation based on empirical experience.

7.4.4 Thermal Shock

Theoretically, thermal shock is thermal cycling with an infinite rate of temperature change or a zero transition time between temperature extremes. In practice, thermal cycling refers to products sitting in a single chamber that changes temperature using flows of hot and cold air. In contrast, thermal shock uses a hot box and a cold box, with the product moving between them manually or on a conveyor, as shown in Figure 7-5. The technique can be a cost-effective way to screen for component-level defects, especially on ICs, where only a high rate of temperature change can expose latent defects. The *Handbook* cautions, however, that a thermal-shock-screen's severe rates of change may cause needless damage, particularly on complex assemblies containing many components other than ICs.

Thermal shock has several other drawbacks. Normally, one box is always empty, which effectively halves process capacity. The act of transferring boards between boxes can introduce problems. People who handle boards can drop them or subject them to electrostatic discharge. Mechanical conveyors can fail. The logis-

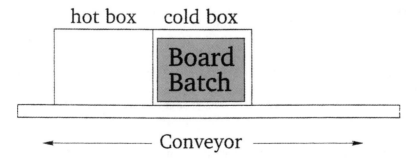

Figure 7-5 Thermal shock uses a hot box and a cold box, with the product moving between them manually or on a conveyor. (Schlagheck, Jerry, 1988. *Methodology and Techniques of Environmental-Stress Screening*. ESSC, Cincinnati, Ohio.)

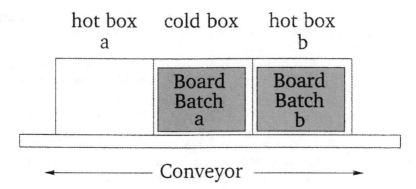

Figure 7-6 To minimize thermal shock costs, Schlagheck (1988) advocates a three-chamber approach. (Schlagheck, Jerry, 1988. *Methodology and Techniques of Environmental-Stress Screening.* ESSC, Cincinnati, Ohio.)

tics of transfer may make powering boards or monitoring failures in real time very difficult, compromising data collection and analysis.

Maintaining a low-temperature box is more expensive than keeping an identical box at high temperature. Therefore, to minimize thermal-shock costs, Schlagheck (1988) advocates a three-chamber approach, as shown in Figure 7-6. This variation processes two board batches at the same time. Batch *a* occupies hot box *a* while batch *b* resides in the cold box. When both batches reach equilibrium, a conveyor moves them so that batch *a* now occupies the cold box, while batch *b* enters hot box *b*. The cold box is always occupied except during material transfer, permitting a smaller box for the same throughput, which translates to lower refrigeration costs and therefore lower overall screen costs.

7.4.5   Mechanical Shock and Vibration

Mechanical shock subjects a board or system to a single, nearly instantaneous displacement. Often called the "drop test," it simulates the effect of proximity to an explosion or to being dropped during shipping or operation. Manufacturers of portable instruments, notebook computers, personal digital assistants (PDAs), and other hand-held products often require that at least some of their products endure this particular indignity. Examples include the "drop out of an airplane's overhead compartment" from the common notebook-computer "torture test." If performed only on samples, this step verifies that the product's design will endure rugged use. As a true screen, mechanical shock detects loose components, inadequate solder joints, and other problems.

Shock screening is often a manual procedure and is therefore relatively inexpensive. It is difficult to control, however, and stress magnitudes are generally difficult to measure. Manufacturers usually apply the technique only to products that may experience such severe conditions during normal use.

Vibration applies repetitive mechanical stresses to a board or system at one or more frequencies. There are three basic types: random, sine-vibration fixed frequency, and sine-vibration swept frequency.

Random vibration stresses products over a wide frequency range, usually 20 Hz to 20 kHz. The screen simultaneously excites all resonant frequencies within that range.

Apparatus includes an electrodynamic shaker and a closed-loop digital control system. Fixturing between the shaker and the board must be extremely rigid to transfer stresses properly and to ensure screen accuracy and repeatability.

The shaker may vibrate along one or more rotational axes. Stressing more than one axis generally reveals more faults. Empirical evidence suggests, however, that whether the shaker vibrates along the axes consecutively or simultaneously has little effect on the screen's overall effectiveness. Simultaneous vibration increases shaker cost; consecutive vibration requires longer screen execution times, with the expected impact on throughput and work-in-process inventory.

Random vibration exposes loose solder joints, improper component-to-board bonding, and shorts. It often takes much less time than other techniques, quickly aggravating many intermittent failures. Smithson (1990) reports a case where a shaker offering six uncorrelated, noncoherent, independent vibration axes uncovered 90 percent of all assembly-level defects in five minutes. He contends that, like thermal cycling, vibration screening allows full automation. Shakers can be less expensive than thermal chambers on a cost-per-defect-detected basis. In addition, if vibration occurs during diagnostic testing, it does not even add a process step. According to the *Handbook*, random vibration is the most effective vibration technique and uncovers more failures than any other screening method except temperature cycling.

On the downside, shakers for random vibration generally cost more than comparable equipment for other vibration techniques. Installation, control, and maintenance may be expensive as well. A board undergoing random vibration also experiences higher stresses at the center than at the edges where it attaches to the fixture. Therefore, to avoid overstressing components in the center, specified stress levels might not adequately precipitate faults near the board edge.

In *sine-vibration, fixed-frequency* screening, a mechanical shaker applies stresses as sine waves at a single frequency, usually 60 Hz or less. The technique is less expensive and easier to control than the random approach, but it also uncovers fewer defects.

Sine-vibration, swept frequency involves a hydraulic shaker operating at multiple frequencies up to 500 Hz in a predetermined pattern. Its cost and effectiveness are similar to the fixed-frequency case.

Some companies, not wanting to subject every product to a vibration screen because of time or cost, employ this approach as a process monitor. That is, they vibrate product samples from manufacturing and carefully analyze any boards that fail. Failure mechanisms may indicate correctable process problems. Success of this variation depends on adequate testing procedures after the screen.

7.4.6 Other Techniques

Electrical stress exercises board circuits and stimulates semiconductor-junction temperatures. It consists of power cycling, which turns product power on and off, and voltage margining, which varies board voltages above and below nominal limits, and may occur during a test step. A functional test, for example, may run 5-V power supplies at 5.2 V and 12-V supplies at 12.5 V to increase fault coverage. The approach is inexpensive and easy to control but does not uncover many defects. Some manufacturers combine electrical stress with other screen conditions during functional test to increase their effectiveness. Finding soft failures may require power on the board.

Physical screening approaches include *hermeticity*, which detects leakage of gases into and out of device packages, boards, and systems containing conformal coatings. Hermeticity screening also verifies behavior at high altitudes and susceptibility to attack by particulates such as sand and dust. Manufacturers whose products are destined for unusually rugged environments may want to subject them to this screen.

Radiation hardness looks for failures from individual radioactive particles, referred to as "single-event upsets." It also examines circuit reliability in radioactive environments. Some military products must undergo this screen before certification.

Corrosion screening verifies resistance to chemical contamination, including salt sprays and other hostile conditions. Again, devices used in military applications, such as naval shipboard service, may require this screen.

7.4.7 Combined Screens

Figure 7-7 contains a bar graph from the *Handbook*, indicating the relative effectiveness of various screening techniques. The Venn diagram in Figure 7-8 shows that each approach produces some unique failures, whereas in other areas their fault coverage overlaps. Therefore, combining screens allows one to find the largest possible number of problems and permits shipping the most reliable products. (Note that Figure 7-8 is representational only and does not imply the different techniques' relative effectiveness in any actual manufacturing operation.)

Companies opting for a single screen generally choose temperature cycling. In addition to finding the most failures, it subjects boards to uniform stress, which in turn makes it more controllable than alternatives. Cycling also offers better flexibility in implementation, and adjusting screening conditions to meet the needs of specific situations is easier than with other methods.

Evidence indicates that temperature cycling and random vibration together produce the best total fault coverage for a two-step process. This common combination has earned ESS its nickname "shake and bake." The freedom to enhance an ESS program's comprehensiveness depends on budgets, floor space, personnel availability and skills, allowable work-in-process inventory levels, and allowable manufacturing cycle lengths.

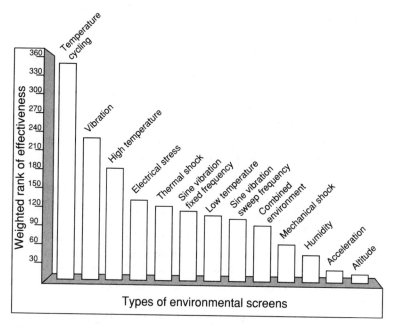

Figure 7-7 Relative effectiveness of various screening techniques. (*Environmental Stress-Screening Handbook*, 1988, Thermotron Industries, Holland, Michigan.)

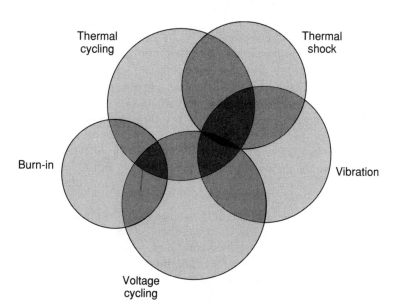

Figure 7-8 Each screening approach produces some unique failures, while in other areas their fault coverage overlaps.

7.5 Failure Analysis

What a manufacturer does with ESS information determines the screening program's success. Analysis of that information indicates whether failures result from vendor problems, handling, assembly operations, or other process steps.

Also important is where you place ESS and test steps. A couple of years ago, a large-system manufacturer farmed out board assembly and process test to a contractor. Incoming boards underwent a temperature-cycling step before functional test. The 74 percent board-yield from the functional test represented considerable cause for concern. However, because testing was not performed before temperature cycling as well as after, the manufacturer could not determine if the failures were introduced by the CM during assembly or handling, or by the temperature-cycling step itself. Lack of this vital piece of the puzzle precludes any kind of process feedback to reduce future failures.

For each ESS-related failure, manufacturers should determine failure symptoms, actual environmental conditions, cycle number (if applicable), time into that cycle or time into the screen, and, based on a physical examination of the assembly, the failure's cause. Correcting problems at the design or process level can eliminate most ESS failures altogether.

As a product matures, applying ESS conditions to only a sample of products becomes more practical. At that point, ESS ceases to be a true screen. Instead, it represents a monitoring technique for ensuring that manufacturing processes remain under control. This change of procedure frees ESS capacity that managers can then assign to other products.

7.6 ESS Costs

Figure 7-9 outlines cost savings that could result from adding an electrodynamic shaker to a 40,000-board assembly and test process. In this example, the shaker, digital-control system, and fixture cost $125,000 depreciated over 3 years, or $41,667 per year. Screening occurs in one shift. Operating costs total $50,000 per year. Therefore, screening costs $2.29 per board. If test after screening reveals problems on 3 percent of all boards and diagnosis-and-repair averages $0.30 per board over the entire production run, total screening cost is $2.59 per board.

In contrast, not screening means that 3 percent of the boards will fail in the field. If correcting a field failure costs $1000, total cost is $1,200,000, or an average of $30.00 per board. Therefore, as the figure shows, this one vibration screen saves $1,108,400. Of course, much lower field-failure rates or field-repair costs will significantly reduce this screen's advantage.

Manufacturers must diligently reevaluate screening costs and benefits on a regular basis to be sure that continuing the program is justified. High-yield products that rarely fail in the field may not require ESS at all.

Screened Boards

Capital Costs $41,667
Operating Costs $50,000

Total Costs $91,667
Cost per Board $ 2.29
                      plus diagnostic and rework costs

Unscreened Boards

Defects on 3% of boards
Average field repair costs $1,000/board

Field Failures $40,000 \times 0.03 = 1,200/\text{yr}.$
Field Repair Cost $1,200 \times \$1,000 = \$1,200,000$
Cost per Board $30.00

Savings on Screened Boards

Savings per Board $\$30-2.29 = \$27.71$
Annual Savings $\$27.71 \times 40,000$
                           $= \$1,108,400$

Figure 7-9   Cost savings that could result from adding an electrodynamic shaker to a 40,000-board assembly and test process. (Scheiber, Stephen F. 1989. "The New Face of Environmental Test, Part 2," *Test & Measurement World*, Newton, Massachusetts.)

7.7 To Screen or Not to Screen

Despite its huge potential benefits, implementing a comprehensive ESS program is an expensive, time-consuming, and operationally awkward endeavor. It requires precious factory floor space and increases both cycle time and work-in-process. Also, many of the fault classes it finds—including inadequate or missing solder—will yield to the esthetically more acceptable inspection. So is ESS worth the effort?

Like many questions in creating a board-test strategy, the answer is a firm "it depends." Again, you have to consider the product, its operating environment, and the consequences of failure.

ESS has become routine, for example, in the automotive industry. Automobiles provide the most hostile common environment that electronics must endure—searing heat, freezing cold, voltage spikes, and bone-crushing vibrations. If boards designed for that application cannot withstand those conditions, they will frequently fail in the field. And few failures annoy customers as much as when their cars break down.

By contrast, desktop computers destined for office use live in clean, relatively controlled environments. A board that would fail when a car hit a pothole will likely function for the life of the PC. However, if the PC is destined for the factory floor in, say, a chemical plant, the conditions under which it must survive are different again.

While a desktop computer generally stays put for its useful life, a notebook machine must endure constant handling, changing environmental conditions (being left in a car trunk, for example), spilled coffee, and other indignities.

Each of these situations suggests an array of test, inspection, and ESS steps. The test engineer must evaluate their relative economic and technical benefits to determine the best mix of techniques in each case.

7.8 Implementation Realities

Schlagheck (1988) observes that, because few civilian guidelines exist, manufacturers establishing ESS programs often rely on some "magic" reliability target, rather than on sound engineering principles and practices. For military and other high-reliability applications, official and semi-official specifications exist. Rather than following these standards blindly, however, Schlagheck recommends tailoring every ESS program to meet its product's specific needs, if possible.

In addition, reference documents leave a number of unanswered questions. For example, what thermal and physical conditions will adequately stress boards and systems without damaging them? For temperature cycling, what are minimum and maximum rates of temperature change? Should temperature measurements occur at a chamber's airflow input, output, or on ICs and assemblies under stress? How many axes of vibration provide the best fault coverage? Should vibration and thermal cycling occur consecutively or simultaneously? What is the appropriate thermal or vibration profile?

Victor (1989) defines the goal of an ESS program as allowing the product to exhibit a user-specified MTBF (S), which is lower than the MTBF (P) that an engineer would predict by combining MTBFs of components, bare boards, and other constituent parts. The two numbers are equal only when the screen is perfect and the product has no design flaws. The wider the disparity between S and P, the smaller the percentage of total latent faults that the screen must find and, therefore, the more residual faults the customer will accept. Victor establishes an adjusted target MTBF (A) that is somewhat more stringent than the customer requirement to ensure an adequate screen:

$$A = S + 0.25(P - S) \qquad \text{(Eq. 7–3)}$$

From excess MTBF factor $X = A/P$, the allowable number of residual defects in the final product R, and the 2σ number of actual defects per board or assembly before screening (D), Victor calculates a screen strength S_s as:

$$S_s = 1 - \frac{R}{D} \qquad \text{(Eq. 7–4a)}$$

$$R = \frac{1}{X} - 1 \qquad \text{(Eq. 7–4b)}$$

For a perfect screen, $S_s = 1$. Consider, for example, an assembly for which the predicted MTBF is 2240 hours, and the customer specifies 2000 hours. In this case,

$A = 2000 + 0.25 \ (2240 - 2000)$

$A = 2060$

$R = 2240/2060 - 1 = 0.09$ allowable faults per board

If each board contains an average of 1 fault, Victor's required screen strength would be 0.91. A reliability engineer then designs an ESS program with that level in mind.

7.9 Long-Term Effects

Many manufacturers have expressed concern that ESS stresses unacceptably shorten product life. Steinberg (1989) developed a model that specifically addresses this issue. He contends that the number of fatigue cycles (N) to circuit failure depends on stress levels S in psi (pounds per square inch) and an empirical coefficient b. Figure 7-10 shows failure curves for 63%-tin, 37%-lead solder during thermal cycling and during vibration screening at room temperature.

Note two significant characteristics of this figure. When plotted on a log-log scale, most of each curve is linear. In addition, for a given stress level, vibration produces fewer failures than thermal cycling. Solder joints fatigue sooner with slow stress changes, as in thermal cycling, than when stress changes more rapidly, as with mechanical vibration or shock.

Steinberg describes the linear portion of the relationships as follows:

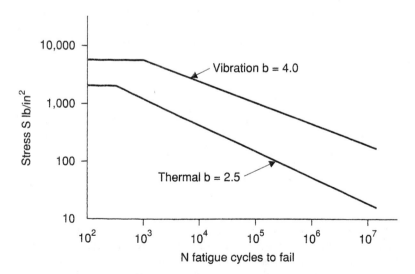

Figure 7-10   Failure curves for 63%-tin, 37%-lead solder during thermal cycling and during vibration-screening at room temperature. (Steinberg, Dave S. 1989. "Fatigue Life in Temperature-Cycling Environments," Nepcon West, Reed Exhibitions, Stamford, Connecticut.)

$$N_1 S_1{}^b = N_2 S_2{}^b \qquad \text{(Eq. 7–5)}$$

In linear systems, the number of cycles to failure is proportional to the time t, whereas stress levels depend on acceleration level G and perpendicular displacement Z, producing the following equations:

$$t_1 G_1{}^b = t_2 G_2{}^b \qquad \text{(Eq. 7–6)}$$

$$N_1 Z_1{}^b = N_2 Z_2{}^b \qquad \text{(Eq. 7–7)}$$

$$N_1 G_1{}^b = N_2 G_2{}^b \qquad \text{(Eq. 7–8)}$$

Using these equations, engineers can determine if a board design will withstand a particular set of environmental-screening stresses.

Steinberg's model assumes that solder joints represent the most likely ESS-related failure, then attempts to calculate whether a specific stress will unacceptably increase the likelihood of solder failure within an anticipated 15-year service life. As a rule of thumb, he recommends keeping stress levels under 300 psi to 400 psi, especially with products aimed at military and other high-reliability applications.

Evidence suggests that briefly higher stress levels during vibration screening will not unduly compromise solder-joint integrity. At the same time, however, extended vibration periods can quickly subject boards to many millions of stress cycles, which may cause fatigue failures. During long exposure to vibration, Steinberg recommends keeping stresses below 400 psi to prevent solder cracking.

As discussed, stresses during thermal cycling result largely from differences in thermal coefficients of expansion of the various materials on the board. Counteracting those stresses by reducing the stiffness, or *spring rate*, of lead wires also reduces corresponding stresses on solder joints.

Thermal expansion forces (P) depend on the relative displacements of the materials Y and the spring constant K.

$$P = KY \qquad \text{(Eq. 7–9)}$$

Magnitude of the spring constant depends on the cause of the stress. During bending, K relates to the modulus of elasticity of the lead material E, the cross-sectional area moment of inertia I, and the wire length L:

$$K = \frac{EI}{L^3} \qquad \text{(Eq. 7–10)}$$

Because of the cubic relationship between K and L, Steinberg recommends increasing wire lengths, such as by looping the wires, which reduces stress very rapidly. *Coining* the wire (pressing round wires into thin flat strips) has a similar effect, because moment of inertia I is a cubic function of the wire's height.

When screening stresses result from tension, K relates to cross-sectional area A and length L as follows:

$$K = \frac{AE}{L} \qquad\qquad \text{(Eq. 7–11)}$$

Again, lengthening the wires will cut the spring constant but much more slowly than in the bending case.

These equations permit calculating the total stress that each component on a board or system experiences during various ESS steps. By adding these stresses to the stresses that the product sees during its service life, a manufacturing engineer can determine the likelihood that any board component will fail. If analysis indicates that each component (including board traces and other circuit elements) will survive the anticipated service, then the board or system will likely survive as well.

Of course, this analysis breaks down if the devices in question are surface-mounted. In that case, only solder paste holds devices on the board. "Give" in surface-mount devices is extremely limited. Stresses that would not bother a through-hole board could prove disastrous.

Actual calculation examples are beyond the scope of this text. Steinberg and other sources offer such examples. In general, experience has shown that a carefully designed ESS program that takes the product's total stress tolerance and likely service into account will not compromise its useful life.

7.10   Case Studies

Thermotron's *Environmental Stress-Screening Handbook* (1988) offers several case studies to illustrate ESS implementations.

7.10.1   Analogic

One Analogic product was failing too often in the field even after conventional burn-in. Replacing burn-in with a thermal-cycling step cut the failure rate in half, uncovering previously unseen IC and display faults. In addition, thermal cycling took less time than burn-in had, reducing overall production time. Based on these results, the company installed thermal-cycling systems on six more production lines.

7.10.2   Bendix

Bendix was experiencing 23.5 percent field failures of its automotive fuel-injector systems. Such failures disable the vehicles, and repair is expensive to vendors and customers alike. Because of the product's complexity (420 components and 1700 soldered joints), the company selected temperature cycling over burn-in or other more conventional techniques. This screen uncovered solder, workmanship, and device defects—all of which generated soft failures—and reduced the field-failure rate to only 8 percent.

7.10.3 Hewlett-Packard (now Agilent Technologies)

Replacing a burn-in system with thermal cycling on the 9826A production line reduced screening time from 2 to 5 days to only a few hours. The increased throughput meant lower unit costs and lower costs for troubleshooting and line repairs. The new process also uncovered two to three times as many defects. Warranty field repairs fell by 50 percent. Altogether, over a 5-year period the company saved more than $1.5 million.

7.11 Summary

Environmental-stress screening attempts to accelerate failures that would otherwise occur during a product's early months in the field so that normal test procedures can find them before the product leaves the factory. To be a true screen, the process must apply to all products, rather than samples. Although ESS is not a test in itself, subsequent testing will reveal problems that the screen has precipitated.

Levels of complexity range from static screens, which subject boards and systems to a constant stress, to full functional methods, where circuits behave as they would during normal operation. Specific techniques include conventional burn-in, the less common but more effective temperature cycling, thermal and mechanical shock, vibration, electrical stress, hermeticity, and corrosion and chemical sensitivity methods.

Although ESS is not inexpensive, its proponents insist that its benefits justify the cost. One drawback to implementation is the lack of coherent guidelines and standards as to what constitutes an adequate screen and how to determine when a product is free of screenable defects.

Some authorities raise the question of whether ESS significantly shortens a product's useful life. Current best evidence indicates that a carefully designed program has no such deleterious effect.

CHAPTER **8**

Evaluating Real Tester Speeds

Anyone creating a board-test strategy must evaluate available equipment to ensure that it can adequately execute the desired tests. To accomplish this, many test engineers and managers merely compare test-equipment specifications to comparable board and system performance numbers. Unfortunately, this approach involves more blind faith than solid engineering.

Tester specifications represent a vendor's best-case scenario. Singly, they accurately characterize tester capability. In actual applications, however, individual factors interact, effectively degrading overall equipment behavior. In addition, design of the board under test introduces its own limitations, as does the board-to-tester interface. Test professionals must consider these issues both when assessing existing equipment capabilities in the factory and when choosing additional equipment for a proposed new strategy.

Consider an analog circuit containing a sizable capacitance. Even a fast tester must wait for the circuit to settle before taking a DC-voltage measurement or else take numerous measurements during the transition and extrapolate to a final value. Either technique produces slower-than-advertised test speeds.

Digital-test realities are somewhat subtler. A 100-MHz tester can no more achieve that exalted performance level in actual use than an off-the-showroom-floor automobile can reach 120 mph without spinning out of control.

A tester's *maximum pattern rate* describes how fast it can deliver signals to the unit under test (UUT). This number ignores the effects of other tester specifications and activities such as verifying outputs or changing stimulus data or direction on the fly.

In contrast, *effective pattern rate* takes into account all tester specifications and fixture interactions, as well as such UUT characteristics as maximum speeds and capacitances. This quantity defines the fastest test consistent with repeatability and control and may be considerably lower than even the UUT's top speed.

Clock rate is the speed at which a tester can produce clock cycles containing two or more transitions each. *Data rate* expresses the same information but includes only one transition per cycle. In Figure 8-1, clock rate is half the data rate. Multiphase clocks increase this disparity. Many vendors prefer to quote their

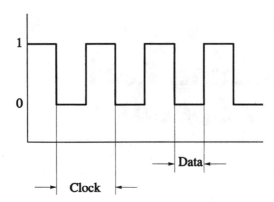

Figure 8-1 Clock rate and data rate.

specifications as data rates because the higher numbers sound better. An unwary customer might assume that one tester's 80-MHz data rate is superior to another's 50-MHz clock rate, when reality is the other way around.

8.1 Resolution and Skew

Clock rates and data rates consider no encumbrances on test speeds. In fact, numerous factors interfere with a tester's ability to run that fast. *Skew*, for example, represents the error around clock-signal boundaries, as shown in Figure 8-2. To avoid overlapping test stimuli, programmers must apply driver and detector edges outside skew bands, consuming much of the available pattern period and thereby reducing the amount of information that each pattern can contain.

Combinational UUTs require only that detectors wait until the last stimulus skew (and any associated propagation delay) expires before measuring the response. Sequential circuits present a much more serious problem. If a programmer calls for a driver or detector signal at time t on a tester with skew s, the signal could occur as early as $t - s$ or as late as $t + s$. Therefore, programmers must place an edge no less than $2s$ after another edge on which it depends. Allowing for skew on test patterns that feature a large number of separate edges is very complicated.

Resolution defines the minimum increment by which a programmer can move or place an edge in a pattern cycle. A 10-ns resolution, for example, permits specifying tester edges at 0 ns, 10 ns, 20 ns, 30 ns, and so on, but prohibits calling for an edge at, say, 25 ns. In Figure 8-3, suppose that tester resolution is 10 ns and the UUT requires asserting tester edge 2 no less than 12 ns after edge 1. The earliest that the tester can place edge 2 is 20 ns after edge 1, introducing 8 ns of resolution "error."

Although coarse resolution will not affect slow testers enough to cause significant problems, high-speed machines featuring pattern periods down to tens of

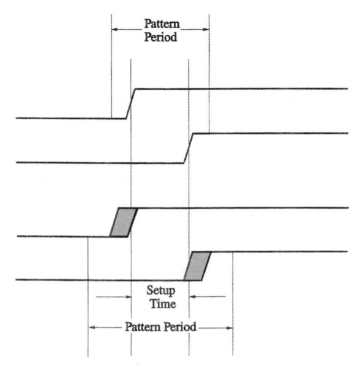

Figure 8-2 *Skew* represents the error around the clock-signal boundaries. (Arena, John and Stephen Cohen. 1989. "Tester Specs vs. Realistic Tester Speeds, Part I," *Test & Measurement World*, Newton, Massachusetts.)

nanoseconds are much more sensitive. On the other hand, fine resolution may provide a better match between tester and UUT speeds, but too fine a resolution may not sufficiently justify its necessarily higher cost. Arena and Cohen (1989) recommend tester resolutions no less than one-tenth of the skew.

Some older testers also require a *dead time*—a specified period during a pattern cycle when the tester cannot apply signals or measure responses. The demands of modern high-performance testing have forced tester designers to eliminate dead time on most of today's high-speed machines.

Figure 8-4 combines these factors to show their collective impact on actual speeds during a flip-flop test. This tester features driver skew of ±30 ns, receiver skew of ±20 ns, 10-ns resolution, and a 10-ns-wide measurement pulse. The flip-flop under test requires 5 ns of setup between data and clock signals and a 22-ns propagation delay. The tester must apply a data signal, wait for the setup time, clock the data, wait for propagation, and measure the device's outputs. On ideal equipment, the test would take 37 ns—setup time plus propagation delay plus the measurement pulse width.

The actual tester shows a somewhat different result. Because of driver skew, asserting the *D* line at time 0 means that the data will first be active somewhere

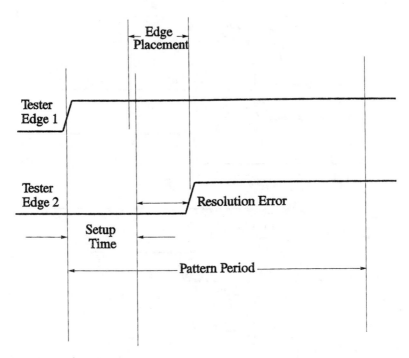

Figure 8-3 Example of tester resolution. (Arena, John and Stephen Cohen. 1989. "Tester Specs vs. Realistic Tester Speeds, Part I," *Test & Measurement World*, Newton, Massachusetts.)

between −30 ns and +30 ns. A test programmer must assume a worst-case scenario. Adding flip-flop setup time, the tester can apply a clock no earlier than +35 ns. Again considering the 30-ns skew, the program must specify the clock for at least +65 ns. The tester's 10-ns resolution permits the first available clock-signal location at +70 ns. Therefore, the clock could assert as late as +100 ns. Adding the propagation delay, the measurement strobe must not come any earlier than +122 ns. The 20-ns receiver skew requires placing the strobe no earlier than +142 ns—+150 ns with the tester's resolution. Adding the width of the strobe and the worst-case skew on the other end, the test could stretch out to +180 ns. Therefore, this simple 37-ns test could take as long as 180 ns + 30 ns, or 210 ns.

8.2 Voltage vs. Time

Another consideration when evaluating apparent test-system capability is the speed with which digital drivers accomplish transitions between logic states. *Slew rate* expresses that transition in volts per nanosecond. If a tester offers a constant slew rate, the total time required for a particular logic swing depends on the difference between end-point voltages.

Resolution	10 ns
Measurement window	10 ns
ATE driver skew	±30 ns
Flip-flop setup	5 ns
Flip-flop prop delay	22 ns
ATE receiver skew	±20 ns

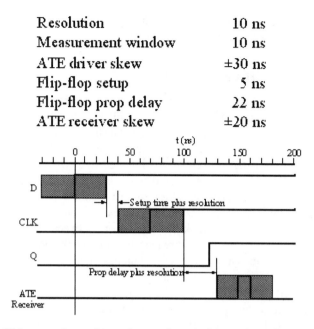

Figure 8-4 This example combines skew and resolution to show their impact on actual speeds during a flip-flop test. On ideal equipment, the test would take 37 ns—setup time plus propagation delay plus the measurement pulse width. Under these less-than-ideal conditions, the test could take as long as 210 ns (180 ns + 30 ns). (Arena, John and Stephen Cohen. "Tester Specs vs. Realistic Tester Speeds, Part I," *Test & Measurement World*, 1989.)

In contrast *rise time* defines how long a tester driver takes to travel from 20 percent to 80 percent of its full logic swing. Therefore, the time for a logic swing on a constant rise-time tester is independent of the size of the swing.

Because most UUT digital logic features fast edge speeds, many engineers consider this criterion the only basis for comparing test systems. Fast edges reduce the likelihood that extraneous noise in a circuit will inappropriately flip sequential logic states, while minimizing device-threshold input errors that can compromise timing accuracy.

Arena and Cohen contend, however, that fast tester edges increase transmission-line effects, such as signal reflections and overshoot, which reduce test accuracy by degrading waveform integrity and otherwise interfering with normal circuit behavior. The optimum test system, therefore, must establish a proper balance between edge speeds and transmission-line effects.

Some systems address this issue by permitting variable slew rates under program control. Although this solution increases test flexibility, a lack of objective guidelines for setting the rates complicates test-programming efforts considerably.

In addition, signal speed at the tester wins only part of the battle. The path between tester drivers and the UUT must faithfully deliver input signals and carry back responses. Fixtures must ensure clean edges, and tester drivers must absorb reflections from high-impedance UUT inputs.

8.3   Other Uncertainties

Digital-device input thresholds also exhibit a margin of error. Suppose that a device specification indicates that outputs will switch from 0 to 1 when input voltage reaches 1.6 V. Some devices will switch at a slightly lower voltage, and others will not switch until the input voltage is higher. Because edge speed is finite, this error adds to the tester-driver skew and is referred to as *input-threshold skew*.

Fixture architecture and associated wires contribute additional delays that depend on the roundtrip distance between tester and UUT. Clamshell designs and the need to accommodate tester multiplexing add significantly to fixture wirelengths. Recent tester designs keep wires short between drivers and receivers. Unfortunately, in some cases, total wire lengths can still exceed 2 to 5 *feet*. Because a signal's travel time totals around 1 ns per 9 inches of straight wire, these distances can unacceptably extend propagation delays.

On clamshell fixtures, wire lengths to nodes on the top and bottom of the board under test are often quite different. Fast sequential circuits may experience performance anomalies because theoretically simultaneous signals do not reach target board inputs at the same time.

Arena and Cohen contend that objectively evaluating the aggregate effect of machine specifications on test processes may be difficult. The only observable evidence that a test exceeds tester abilities may be inadequate production yields.

In assessing the impact of tester specifications on individual test situations, engineers sometimes assume better than worst case for skew and other parameters. Empirical evidence with a particular tester or class of UUT may encourage this approach. In circumstances where a tester's specifications indicate that its ability to meet test conditions is questionable—and when no alternative exists—basing test-solution decisions on trial-run results may be unavoidable. Test professionals should remember, however, that two testers of the same type and one tester at two different times might not perform identically. Generally speaking, erring on the conservative side represents the safest course.

At the same time, a tester that cannot meet objective UUT performance criteria may still perform adequately. Most UUT parameters represent ideal, rather than minimum, conditions. Some years ago, a tester manufacturer experimented on a board based on a 6809 microprocessor to see how slowly a test had to run before the board failed. Although the written specification called for 1-MHz clock rates (admittedly very slow by today's standards), the board performed acceptably down to 25 *kHz*. Readers should not take this incident to endorse in any way executing tests that far below specifications. It merely points out that an appropriate test passes good boards and fails bad ones. With device speeds on many of today's boards reaching 1 GHz and beyond, finding testers to match that level of performance may be less important than finding testers—at whatever speed—that can catch and diagnose faults accurately.

Also remember that on many boards, internal device speeds far exceed I/O speeds between devices. A 1 GHz Pentium microprocessor, for example, may communicate over a 133 MHz bus. For testing purposes, bus speed is more important.

8.4   Impact of Test-Method Choices

In addition to specification interactions, actual tester speeds depend on test-method choices. That is, a functional tester and an in-circuit tester may theoretically run at the same speed, but waveform integrity at that speed is generally very different. For example, *overdrive delay*, which is a function of an in-circuit test system's driver and fixture electronics, interferes with the tester's ability to deliver test-patterns to the UUT. For many typical devices, delay δ may be expressed as:

$$\delta = \frac{L}{R} \qquad\qquad \text{(Eq. 8–1)}$$

where L represents the fixture inductance and R is the series resistance. Corresponding time-dependent voltage response at the UUT node is given by:

$$V(t) = V_{\max} \left(1 - e^{-t/\delta}\right) \qquad\qquad \text{(Eq. 8–2)}$$

Unfortunately, modern complex devices often do not conform to these simple models. First, rise times can vary drastically with overdriven-device input impedance, as Figure 8-5 shows. A lower impedance means that logic transitions require more current. Overdriving some advanced-technology devices can require currents

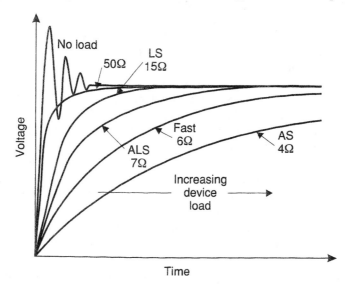

Figure 8-5   Risetimes can vary drastically with overdriven-device input impedance. (Arena, John and Stephen Cohen. 1990a. "Tester Specs vs. Realistic Tester Speeds, Part II," *Test & Measurement World*, Newton, Massachusetts.)

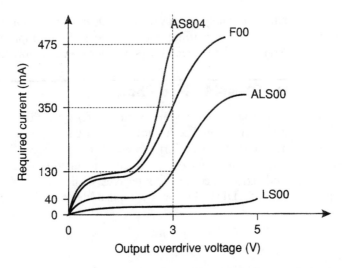

Figure 8-6 Some typical devices whose input impedance changes during testing. (Arena, John and Stephen Cohen. 1990a. "Tester Specs vs. Realistic Tester Speeds, Part II," *Test & Measurement World*, Newton, Massachusetts.)

of 1 A or more, increasing the likelihood that testing will cause secondary damage. Low output impedances and parasitic fixture inductances also lengthen delays from tester drivers to UUTs and back to tester receivers.

The input impedance itself may change during testing. Figure 8-6 offers some typical examples. In addition, one device's impedance may be two or more times greater than the impedance of another device of the same logic type, depending on internal resistor values and other design or production factors.

Arena and Cohen (1990a) offer the sample delay-time calculation in Figure 8-7. In this case, an in-circuit tester drives a FAST device whose impedance varies between 2Ω and 10Ω. With a 3-foot fixture wire featuring a 1-µH inductance, the signal takes between 70 ns and 347 ns to reach its logic threshold. Shortening wire lengths to 1 foot reduces the overdrive delay to between 23 ns and 116 ns. Even at this level, the delay interferes with tester performance and signal quality. This fact constitutes the primary reason that practical in-circuit test speeds cannot keep pace with speeds of the board under test.

A functional tester's low output impedance generally drives high impedances at the board edge, often producing overshoot, reflections, and ringing. These phenomena can result from fixture-path length, tester-driver to fixture-path impedance mismatches, fast signal edge speeds, and UUT impedances that vary with time or device type.

Arena and Cohen anticipate reflections during testing if propagation between tester and UUT takes longer than 20 percent of signal rise time. When fixture wiring contains that much of the signal's wave front, the fixture acts like a classical transmission line.

From Equations 8-1 and 8-2

$$t = \left(\frac{L}{R}\right) \ln\left(\frac{V}{V_{max}}\right)$$

Logic threshold (V) is 2 volts
Logic swing (V_{max}) is 4 volts
Fixture-wire inductance (L) is 1μH
Overdriven impedance (R) is 2Ω

$$t = \left(\frac{1\times10^{-6}}{2}\right) \ln\left(\frac{2}{4}\right) = 347 \text{ ns}$$

Figure 8-7   Sample delay-time calculation. (Arena, John and Stephen Cohen. 1990a. "Tester specs vs. Realistic Tester Speeds, Part II," *Test & Measurement World*, Newton, Massachusetts.)

Reflections occur when logic-transition energy bounces off a high-impedance UUT node, then off the low-impedance tester channel, and back again, producing oscillating waveforms of decreasing amplitude. If the extraneous signals cross an input's logic threshold, this effect can create glitches, double-clocking, and other manifestations that render test results unreliable and unrepeatable. The researchers suggest that keeping reflection overshoot at less than 10 percent will likely prevent the associated test problems.

As testers and UUTs get faster, however, conforming to this recommendation is becoming more difficult. At one time, testing a typical digital circuit involved functional-driver rise times in the 20-ns to 50-ns range. Normal care in fixture design sufficed to meet the less than 10 percent criterion. Unfortunately, with modern circuits commonly running at 500 MHz and above (which translates into pattern periods of less than 2 ns), even very short fixture wires can permit impedance mismatches and their accompanying undesirable behaviors.

Therefore, Arena and Cohen insist on matching line terminations to fixture impedances to provide a controlled-impedance path between tester and UUT. One solution adds such terminations directly to fixtures. As an alternative, some tester manufacturers have designed a *terminated driver*, like the one in Figure 8-8. Here, a source resistor whose value equals the line impedance absorbs reflections between driver and load. According to the researchers, this arrangement permits slew rates of 1 ns/V to 2 ns/V—and therefore higher-speed testing. The version in the figure also allows bypassing the resistor to reduce DC errors when driving large loads, although this action defeats the reflection protection. Using this tester design, only fixture-wiring impedance variations contribute to signal distortion.

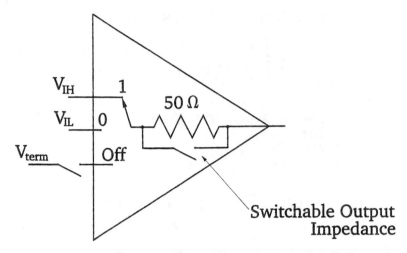

Figure 8-8 To match fixture impedances to line terminations, some tester manufacturers have designed a *terminated driver*. A source resistor equal to the line impedance absorbs reflections between driver and load. V_{IH} is the input-HIGH voltage, V_{IL} is the input-LOW voltage, V_{term} is the termination voltage. (Arena, John and Stephen Cohen. 1990a. "Tester Specs vs. Realistic Tester Speeds, Part II," *Test & Measurement World*, Newton, Massachusetts.)

Situations where board speed exceeds even a fast tester's capabilities introduce their own reflections. Resulting ringing further reduces test speeds, so that terminated tester *receivers* may also become necessary.

Arena and Cohen (1990b) offer additional sample calculations for determining whether a particular tester's actual performance will meet an electronics manufacturer's needs. Test engineers and managers should consider these issues carefully before settling on a specific test strategy and its associated tactics.

8.5 Summary

Vendors quote tester specifications in isolation, under ideal conditions. Because specifications interact, however, performance is generally lower than expected. In addition, fixture architectures can compromise waveform integrity and test timing, and the design of the board under test can make achieving the tester's best performance very difficult.

Some specifications express the same information in different ways. For example, *clock rate* is the speed at which a tester can produce clock cycles containing two or more transitions each. *Data rate*, on the other hand, includes only one transition per cycle. *Slew rate* defines digital data transition speeds in volts per nanosecond. *Rise time* defines how long a tester driver takes to travel from 20 percent to 80 percent of its full logic swing. Therefore, constant-slew-rate testers and constant-rise-time testers do not behave in the same way.

Test-method choices also affect tester speeds. In-circuit beds-of-nails contribute significant capacitance to test environments, so that square waves may show distinctly rounded edges. In contrast, functional testers can run closer to board-rated speeds with reasonable accuracy. Engineers and managers must consider these issues when evaluating existing test equipment and when selecting new equipment for a proposed test strategy.

CHAPTER 9

Test-Program Development and Simulation

Test-program development can represent the single biggest differentiator between two otherwise equivalent test strategies. The strategies may offer equivalent fault coverage, as well as comparable costs for equipment, operation, fixturing, and maintenance. Nevertheless, some test steps may permit largely automatic test generation by a person with only basic familiarity with board technology and layout, while others require manual trial-and-error programming by a test expert. The better the board is designed for test, the smaller this disparity will likely be.

Whatever strategy you choose, efficient test programming plays a large role in its implementation. A strategy's effectiveness can never exceed the quality of its test programs. At the same time, a bad program will compromise even the best test choices.

9.1 The Program-Generation Process

Figure 9-1 shows an idealized test-program-generation process. First, engineers subject the product design to a test-requirements analysis. If a company practices design-for-testability, either by itself or as part of a concurrent-engineering program, test engineers and design engineers perform this analysis together early in the product-development cycle, before the design is "set in stone." The more traditional approach has designers handing products to test engineers as *faits accomplis*. In the latter situation, subsequent design changes can affect the entire process, and may require starting again.

Test-requirements analysis examines test issues that relate to product design. The test requirements *document* takes a test point of view. Based on this document and an assessment of existing and potential test capabilities, engineers create a *test design* that specifies classes of test, likely fault coverages, and specific equipment needed. The document should also outline diagnostic techniques to facilitate board repair.

Once the test design is set, test-program generation can begin. Engineers can also define and build any necessary bed-of-nails or edge-connector test interfaces at this time. Test-program documentation from this stage should reflect the actual,

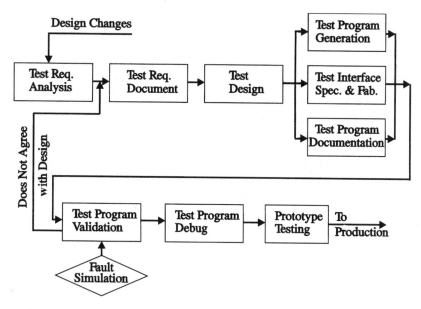

Figure 9-1 An idealized test-program generation process. (Racal-Dana. 1990. Sales Presentations, MRT, Winchester, Massachusetts.)

rather than the requested, program. Documentation should include program scope and fault coverage (noting where they differ from the test design), as well as operator instructions, expected test results, and how best to analyze results to improve product quality, monitor processes, and perform statistical process control. Engineers who prepare the documents should thoroughly understand both test designs and actual programs. In an ideal world, these three steps would occur simultaneously.

Many companies specify fixture construction and generate first-pass test programs based entirely on CAE information. Again, design changes at this stage will interfere with the process. If the test strategy includes either a shorts-and-opens tester or an MDA, manufacturers may try to build fixtures *before* program generation to permit self-learning as soon as actual boards become available. Because test documentation consumes scarce engineering resources, some companies opt for a less comprehensive version or forego this step entirely.

Inspection also allows learning programs from known-good boards or simulations. Nevertheless, most boards change rapidly during the last stages before release. A self-learned program is only as accurate as the board or representation from which it was generated. Execute that step too early and board changes make the effort worthless. Wait too long and you endanger time-to-market goals.

Test engineers must next validate programs against design requirements, using a combination of fault simulation and manual analysis. Automatic tools can help on primarily digital boards and on boards where digital circuitry is fairly well segregated. On heavily analog boards and on boards containing considerable mixed circuitry, such as A/Ds, D/As, and modems, however, manual approaches still provide the best results.

The validation step is theoretical and relates only to how well programs agree with corresponding designs. If they do not—assuming that original test-requirements analyses were correct—the entire process must begin again from the test-requirements-document step. Of course, no matter how carefully you verify a program, the only true test of its success is how well it passes good boards and fails bad ones in production.

Program validation may also reveal some conventional bugs, such as incorrectly specified test limits, incorrect timing sequences, and syntax errors. Applying debugged programs to board prototypes can reveal additional anomalies and may expose potential manufacturing problems that would otherwise surface only after production ramp-up.

This procedure certainly presents an ideal task flow. In practice, companies rarely execute all the steps in this way, combining or eliminating some steps to save time or money. Doing so, however, may reduce the resulting test strategy's success.

Managers find documentation in particular difficult to justify. Nevertheless, to support rapidly changing products, personnel turnover, and other disruptions requires clearly knowing the test-and-inspection steps in the accepted strategy.

9.2   Cutting Test-Programming Time and Costs

Strategic choices significantly affect test-program-development time, effectiveness, and cost. For example, creating a functional test yielding better than 99 percent good boards requires more elaborate and expensive tools and takes longer than specifying that same test for 95 percent yields and removing the remaining faults at the system level. Following the Pareto rule, the last 20 percent of any job takes 80 percent of the total effort. Of course, detecting extra faults at system test increases costs at that stage.

Similarly, prescreening functional test with an inspection station may reduce time and costs for functional test development but requires additional capital equipment. An in-circuit step requires both capital equipment and a complete bed-of-nails fixture. The correct strategy depends on whether the savings at one step compensate for additional spending at another. Also, simplifying a later test step requires taking advantage of the results at earlier steps. If you find shorts during an optical or x-ray inspection, avoid looking for those same faults again. Similarly, many test strategies ignore carefully created self-tests other than to ensure that they exist and that they function properly. With self-test fault coverage exceeding 75 percent in some cases, aiming subsequent tests only at the remaining 25 percent will cut both development and execution times considerably.

Tester choices also affect test-development costs. Programming a large full-function tester takes longer than creating a similar test for a less-expensive model. In this case, higher speeds, throughputs, and fault coverages can justify the extra effort.

Test-development language alternatives exist as well. Programming directly in C or BASIC may take longer than doing so in an object-oriented or "drag-and-drop" environment, where a mouse-driven interface guides a test engineer through the development steps, generating actual program code in response to user selections. Situations that permit teaching the tester about good-board and bad-board responses to test stimuli instead of using formal programming can also significantly cut time and costs.

At the same time, formal programming gives designers and test engineers more control over the resulting tests. Successful program development requires comprehensive and controllable documentation. Tracking changes in the product, the process, or the involved personnel must proceed with as few glitches as possible. A new person developing or revising a program must clearly understand the previous person's efforts. Formal coding includes code-development documents and standards, comments in the source code (in theory, anyway), and some kind of established revision-control system. With less-formal, menu-driven "point-and-click" programs, tracing development steps is much more difficult.

Some code generators learn programmer's steps directly. This can create enormous inefficiencies if the programmer is experimenting or trying several alternatives before settling on tests that work. This situation demands either optimizing the program after initial development or teaching the program to the tester for production only in its final form.

Software tools can provide automatic tracking of product revisions, code revisions, and configuration management. Web-browser technology adds considerable power to this process. Test-development engineers may not work in the same facility, so the best available expert for each program section can participate, no matter where he or she lives. Tester manufacturers and their customers can work together to ensure adherence to common goals. Principals can create multimedia training tools—such as hyperlinks to video clips—for people responsible for day-to-day operations. Nevertheless, engineers and managers must rigorously maintain these tools to keep them current.

For formal programming, automated tools such as fault simulators and test-program generators can reduce the code-generation task. Other aids that include word processors can help in the later stages of debugging and verification—for example, by permitting easy modification to detect extra faults. A computer-bound syntax checker prevents numerous runtime delays by ensuring that program statements are valid before a test engineer first installs the program on the tester itself.

Some test programs are compiled; others are interpreted. A *compiler* translates a completed program as a block into machine language (called an "object program") at the end of the programming process prior to execution. An *interpreter* translates the program at runtime, one line at a time. Because of this extra runtime step, interpreted programs generally run more slowly than the compiled variety. For debugging, however, interpreters can be much more convenient. Making even a small change in a conventional compiled program requires complete recompilation, a process that could take several minutes, before the programmer can assess the change's impact. Because an interpreted program enters the tester in source-code form, it is immediately ready for execution following any revision. C is a compiled language. Most versions of BASIC are interpreted, although BASIC compilers exist as well.

An *incremental compiler* offers one compromise between the runtime speed of compiled code and the debugging convenience of an interpreter. This variation decompiles individual routines within a large test program to allow debugging those routines or examining intermediate results. During debugging, this approach either interprets code in the program segment or compiles it separately, which takes much less time than compiling the entire program. Complete recompilation occurs only at the end of a debugging session.

In any of these situations, a clear test plan and a thorough understanding of the target tester's capabilities will improve development-process efficiency. In addition, UUT characteristics affect development time and cost. Obviously, creating tests for more complex boards generally takes longer than for simpler ones. On the other hand, the per board test-development cost for even a very complex board that is manufactured at high volume may be much less than for a simpler board with lower volumes. Also, resource availability remains a factor. A high-mix operation (such as a CM) must generate many programs, reducing the time, money, and personnel available for each one.

Again, taking advantage of self-test circuitry can significantly reduce the number of failure mechanisms that more conventional test development must cover. Inspection coverage should also reduce the test burden. In addition, complexity-reduction factors, such as boundary-scan design and other forms of design-for-testability, significantly improve programming productivity.

Development time and costs depend on UUT logical and mechanical maturity as well. Logical maturity determines how easily an engineer can create a functional test program. A mechanically stable board—where the sizes and locations of individual components are fixed—permits easier bed-of-nails construction and minimizes the number of fixture revisions, facilitating MDA and in-circuit test development. Trying to create a program for an immature product is like shooting at a moving target.

Efficient programming demands comprehensive documentation of UUT design, construction, and production processes. The cycle is much shorter if design data and fault-simulation data reside in-house. If you need cooperation from outside designers and other vendors, creating and evaluating a test will be both more difficult and more time-consuming.

On the ATE side, if a board design confines testing to the latest available tester generations, users should plan for delivery delays, tester hardware and software bugs, frequent updates, and other disruptions. On ATE generations even a year old, vendors have had time to work out the kinks, and test development will generally proceed more smoothly. If a vendor's tester is VXI-based or similarly modular, bug-fixes, upgrades, and enhancements proceed much more easily, again shortening test development and making the process more efficient. If the strategy includes only one tester of a particular type, as is often the case with large equipment, availability of offline development tools and machine time for final program checkout can substantially influence development schedules and costs.

It is also important to assess the level of compatibility between testers and UUTs. That is, it is difficult to create a bed-of-nails test for a board containing hundreds of surface-mounted components with no test pads.

In addition to target fault coverage, test programmers must decide to what extent they wish to isolate faults to failing components. Program development for a go/no-go-type test is significantly simpler and less expensive than for a test that identifies actual failures, especially at the functional level. Inspection and bed-of-nails techniques theoretically examine one component at a time anyway, so a failure should automatically indict the faulty part. On a board containing primarily large ASICs, identifying offending devices is fairly straightforward. If the board contains hundreds of surface-mounted components—a disk-drive controller board or laptop-computer motherboard, for example—functional fault diagnosis by guided-fault isolation to the component level may be nearly impossible, and fault-dictionary isolation beyond a board section or functional unit could require lengthy simulation times.

Another important consideration is whether the board will be repaired or scrapped. Fixing multichip modules and some small boards can cost more than they are worth. Locomotive-engine controller boards worth many thousands of dollars and manufactured one at a time require fault isolation and repair even if the process is slow and painful.

Some self-tests require that certain board functions work correctly before executing. Test developers must take this fact into consideration in ordering individual test steps. As with determining whether boundary-scan circuitry works before executing a boundary-scan test, this requirement simply adds a step to the test hierarchy. Such a board has more ways to fail than a similar board without self-test has, and the board's overall yield through manufacturing may, therefore, be lower. Nevertheless, self-test can drastically cut test-program development times, and the quality of the target system is generally at least as high as for a more conventional product.

Extensive test-program validation can add significantly to schedules and budgets. How long do you play with a program before you can declare with confidence that it works adequately? In many cases, the answer to this question depends on what tools are available. In some cases, you can compare test-program results against a manual test or previous-generation test. For many digital boards, you can judge the test program against a simulation. In any situation, UUT samples must be available for the final checkout.

With sufficient bed-of-nails access, an MDA test should provide next-stage yields of 80 to 90 percent, depending on the board's actual fault spectrum. A full in-circuit test can offer 85 to about 95 percent, and functional or combinational testers can approach 100 percent, again depending on board design and actual fault spectrum. Inspection results can vary widely. Some manufacturers claim 80 percent fault identification with simple post-paste inspection. AOI effectiveness depends heavily on node visibility. Proponents of x-ray inspection claim 99 percent or better detection of solder faults.

Heavily analog boards will frustrate functional-test methods. High-speed digital circuitry and access problems will reduce the effectiveness of MDAs and in-circuit tests. One application of combinational testers is to perform in-circuit test on analog portions of the board and functional tests on digital sections. Hybrid devices such as A/Ds may require applying both techniques simultaneously.

Conventional test methods evaluate a board's hardware condition. For some boards, a form of emulation that executes target-system software may provide better results. Software for these boards is usually available at the same time as the hardware, because it forms part of the delivered product. An instrument-controller board—for a logic analyzer or waveform generator, for example—has relatively few functions. Executing those functions and observing their responses confirm the board's behavior. Other types of boards that may respond to this approach include PC motherboards and automotive electronics.

9.3 Simulation vs. Prototyping

An important part of test planning is comprehensive design verification. If the design is not correct, test generation requires more steps, and the resulting program will not likely yield the highest-quality product.

Traditional board design involved simulation to a certain point, followed by numerous generations of physical prototypes to ensure that the actual board would work as designers intended. Practical prototyping of today's circuits ranges from difficult to impossible. To alleviate this problem, designers are resorting to more simulation iterations and limiting prototype generations before production sampling.

Replacing prototypes with simulations offers numerous advantages. A board-design simulation can include the newest components and technologies by incorporating their logical representations, even if physical devices do not yet exist. This statement is particularly true for designs containing ASICs and other custom components, because these parts often arrive only barely before manufacturing begins. In addition, if design verification in simulation indicates that the board will not work properly because of custom-device design decisions or device selections, engineers can propose changes to fabrication houses before full-scale production begins—at considerable savings.

Board prototypes often do not work like final versions because wire-wraps and other prototyping techniques, by the laws of physics, cannot provide the performance of printed circuits. Similarly, breadboards lack the speed and signal fidelity of the much smaller ASICs and other complex devices that they emulate. As a result, a "real-board" prototype is not as close to final-product behavior as the "virtual" board that a simulation creates.

Prototyping generally involves three to five iterations, not counting minor changes such as cuts and jumps. The resulting hardware offers limited access for analysis and generally does not include a test program. Some companies do not even begin to write test programs until at least the third or fourth prototype. In addition, because the prototype is manufactured in a lot of one on a laboratory bench and does not physically resemble the final product, it offers little insight into manufacturability or expected product quality.

Performing design verification in simulation permits examining many variations at little cost in time or money. Ideally, simulation reduces the need for prototyping to at most a single prototyping step. Final designs are cleaner and more manufacturable, and resulting products require fewer engineering changes before maturity. Primarily digital boards can allow skipping the prototype stage altogether, moving directly from simulation to preproduction boards. For heavily analog boards, limiting development to a single prototype step may be unattainable, but there will still be fewer such steps than without the simulation-based analysis.

Prototype testing requires equipment and physical probes or fixtures, along with skilled operators or technicians. Test programs, when they exist, most often are unique to the prototyping step and have little value to the manufacturing portion of the process. On the other hand, a design engineer can often verify simulated behavior using the same tools that permitted creating the simulation in the first place, eliminating most test programming, additional equipment, and diagnostic errors such as misprobes.

Available design tools can identify subtle, complex design errors in simulation that may not even manifest themselves in a prototype. For correcting problems, instructing software to move a connection or change a component is faster and far less expensive than cutting and jumpering prototype wires or physically replacing parts. If a board contains multiple layers, cutting and jumpering real traces may be impractical, if not completely impossible. Simulation permits probing blind and buried nodes and vias, even on the densest surface-mount designs. Because much test-program debugging can occur offline, simulation also frees valuable tester time for production.

Figure 9-2 shows a typical design process without simulation. The board undergoes a logic design, followed by a physical design, a manufacturability design, and a test design. At each step, any problems that arise trigger design iterations that may take the process back to the preceding step or all the way back to the beginning. Notice that prototyping occurs after physical design.

In Figure 9-3, a simulation occurs at each stage. Logic simulation assures that the physical design's logical behavior will be correct. Timing simulation handles practical implementation of designers' intentions. A prototype at this point is unlikely to show either logic or timing problems. Design changes resulting from manufacturability analyses may require returning to physical design before proceeding to actual product. As discussed, fault simulation ensures that testing will uncover any problems before a product ships to customers.

9.4 Design for Testability

Just as design-for-testability and related techniques make testing possible, they also reduce the test-programming effort. Individual components must be

Design Process

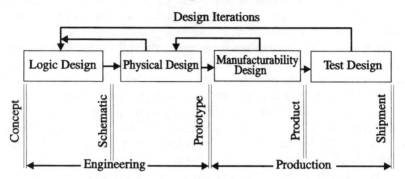

Figure 9-2   A typical design process without simulation. (Racal-Dana. 1990. Sales Presentations, MRT, Winchester, Massachusetts.)

Simulation-Based Design Process

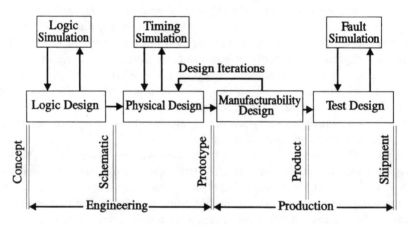

Figure 9-3   A design process showing a simulation at each stage. (Racal-Dana. 1990. Sales Presentations, MRT, Winchester, Massachusetts.)

testable to ensure both that they work and that they function on the target board. Ensuring testability may involve specifying that devices incorporate boundary-scan or other techniques that improve visibility from package pins or the board edge.

As mentioned in earlier chapters, it is particularly important that complex components and custom devices such as ASICs not only be testable but also be test*ed* before board assembly. If a manufacturing engineer can assume that a device worked at one time, the number of possible failure mechanisms at the board level is quite limited, which limits test complexity as well.

Just because a component is testable does not necessarily mean that its test is *available* for use during board production. An ASIC vendor who generates all production tests, even in cooperation with the board-manufacturer's design-engineering team, may not willingly share them. If this problem arises, the simplest recommendation is to find another vendor. Providing device-level programs both for device verification and board test should be a condition of contract for ASIC fabrication.

Board layout should, as much as possible, provide node access, both for bed-of-nails testing and guided probing for functional-test fault diagnostics. Long digital logic chains must be easily initializable because complicated initialization sequences add to programming burdens, as well as to test times.

Designing for testability also means partitioning boards into functional-simulation units, limiting functional-unit size, isolating analog circuitry (if possible) to allow digital functional testing, and permitting access to analog circuitry for instrument-based analysis. In addition, there should be a convenient means to break long feedback loops. Incorporating more circuitry onto ASICs often limits board-test complexity and therefore reduces the effort of creating the test. For whatever test methods are selected, inputs must be controllable. Good and faulty output signals must be observable and distinguishable from one another.

Incorporating DFT and other concurrent-engineering techniques into board designs from the outset can spell the difference between boards whose high quality is achievable and boards where quality is always suspect. Maintainability and repairability are as important to customers as testability is to manufacturers.

9.5 Summary

The relative effectiveness of two otherwise equivalent test strategies depends on ease of test-program development. For example, a go/no-go test requires less effort than does a test that includes fault diagnostics to the failing component. Similarly, the higher the fault coverage, the greater the necessary programming effort. Test-method, test-language, and tester choices also significantly affect schedules, costs, and results.

An ideal development process includes a test-requirements analysis and a test-requirements document, then simultaneous test generation, fixture construction, and test documentation, followed by program validation against the product design and the test design before production ramp-up. Companies may combine or eliminate some of these steps to save time or money.

Verifying board designs in simulation offers distinct advantages over the more traditional prototyping approach. With today's complex circuits, wire-wrapped boards and breadboard ASICs do not permit the speed and signal integrity of printed-circuit and silicon-bound alternatives. Simulation allows verification of the newest technologies even before actual hardware is available. Modification of a simulated model to see how the circuit will behave is relatively painless, encouraging design optimization before production begins, thereby minimizing changes later.

Design-for-testability techniques also reduce program-generation efforts. Components should be testable and tested, and test programs should be available for application at the board level. Board layout should permit node access for bed-of-nails testing and guided probing, wherever possible. Inputs should be controllable, and both good and faulty outputs should be observable. Test planning must last the product's lifetime from conception until the last one in the field dies.

CHAPTER 10

Test-Strategy Economics

To this point, the discussion of test-strategy building blocks and alternatives has centered around their relative technical merit. Test problems, however, permit numerous technically "correct" solutions, each of which also has economic consequences. An option that offers the most sophisticated tests and the highest yields (the biggest bang) may require too large an investment, too large a support budget, or too large a programming effort. On the other hand, the most economical choice (the fewest bucks) may not provide adequate test coverage or fault diagnosis.

Naturally, every test step has to meet minimum technical standards to have any economic value. On the other hand, unnecessary test capabilities cost money for no direct benefit. In planning test purchases, overbuying in anticipation of your needs for the next year or two makes sense. Beyond that, by the time you need the extra capabilities, new testers will be available that will provide them much better than current offerings—usually at lower prices.

In addition, many vendors offer some enhancements as field upgrades. Testers based on VXI, for example, permit a wide range of such upgrades with only minimum impact on production. Buying equipment without those features reduces initial costs and serves as a hedge that you will not need them during the equipment's lifetime.

Careful economic analysis of technically acceptable alternatives can represent the difference between choosing an inadequate strategy, an expensive albatross, or an efficient, cost-effective solution. The final "best" strategy may not perform any one test step in the least-expensive way. Rather, it minimizes *overall* test costs. Spending more at one test stage can avoid costs later. Comprehensive production testing, for example, can reduce field-service and warranty-repair costs.

This chapter explores economic implications of the various stages of a test-and-manufacturing strategy and how test-strategy choices affect them. Included will be specific techniques to help managers evaluate test alternatives. Some of the most powerful of these techniques fall into the category of "break-even analysis," which quantitatively assesses costs and benefits.

Unfortunately, economic analysis is far from an exact science. Even when the models themselves can be considered accurate, drawing quantitative conclusions based on available data is dangerous at best. The tools presented here cannot guarantee that you will make the most cost-effective choices at all levels. They can, however, help eliminate the worst alternatives quickly and indicate which strategies belong on the "short list" for further consideration.

In addition, initiating a test strategy is an expensive undertaking. Corporate managers often demand concrete accounting of anticipated cost-savings and other benefits of adopting any proposed program. The techniques presented here can provide some of that information, helping test engineers and managers justify their recommendations to their superiors.

Experience in seminars and courses indicates that this material represents the least familiar territory in the entire book. Because the existence of any organization depends on its profitability, however, it may be the most important. Not long ago, I presented the justification process to an engineer with 30 years' experience. He looked at me incredulously and responded, "But if I need a piece of equipment to get my job done, why do I have to go through this rigmarole?" Even *he* felt that a technical argument would be sufficient.

Interested readers can get a much more complete presentation in Brendan Davis's book, *The Economics of Automatic Testing,* second edition (1994). Since the book's first edition was published in 1982, it has been the definitive source on this subject.

10.1 Manufacturing Costs

As Figure 10-1 shows, the cost to manufacture one printed-circuit board is the total of costs incurred during various process stages. A manufacturer has to buy materials and supplies, expend labor, occupy plant facilities, and use equip-

$$C_m = C_s + C_l + C_f + C_d + C_t$$

where

C_m = Manufacturing cost of one board

C_s = Supplies or material cost

C_l = Amortized labor cost

C_f = Amortized plant facilities cost

C_d = Equipment depreciation cost

C_t = Test cost

Only C_t depends on whether the board is good or bad

Figure 10-1   The cost to manufacture one board is the total of costs incurred during various stages of the production process.

ment. Supply and material costs depend on the particular board's design. Labor and plant-facility costs reflect the percentage of total manufacturing capacity that the board consumes. Similarly, equipment-utilization costs depend on the equipment's total capacity as well as on its anticipated life. In this calculation, only test costs depend on whether a board is good or bad.

A test-cost breakdown resembles the overall cost picture in a microcosm. Test installations include test equipment, test-program development, and repair materials, as well as operators, technicians, and supervisors.

Figure 10-2 computes the cost of producing a *good* board simply by dividing manufacturing cost C_m by the yield, producing the larger C_g. This calculation assumes that bad boards have no economic value; that is, that you throw bad boards away rather than repair them.

Another equally valid approach would total the costs for diagnosing, repairing, and retesting all bad boards, divide by the total number of boards, and add the result to the initial per board cost. Organizations with large board lots and fairly low failure rates will find that the difference between these two methods is small. Remembering that all economic analysis is inexact at best, engineers in that case should select whichever alternative is more appropriate.

At the same time, where yields are less than optimum, scrap can overwhelm other test costs. Reducing or eliminating scrap often represents the most potent economic justification for creating the most effective possible test strategy.

10.2   Test-Cost Breakdown

How much does testing actually cost, and what do those costs include? Consider a board worth $600 at system sale. That number does not represent the manufacturing cost, but rather an appropriate percentage of the system's selling price. Assuming a typical *gross margin* (selling price minus manufacturing costs) of 60 percent, total manufacturing cost is $240. If testing consumes one-third of that number (test costs generally range from one-third to one-half of total manu-

$$C_g = \frac{C_m}{Y}$$

where C_g is the cost to manufacture
one good board, and
Y is the yield percentage

Figure 10-2   The cost of producing a *good* board is calculated simply by dividing manufacturing cost C_m by the yield, producing the larger C_g.

Test cost of $80 per board includes

* Startup costs
    – Tester-evaluation costs
    – Capital-acquisition costs
    – Test-program-development costs
    – Training costs
* Facilities costs
* Site-preparation costs
* Labor
* Board repair
* Tester maintenance and repair
* Spares
* Downtime

Figure 10-3   Some of the cost components included in test cost.

facturing costs), then $80 or 13.3 percent of the board's selling price goes simply for test. Figure 10-3 shows some of this cost's components.

Every dollar that you eliminate from test costs (assuming equivalent product quality) goes directly to the bottom line. Selling prices do not depend on how much testing a production process contains. No wonder managers are so eager to eliminate test! ("If people did their jobs properly the first time, we would not need to test at all.")

10.2.1 Startup Costs

Most *startup costs* are *sunk costs*—that is, you have to spend the money, even if you then abandon part or all of the test strategy before it tests a single board. *Strategy-evaluation costs*, for example, accrue while you consider the relative merits of various alternatives. *Tester-evaluation costs* support examining and selecting specific equipment for each stage in the chosen strategy. In each case, the more alternatives, the higher the evaluation costs. Therefore, you must carefully consider how many merit thorough investigation. Even if the next choice might provide lower operational costs, is the difference worth spending additional time and effort to evaluate it?

These costs include such items as travel expenses to visit tester manufacturers and trade shows, salaries and lost productivity for those trips, meetings, time to prepare and present management reports, and other activities that directly relate to tester and strategy selection. To burden all products fairly, distribute evaluation costs among the total number of boards that will follow the chosen strategy throughout its life.

Note that these costs matter in budgeting the test-strategy implementation. When deciding economic merit among the alternatives under investigation, however, they become irrelevant. That is, they do not change regardless of your final decision, and therefore subtract out of any cost-comparison calculations.

Once you have selected specific vendors (or in-house project teams) and tester types, you incur *capital acquisition costs* to put the equipment in place. For vendor-supplied equipment, these costs include tester and instrument prices and prices for mounting racks, computer engines, and general-purpose support software. Calculations at this stage should also consider the cost of money—the cost of borrowing money or the benefits lost by applying available cash to this rather than another project. This *opportunity cost* will be discussed in greater detail later in this chapter.

Getting a site ready for tester delivery involves *site-preparation* costs. These can include construction costs for a dropped ceiling or raised floor, putting in three-phase power, and other accommodations. In addition, site preparation incurs personnel-related costs for the construction, as well as costs for machine acceptance and similar activities. Production lost during this time and necessary labor overtime to meet commitments also cost money.

Test-program-development costs depend on a board's complexity and its physical design. Developing a program (and any associated fixture) to test a particular board on a specific piece of equipment adds to the cost of manufacturing that board. The per board burden is substantially less for high-production-level boards than for boards with smaller production runs. Therefore, amortize each board's programming costs across only that board's run.

Many companies spread test-development costs evenly among all boards crossing a piece of equipment. Unfortunately, this technique makes gauging the economic benefits of each program much more difficult. Of course, in a small facility where only one tester is available for a particular test step, a manager's strategic options are limited. In that case, the simpler programming-cost calculation may suffice.

Include programmer time for initial development, debugging, and documentation in the cost calculation. In addition, debugging and startup demand tester time, which may interfere with production targets. Again, lost production represents a cost. Delays can affect existing products at each test step, as well as the learning curve to ramp-up a new product.

Upgrading and enhancing a test program after initial implementation can also involve significant costs. Sometimes the product does not change—the programmer merely improves test performance. In other cases, engineering change orders (ECOs) and other product modifications demand corresponding program changes.

Every strategy startup incurs costs for *training*. The amount of training depends on people's familiarity with the strategic steps. For example, when adding in-circuit test to a functional-test-only facility or adding x-ray inspection to a department that previously used only ATE, the ramp-up will be bumpy as people learn new capabilities and procedures for test, diagnosis, and repair. On the other hand, a test department replacing a less-capable in-circuit tester with a more capable machine may experience little difference in day-to-day operation, except that the new tester will find some faults that in the older configuration fell out at a later test step. In that case, the level and cost of extra training may be minimal.

Changing tester vendors within a test approach or introducing home-grown equipment featuring a unique software environment to replace a commercial system generally does not require training personnel on overall procedures and capabilities. People must learn new operations, however, such as interface languages and step-by-step test execution.

Training costs cover formal courses, which may be included in tester prices, as well as travel expenses and the cost of space for in-house training facilities. Amortization of training costs depends on the type of training. Distribute costs for programmer training over the board types for which that person is responsible, then over the entire production runs for those types. For example, training a programmer for $5000 to write programs for five boards means burdening each board's production run with an extra $1000. Dividing that number by the number of boards in the run provides the burden per board.

Amortize tester-operator training over the total number of boards that will cross the tester under that person's supervision. Factor in personnel turnover rates, because new people require extra training. In some cases, the training burden may be too small to consider separately. If that is true, add it to the tester's price in the economic model, or ignore it and move on.

10.2.2 Operating Costs

A number of costs fall into the *operating costs* category. *Facilities costs*—representing space for test and repair areas, including people space and directly related office space—are *fixed*. That is, the total cost of a manufacturing and test facility does not change with production levels. As production levels fall, per board costs go up.

Assuming a fully utilized factory, the facilities cost for a particular project represents the portion of total floor space that the project occupies. Because eliminating a project from the factory floor does not automatically shrink the factory, facilities costs must include an allocated portion of empty space as well. These costs include rents or mortgage payments, utilities, and security and other plantwide services. Company managers generally establish a standard cost per square foot, then charge each department on that basis. Rather than calculate these costs individually for different test steps and test strategies, managers can estimate a cost per test station by dividing the aggregate cost for the entire test operation by the number of stations.

Labor costs cover salaries and benefits for all participants in test activities. Obviously, operators, repair technicians, and people who ferry boards from one test station to the next qualify. The costs must also include an allocated portion of time for indirect employees, such as supervisors, managers, administrative assistants, and clerical people.

This number can be *burdened*, *partially burdened*, or *unburdened*. Unburdened labor costs include only direct employees' salaries. Accurate calculations require adding costs separately for indirect employees and for benefits such as vacations, health insurance, and holidays. Partially burdened labor includes these employee benefits. Fully burdened labor covers the facilities costs, test-equipment amortization, a portion of indirect salaries, and so on. Which version you use will depend on the purpose of the economic exercise. Adding a strategy to an existing facility does not materially change the facility's overall costs for mortgage, security, utilities, and insurance. Therefore, the new strategy's *added* cost can exclude these factors, and partially burdened labor suffices. If implementing the new strategy will require adding clerical and other indirect workers, however, you must add their costs to the justification calculation.

Labor-cost calculations depend primarily on how a company's controller's office handles the numbers. Salary benefits alone generally add about 30 percent. It is not unusual for fully burdened rates to be two or three times actual salaries, or higher.

Repair costs account for extra equipment—logic analyzers, oscilloscopes, multimeters, even spare testers at the repair station—as well as people who fix faulty boards. Labor cost for technicians is generally higher than for test operators, so per person repair costs will be higher as well.

There are several ways to calculate this cost. Perhaps the simplest method is to divide the overall operating cost of a repair department by the total number of faulty boards that pass through that department. This assumes that the repair cost for all bad boards is the same. Consider two boards with comparable production levels. If board number 1 produces a 90 percent yield, and board number 2 produces a 95 percent yield, the repair-cost burden for board number 1 is twice as high. As an alternative, dividing operating costs by the number of *faults* handled by the department allows boards with more than one fault to bear a greater burden.

Another approach uses a standard repair cost per hour and standard repair times for particular board types or fault types. Because diagnosis and repair take longer for a complex fault than a simpler one, the corresponding cost is higher as well. Repair costs include the price of consumed components, also at standard prices.

Scrap costs account for the value, including labor and materials, that a scrapped board contains. If a company does not repair faulty boards, repair costs are zero, but scrap costs will be quite high.

Consider a situation where 5 percent of our $600 boards fail and we do no repair. The additional production cost is $12.00 per board (5 percent of the $240 manufacturing cost). In contrast, suppose diagnosis and repair cost $50 per faulty board, and one bad board out of 10 must be scrapped anyway. In a lot of 200 boards, the repair cost is 10 times $50 (because even the eventually scrapped board must undergo the diagnostic steps) plus $240 for the scrap, or $740. The cost per manufactured board is $3.70. The $8.30 difference represents 1.38 percentage points in gross margin (and therefore profit). If the company was earning 10 percent before taxes, repairing rather than scrapping bad boards increases profits by 13.8 percent.

This oversimplified example ignores the fact that scrapping bad boards requires raising production levels to ship the same number of products out the door. Including the resulting increased costs would increase the economic advantage of the repair strategy. As with all economic analyses, this one uses estimates and assumptions to reduce the number of variables and make a solution possible. Whenever you do that, however, try to cast all such simplifications so that they reinforce, rather than mask, the correct decision.

10.2.3   *Maintenance and Repair*

Another necessary cost covers maintenance and repair of the test equipment itself. Most companies estimate costs for preventive and catastrophic maintenance at either 1 percent per month or 10 percent per year of the equipment purchase price. This number is easy to calculate and represents what many vendors and third-party houses charge for service contracts. Of course, if you have a contract or otherwise know that your costs are substantially different, base your own estimates on your experience.

Allowance for tester spares depends on the nature of the equipment, how much downtime you expect and can tolerate, and who performs repairs. Spares inventory for a large cadre of similar or identical machines will be fairly modest, because the parts will fit any machine that goes down. Machines with many duplicate parts also minimize spares requirements. For example, in large testers, a few unique printed-circuit boards often execute actual measurements, whereas test points, including pin drivers and switching technology, reside on a number of identical boards that plug into the tester chassis. Even if the tester contains 75 such boards, you need only a couple of spares. Similarly, if much of your equipment is PC-based, you can generally find a spare PC when an engine goes down. In that case, it may not be necessary to keep a computer in inventory specifically to serve as a spare.

If, on the other hand, you employ a wide variety of unique machines and you cannot tolerate the downtime required for repair or to obtain a loaner from a vendor or rental company, then you must keep a large stock of spare parts or even spare machines. This situation often occurs when production volumes are fairly low and the strategy relies heavily on rack-and-stack equipment. Such a test operation is unlikely to have more than one of any instrument. If lost production resulting from a down test system has severe consequences, you may need a spare of every instrument or a service on call that can provide one at a moment's notice.

A general rule of thumb for estimating spares' costs is 10 to 50 percent of the total capital cost for one of each equipment type. A large facility with more than one of each tester type would therefore spend a lower percentage on spares than a smaller operation would. Some maintenance contracts specify a minimum supply of spares, so the spares cost is known.

The spares cost is not a yearly budget but an inventory cost that accompanies test-equipment purchase. The cost of replacing spares is part of the maintenance allocation.

As with all production stoppages, downtime for either scheduled or unscheduled maintenance incurs a cost for lost production or extra labor to replace lost production. What is considered a tolerable level of downtime depends on how close to maximum capacity production normally runs. A facility that cannot afford downtime may want to train in-house people to maintain the equipment, thereby avoiding the inevitable wait before a contract person arrives. On the other hand, both the training and that person's salary represent an additional maintenance cost. Most companies do their own calibration and preventive maintenance. Many companies also opt for an in-house person to do board-swapping and other simple jobs that get equipment up and running quickly. They call in vendors or third parties to repair bad boards and to address other complex problems.

Other recurring costs include material consumed during product board repair, to replace parts either failed during testing or damaged by the repair process. Allowing for *depreciation*, although not technically a cost (no money changes hands), permits planning for relatively painless equipment replacement several years out. The cost of *recruitment* to add staff or to replace people who leave includes search, overtime charges to cover short staff, lost work, and training. Upgrading or adding equipment after implementing a strategy also incurs a training cost.

10.3 Workload Analysis

The first step in estimating any strategy's *benefits* compared with other strategies or not testing requires calculating test capacity at theoretical full production. Numerous factors affect the results. Consider the relationship between good-board and bad-board test times and yield. For bed-of-nails testing, test times are about the same, and the only effect of yield on capacity is that bad boards require retesting after repair. Therefore, if the maximum throughput at a particular test step is 1000 boards per day and the yield is 100 percent, the capacity is 1000 boards per day. If the yield is 90 percent, for every 1000 boards tested, 100 require at least one retest, so the operation can produce no more than 909 good boards per day. At 80 percent yield, the capacity is only 833 boards. As a product matures, its yields tend to rise, increasing effective manufacturing capacity.

For functional testing, the situation can be more severe. If fault diagnosis occurs on the tester, bad-board test times can be many times longer than good-board times. Suppose a bad-board functional test takes 10 times as long as the same test for a good board. With 100 percent yields, the 1000-board capacity remains. Even a 90 percent yield reduces maximum capacity to only 500 boards per day. The 450 good boards occupy 450 good-board equivalents of tester time. The 50 bad boards require 500 good-board equivalents on first test and 50 good-board equivalents after repair.

To maximize throughput, functional testers often perform only go/no-go tests, for which bad-board and good-board test times are about the same. The cost of such a strategy is in higher diagnosis and repair costs at the technician's bench.

Fault-diagnosis technique also helps determine capacity. Guided-fault isolation generally takes considerably longer than one of the alternatives. Unless yields are very high, GFI reduces throughput considerably.

Programming errors introduce an additional burden on throughput. If a bad board passes board-level test, it will likely fail at system test, reducing overall capacity only slightly. On the other hand, a good board that fails enters the bad-board diagnostic process. This scenario results in the "no-fault-found" loop that Figure 10-4 depicts. In a largely automated facility, such a board may travel through this loop many times before someone notices that it is merely consuming production time with no benefit.

Test-strategy design must include the number of times such a false failure may navigate this loop. (Many companies use three.) From in-circuit test or inspection, at that point the board proceeds to the next test step. The strategy assumes that either the board was actually good or that subsequent test steps will pinpoint the problem. A board with an unknown failure from functional or system test generally ends up as scrap.

In addition to actual test-time issues, a capacity calculation must consider *handling times* to transfer boards between test stations or to mount boards on test fixtures and attach appropriate connectors. *Administrative time* for test-floor workers includes reporting test results, filling out time sheets, lunch, breaks, and any other periods that are not directly productive. *Setup* and *teardown* cover changeover between board types, as well as normal procedures that occur at the beginning and the end of each shift. For clean-room employees, it can include dressing time and showers.

Test programming often impinges on production. Some companies require that programmers do most of their work on the test equipment off-shift (which may contribute to higher personnel turnover). When purchasing some of today's

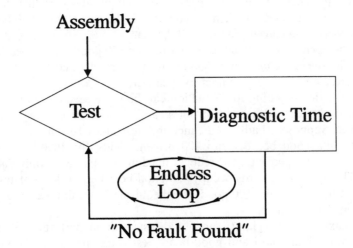

Figure 10-4 A good board that fails enters the bad-board process loop, consuming precious test and diagnosis resources with no ultimate benefit.

less-expensive testers or instrument suites, manufacturers may buy extra equipment for program development. Such extra systems also serve as spares in case a production machine fails, virtually eliminating downtime while reducing effective spares cost.

Calibration and other *preventive maintenance* follow fairly routine schedules. Budgeting time for these activities in the capacity calculation helps improve its accuracy.

The following throughput example shows how strategic decisions can affect test capacity. Assume a 2-minute test time per board, regardless of whether the board is good or bad, which is reasonable for in-circuit test and other bed-of-nails techniques. Yield is 90 percent. Average repair time, including turnaround, is 10 minutes. One retest after repair confirms that the board contained only one fault and that the test process corrected it. Changeover between board types—to change fixtures, retrieve programs, set up log reports, and perform other chores—is 20 minutes. In addition, the tester handles two or more board types (to invoke the changeover time) and lot size is 100 boards. A higher lot size would reduce the percentage of the total-lot test time that changeover consumes.

A normal work year is 250 days—5 days per week, 50 weeks per year with a 2-week shutdown. Assume 6 productive hours per shift. This number excludes setup/teardown, lunch on job, breaks, preventive (scheduled) maintenance, and similar nonproductive activities. Also, by assuming a relatively low number of hours per shift, the capacity calculation is somewhat conservative, leaving room for unanticipated problems. Of course, excessive tester downtime may reduce the number of hours even further, as will downtime elsewhere in the production cycle, such as with a chip shooter or pick-and-place machine. This problem is particularly acute in factories that practice just-in-time inventory control, where there is no stock of work-in-process to keep downstream activities occupied when a stoppage occurs upstream. Anyway, for the analyses in this chapter, a work year contains 1500 productive hours.

Suppose that the tester determines only whether a board is good or bad, with diagnosis and repair offline. Processing one board lot requires testing 110 boards (since 10 failing boards get tested twice) with one board changeover, for a total of 220 + 20 = 240 minutes, or 4 hours per lot. One shift can handle 1500/4, or 375 lots, totaling 37,500 boards in a year.

If diagnosis and repair shift to the test station, each bad board adds 10 minutes to the test time, totaling 100 extra minutes per lot. Processing one lot now takes 240 + 100, or 340 minutes. Full capacity falls to 264 lots, or 26,400 boards, a reduction of 29.6 percent.

10.4 An Order-of-Magnitude Rule Counterexample

Conventional wisdom states that catching faults early in the production cycle is much less expensive than finding them later. Is that always true?

To a point, designing a test strategy for catching faults at the earliest possible level makes sense, minimizing overall test costs, including costs prior to and subsequent to that test step. Increasing coverage beyond that point generally requires additional programming time or greater manual effort for fault diagnosis. In addition, particular test steps carry hard fault-coverage limits. For example, you cannot expect an in-circuit test to find functional-performance problems or an inspection station to identify a bad part. Therefore, a strategy's ultimate fault coverage depends on its constituent elements. Expanding fault coverage beyond those limits requires adding a test step or replacing a test step with a more capable (and more expensive) alternative, such as by trading in an MDA on a full in-circuit tester. In many cases, increasing fault coverage in this way does not generate enough savings to justify the additional cost.

Consider two strategies, one of which requires spending $400,000 more for board-level test equipment than the other. Perhaps one strategy includes only in-circuit testing, whereas the other also incorporates functional test. The less-expensive strategy might consist of a cadre of small testers, while the more-expensive alternative uses a large monolithic machine. For the purpose of discussion, the only relevant issue is that the additional cost buys extra fault coverage, so that fewer faults survive to system-level test.

After adding extra maintenance, supporting labor, cost of money, and other incidentals, the cost difference is $491,000. (These seemingly peculiar numbers come from an actual case.) Assuming a 2.5-year amortization period and a 1500-hour work year, the new strategy must pay back the initial cost in 3750 hours. Therefore, the loaded tester cost differential is $130.67 per hour. Partially loaded board-test operator labor cost is $17.00 per hour, including health insurance, vacation, and other benefits, while a system-test technician costs $55.00 per hour.

As Figure 10-5 shows, the total loaded cost for the extra tester and operator is $2.46 per minute. Assuming an average of 1.5 faults per bad board and an extra diagnostic time of 3 minutes per fault, the average cost to diagnose faults at the board level is $11.06 per board.

Finding the same problems at system test invokes the higher per-hour labor charges. Diagnosing each fault also takes longer, in this case 36.25 minutes. Therefore, the system-test cost is $49.84 per board, assuming that the instruments and other tools that the system-test technician requires already exist in the facility. Introducing the more comprehensive board-test step saves an average of $38.78 per board.

Justifying the additional equipment cost depends on the yield improvement and the production volume. In Figure 10-6, if the new strategy buys a 1 percent improvement, a facility that manufactures 1000 boards per month will save only $388 per month. Over the 2.5-year amortization period, the payback will be only $11,640, clearly insufficient, and finding the faults at system test will be much less expensive. For such a small yield improvement, production volumes would have to exceed about 50,000 boards per month to make the additional investment worthwhile. If, on the other hand, yield improvement is at least 10 percent, board volumes of less than 5000 per month will be enough to justify the change.

As stated previously, other quality factors being equal, the most successful combination of test steps generates the lowest *overall* costs. As in dealing with all

Days/year	250
Hours/day	6
Hours/year	1500
Amortization years	2.5
Hours to payback	3750
Loaded tester cost	$491,000

At Board Test

Loaded ATE operator cost	$ 17.00/hr	$ 0.28/min
Loaded ATE cost	$130.67/hr	$ 2.18/min
Total loaded cost		$ 2.46/min
Avg. faults/board	1.5 faults	
Avg. diagnosis/fault	3 min	$ 7.28/fault
Avg. ATE diag. cost		$11.06/board

At System Test

Loaded test tech. cost	$55.00/hr	$ 0.92/min
Avg. system-test diag.	36.25 min	$33.23/fault
Avg. system-test diag. cost		$49.84/board
Savings with board ATE		$38.78/board

Figure 10-5 In this example, increasing fault coverage at an earlier test stage saves an average of $38.78 per board. (Scheiber, Stephen F. 1992. "Evaluating Test-Strategy Alternatives," *Test & Measurement World*, Newton, Massachusetts.)

aspects of electronics manufacturing (and life, for that matter), slavish adherence to time-honored generalizations such as the order-of-magnitude rule is at best questionable.

10.5 Comparing Test Strategies

In evaluating the relative merits of different test strategies, economic analysis represents only one tool, not the only tool. For example, on first examination, a manufacturer may determine that several small testers provide lower cost than a single monolithic tester performing the same job. An automated facility that relies heavily on board-handling equipment, however, may experience the opposite effect. The small-tester option requires more handling stations, incurs higher software and support costs, and creates higher levels of in-transit damage.

Board Test vs. System Test Costs

% Yield Improvement

Bds/mo.	1%	2.5yrs.	10%	2.5yrs.
1K/mo	$ 388	$ 11,640	$ 3,880	$116,400
5K/mo	$ 1,940	$ 58,200	$ 19,400	$582,000
10K/mo	$ 3,880	$116,400	$ 38,800	$ 1.16M
25K/mo	$ 9,700	$291,000	$ 97,000	$ 2.91M
50K/mo	$ 19,400	$582,000	$194,000	$ 5.82M
100K/mo	$ 38,800	$ 1.16M	$388,000	$ 11.6M
500K/mo	$194,000	$ 5.82M	$ 1.94M	$ 58.2M
1M/mo	$388,000	$ 11.6M	$ 3.88M	$ 116M

Figure 10-6   The aggregate savings during the amortization period for adding the extra test step.

Figure 10-7 considers the relative merits of three test strategies. This discussion will not definitively select one but merely explores some appropriate components of a rational evaluation.

Tester number 1 is a relatively conventional small functional tester, perhaps using emulation technology to measure board performance. Number 2 is an inspection machine that requires no traditional test programming, learning its program from simulators and good boards. The third tester is a large monolithic stimulus/response type. The two small machines are the same price, and number 3 is much more expensive.

Setup cost assumes that the two small testers are relatively easy to install. Installing the larger machine is more difficult but less expensive than the purchase-price-multiple. A facility cost of 2 percent of the purchase price assumes that the testers occupy factory-floor space proportionate to their price.

The programming cost assumes that the large tester offers numerous elaborate test-generation tools but that actual test generation and debugging are long and expensive. Similarly, normal manual test-program development for the emulation machine costs more than the self-learn method does on machine number 2. Costs for edge-connector fixtures for functional test and a tooling-pin bed for the inspection system are identical, and the service cost is 1 percent per month, producing the annual system costs shown in the figure.

The next step in this model separates go/no-go sorting costs from the cost of fault diagnosis. In this case, testing a good board by emulation takes two minutes, twice as long as an equivalent test for either of the other approaches. Sort-cost calculations assume that the salary for a test operator is higher on the more complex monolithic tester than on its smaller siblings. Fault-isolation costs assume that the person who diagnoses failures using either GFI or emulation techniques requires

	1	2	3
	Small ATE	Inspection Station	Large ATE
Purchase price	$ 50,000	$ 50,000	$250,000
Setup cost	2,000	2,000	5,000
Purchase cost/yr	10,000	10,000	50,000
Facility cost	1,000	1,000	5,000
Prog. cost (2/yr)	100,000	40,000	150,000
Fixture cost	5,000	5,000	5,000
Service cost	6,000	6,000	30,000
Annual cost	$124,000	$ 64,000	$245,000
	Small ATE	Inspection Station	Large ATE
Max. boards/year	60,000	120,000	120,000
# boards/year	25,000	25,000	25,000
% utilization	42%	21%	21%
Good board test times (minutes)	2	1	1
Labor cost/hour	$40	$40	$60
Annual sort cost	$33,333	$16,667	$25,000
	Small ATE	Inspection Station	Large ATE
Bad boards (10%)	2,500	2,500	2,500
Diag. time (min)	10	1	5
Labor cost/hour	$70	$40	$70
Faults/board	2	2	2
Fault-detection cost	$58,333	$3,333	$29,167
Total cost	$215,666	$84,000	$299,167

Figure 10-7 The relative merits of three test strategies.

higher skills than the person performing the same task on the inspection system and that the isolation task will take much longer as well.

If all of the individual numbers are accurate, this model selects the inspection system as providing the least-expensive approach. Before adopting that approach, however, you must carefully examine the model's assumptions. For example, large-tester operators are not necessarily more expensive than their small-tester counterparts. In fact, in automated, high-throughput situations, the portion of an operator's time devoted to a single board test may be less than with a smaller machine.

To compare these three testers properly, they must all test all boards and provide equivalent fault coverage. Because large testers often provide higher speeds, can test larger boards, and find more faults, the additional features may justify the price difference. Similarly, tester number 1 will find component and other faults that the inspection station will miss. If the board under test contains slower, older-generation electronics technology, higher speeds may not be necessary. On the other hand, edge-of-technology products may require the superior performance of the higher-priced machine.

If the large tester is already in place on the factory floor and the new board adds to its test load without taxing its capacity, incremental costs may be lower than for introducing an altogether new machine, no matter how inexpensive the new machine might be. In that case, training costs are zero, and there is no perceptible learning curve.

The model ignores the cost of escapes, as well as automated handling and other high-throughput issues. Moreover, because the majority of the stated advantage of the winning tester rests with fault-diagnostic procedures, the numbers change drastically if yields are much higher (or much lower) than the assumed 10 percent or if failing boards will be scrapped and not repaired. If testers 1 and 3 can take advantage of some portion of the board's simulation, their programming costs will fall.

Do not reject a model such as this one simply because it contains assumptions that may not be valid. The analysis is still useful if the person responsible for the final decision understands its limitations. Ultimately, no generalization can point unfailingly to the best test strategy. Each situation is fundamentally different, and test managers must consider all known factors before making a choice.

10.6 Break-Even Analysis

The model in the preceding section shows how to quantify different test strategies' relative costs. Accurately comparing alternatives, however, also requires understanding benefits—actual savings and cost avoidance compared to no test or another strategy, enhanced goodwill (and therefore improved product sales), and so on. The best strategy is not necessarily the lowest-cost alternative but the one providing the best payback for money spent. *Break-even analysis* techniques evaluate test options, including inspection and monitoring processes rather than testing products, to aid in decision making. As with all models, even when the individual numerical calculations are less than accurate, if the relationships are correct, the strategy selection will also be correct.

The following discussion applies several analysis methods to the strategic results in Figure 10-8. The tester costs $250,000 and carries an annual maintenance

Break-even analysis example

* Purchase price = $250,000
* Yearly maintenance = $25,000
* Programming costs
  $30,000 first year
  $40,000 second year
  $10,000 (maintenance) per year after that
* Savings
  $75,000 first year
  $150,000 per year after that
* Depreciation – 5-year straight line
*Tax rate – 35%

Figure 10-8   A typical break-even analysis. (Scheiber, Stephen F. 1992. "Evaluating Test-Strategy Alternatives," *Test & Measurement World*, Newton, Massachusetts.)

charge of 10 percent, or $25,000. The test manager plans two test programs, developing one in each of the first two project years. Programming costs are $30,000 per program, plus $10,000 for program maintenance once the first program is implemented after the first year. Savings are $75,000 per program per year, which translates to $75,000 the first year and $150,000 each year after that, compared with a strategy that does not include the new tester.

Note that all net savings are subject to income tax. The company calculates five-year, straight-line depreciation, and the corporate tax rate is 35 percent. Therefore, the after-tax savings is 65 percent of the total savings. In addition, because depreciation is not a cash outlay, but a tax deduction, the company adds 35 percent of the depreciation to the after-tax figure to get the actual cash benefit of a strategy in each year. The next sections will help clarify these points.

10.6.1   Payback Period

The simplest break-even-analysis method, *payback period*, calculates the cost of equipment, programs, and fixtures, then deducts anticipated savings. Payback is the point when aggregate benefits exceed aggregate expenses, net of depreciation and taxes.

Figure 10-9 shows a payback analysis of our example. The zero-year outlay represents the capital expenditure itself, which is both the starting point and the number that is subject to depreciation. In year 1, the $75,000 savings exceeds the $55,000 in outlays by $20,000, producing a benefit of $13,000 after taxes. The $50,000 income-tax deduction for depreciation reduces the tax bite by 35 percent of $50,000, or $17,500. The total savings of $30,500 reduces the aggregate net savings to –$219,500.

Similarly, the savings in year 2 is $85,000, or $55,250 after taxes. Added to the depreciation benefit, the total savings is $72,750, and the net savings after year

Year	Outlays	Savings	Before Tax Net	After Tax Net	Deprec Benefit	Total Savings	Overall Net
0th	$250,000	0	0	0			($250,000)
1st	$55,000	$75,000	$20,000	$13,000	$17,500	$30,500	($219,500)
2nd	$65,000	$150,000	$85,000	$55,250	$17,500	$72,750	($147,250)
3rd	$35,000	$150,000	$115,000	$74,750	$17,500	$92,250	($55,000)
4th	$35,000	$150,000	$115,000	$74,750	$17,500	$92,250	$37,250

Interpolation:
Payback = 3.60

Figure 10-9 A payback analysis of the example in Figure 10-8. (Scheiber, Stephen F. 1992. "Evaluating Test-Strategy Alternatives," *Test & Measurement World*, Newton, Massachusetts.)

2 is down to –$147,250. In each of the next three years, the after-tax savings is 65 percent of $115,000, or $74,750, and the total savings is $92,250 per year. There is no depreciation benefit after year 5.

Notice that the aggregate net savings becomes positive between the end of year 3 and the end of year 4. Interpolation produces a payback period of 3.60 years.

Most managers adopt this technique because it is simple to calculate, easy to understand, and easy to communicate. It favors projects with fast returns (the "hares") over projects that take longer to pay back but have a higher overall return (the "tortoises") because it ignores savings after the payback period. Test strategies often fall into the "hare" category. Nevertheless, comparing two fast-payback strategies may be difficult because of inaccuracies inherent in the payback calculation. It ignores the time value of money, so it ignores the timing of project benefits. (Actually, this limitation may not be as restrictive as it first appears. In today's fast-changing manufacturing environment, many companies require payback in less than 18 months, even less than a year. In such a short time, the time value of money will have less influence on the analysis than if you were to calculate economic performance over, say, five years.) On the other hand, many people who apply the technique ignore depreciation and tax effects, a much more serious omission.

10.6.2 Accounting Rate of Return

Davis also computes an *accounting rate of return* (*ARR*; also called the *average return on investment*, or *average ROI*). This technique divides the project's average annual savings by the initial investment, thereby estimating the percentage of initial investment that the project returns each year. In many respects, it is the reciprocal of payback period. Like that technique, ARR ignores the time value of money, is easy to calculate, and allows quickly eliminating the worst alternatives. Because it considers benefits that accrue throughout a project's life, however, it more accurately compares strategies where the timing of savings is very different.

Year	Savings
1	$30,500
2	$72,750
3	$92,250
4	$92,250
5	$92,250
Total	$380,000
Average	$ 76,000/year

$$\frac{76,000}{250,000} = 30.4\%$$

Figure 10-10   The average rate of return includes the fifth project year and its $92,250 in additional savings.

In Figure 10-10, the year-by-year savings are the same as with the payback method. This case, however, includes the fifth project year and its $92,250 in additional savings. Therefore, the total savings over the 5-year project life is $380,000, an average of $76,000 per year, which represents 30.4 percent of the initial investment.

10.6.3 The Time Value of Money

Both the payback and ARR techniques ignore the fact that money costs money. Income a year from today is worth less than that same income today. Today's recipient could deposit the money in a bank account or other instrument that earns interest. Also, in many business situations, earnings from one project represent an investment in some future project that can earn more than the going interest rate.

The value of an investment at some future time (FV) depends on its present value (PV), the interest rate per period (i), and the number of periods (n), as follows:

$$FV = PV(1 + i)^n \qquad \text{(Eq. 10–1)}$$

Therefore, the present value of any cost or payback from a project is:

$$PV = \frac{FV}{(1+i)^n} \qquad \text{(Eq. 10–2)}$$

Appendix A contains tables of factors for determining present value and future value that result from either a single payment or an annuity. For example, using the present-value table for an interest rate of 10 percent, $1000 of income today is worth $1000 (obviously). That same income a year from now is worth only $1000(.909), or $909 today. If it comes two years from now, it is worth $826 today. For the income at three years, the present value is $751, and the equivalent if the income does not come for four years is only $683. Standard financial software and many financial calculators can compute these factors automatically.

The economic benefit of different test strategies based on time-value-of-money techniques depends both on the magnitude and the *timing* of payments and savings. Two strategies that produce the same benefits, but not at the same time, are rarely equivalent.

What do you use for an interest rate? A minimum rate of return, called the *opportunity cost*, represents the minimum rate that the money would earn in a savings account or other instrument. If investment cash is unavailable, the opportunity cost is the minimum interest rate required to borrow the money. Most company controllers' offices establish a standard opportunity cost that all managers must adopt when evaluating project investments. This value changes periodically as economic conditions change.

Generally speaking, a manager should reject any project whose return is below this level. When evaluating test-strategy alternatives, however, the realities of maintaining a high-quality manufacturing operation make some testing unavoidable. If analysis shows that the return from all strategies is below the opportunity cost, then the best option is the least expensive choice that produces acceptable product quality, with the cost difference invested at the opportunity-cost rate. In practice, test-strategy returns are usually much higher than opportunity costs, so this problem rarely comes up.

Assuming that cash-flow and opportunity-cost predictions are accurate, these evaluation techniques provide the best means to compare test-strategy options. The chief drawback to these approaches, of course, is that they require accurate cash-flow and opportunity-cost predictions.

10.6.4   Net Present Value

Figure 10-11 contains a timeline of the investments and savings for the example in Figure 10-8. To avoid confusion, the figure presents investments and savings separately, rather than combining them. In keeping with convention, it assumes that all expenditures occur at the beginning of any year and that all savings occur at the end.

The *net present value (NPV)* method assumes an opportunity cost, then calculates the present value of all cash flows that a project will generate throughout its life, including the initial investment. If the total is greater than zero, the project's payback exceeds the opportunity cost, and it is worth doing. A net present value less than zero indicates that investing the money elsewhere would yield a better return. When comparing test strategies, managers also look at the NPV's magnitude or its percentage of the initial project investment.

Figure 10-12 shows the present value for each of the cash flows in Figure 10-11. In practice, netting the flows at each time would simplify the calculation, but they are presented this way for clarity. In this case, the NPV is $57,219, or 22.9 percent of the initial $250,000 investment. (Remember that $55,000 of the year 0 expenditure is for operational costs.)

Income (Thousands)

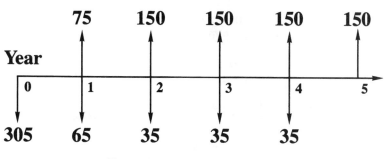

Investment (Thousands)

Figure 10-11 A timeline of the investments and savings for the example in Figure 10-8.

PV(−305,000)at year 0	($305,000)
PV(75,000) at year 1	$68,182
PV(−65,000) at year 1	($59,090)
PV(150,000) at year 2	$123,967
PV(−35,000) at year 2	($28,926)
PV(150,000) at year 3	$112,697
PV(−35,000) at year 3	($26,296)
PV(150,000) at year 4	$102,452
PV(−35,000) at year 4	($23,905)
PV(150,000) at year 5	$93,138

Net present value (NPV)	$57,219
% of initial investment ($250,000)	
	22.9%

Figure 10-12 The present value for each of the cash flows in Figure 10-11. (Scheiber, Stephen F. 1992. "Evaluating Test-Strategy Alternatives," *Test & Measurement World*, Newton, Massachusetts.)

This version of the analysis ignores the effects of taxes and depreciation. Again assuming a 35 percent tax rate and 5-year, straight-line depreciation produces the numbers in Figure 10-13. The flow at year 0 includes the cash benefit from the first $50,000 of depreciation because depreciation does not represent savings that occur over time but, rather, results from an accounting convenience (a tax deduction). Also, a company can begin depreciating an asset as soon as it is placed in service. There is no depreciation benefit at year 5 because the tester is already fully depreciated. Notice that the tax bite has lowered the net-present value

Year	Flows	Net	PV
0	−$250,000 (−$ 55,000)(0.65) (+$ 50,000)(0.35)	−$268,000	−$268,000
1	(+$10,000)(0.65) (+$50,000)(0.35)	+$24,000	+$21,818
2	(+$115,000)(0.65) (+$50,000)(0.35)	+$92,250	+$76,240
3	(+$115,000)(0.65) (+$50,000)(0.35)	+$92,250	+$69,309
4	(+$115,000)(0.65) (+$50,000)(0.35)	+$92,250	+$63,008
5	(+$115,000)(0.65)	+$74,750	+$46,414
NPV			+$ 8,789
NPV%	of inital $250,000		+ 3.52%

Figure 10-13 Numbers from Figure 10-12 assuming a 35% tax rate and five-year, straight-line depreciation.

considerably. Although it is still greater than zero, it now represents only 3.52 percent of the initial investment (above the opportunity cost). In addition to depreciation, many countries offer investment tax credits and other incentives that significantly improve the benefit side of this calculation. When in doubt, check with the company's accounting department or controller's office before completing the analysis.

10.6.5 Internal Rate of Return

The *internal rate of return* technique computes the interest rate that reduces the net present value to exactly zero. In effect, this approach asks the question, "How high would the opportunity cost have to be to make this project not worth doing?" Figure 10-14 shows the NPV for our example at a number of interest rates, starting with 10 percent. Notice that, excluding tax effects, crossover occurs between 15 percent and 16 percent. Interpolation yields an IRR of 15.7406 percent. A computer program designed to calculate the number directly generates 15.7366 percent. In no way do these calculations suggest that the values are significant to six figures. The numbers merely show that, within the limits of reasonable error, it does not matter whether you obtain results by trial and error or by direct calculation.

Ignoring Depreciation and Taxes

NPV at 10%	$ 57,219
NPV at 11%	$ 46,204
NPV at 12%	$ 35,659
NPV at 13%	$ 25,557
NPV at 14%	$ 15,877
NPV at 15%	$ 6,595
NPV at 16%	$ -2,310
Interpolated IRR	15.7406%
Calculated IRR	15.7366%

Including Depreciation and Taxes

NPV at 10%	$ 8,789
NPV at 11%	$ 1,075
NPV at 12%	$ -6,327
Interpolated IRR	11.1452%
Calculated IRR	11.1427%

Figure 10-14   The NPV for our example (excluding and including tax effects) at a number of interest rates.

In a similar manner, the IRR including taxes and depreciation is 11.14 percent. Again, Figure 10-14 includes both interpolated and calculated results for comparison.

10.7   Estimating Cash Flows

As stated, the largest impediment to obtaining accurate economic forecasts is the difficulty of estimating cash flows accurately. Outflows include capital investments, with all of their attendant costs, software and test programming, maintenance, and board repair. Inflows come primarily from cost avoidance compared with not testing and from lost-sale avoidance, as well as from the economic benefits of getting products to market quickly. Customer goodwill from offering high-quality products leads to higher sales and represents an inflow as well. Boards that fail in the field after warranty expiration may generate revenue through the field-service organization, but that hardly represents a desirable situation.

Where do cost numbers for an economic analysis come from? Most companies declare an opportunity cost, based on managers' previous experience with project evaluations and those projects' subsequent success rate. Labor rates, burdens, overheads, and other costs are usually available from the company controller's office. In fact, many companies report those numbers monthly to the management staff, although managers rarely look past their own department's performance against budget.

Vendors can provide equipment costs. Software and test-program-development costs are a bit less certain, but estimating them from other projects generally produces reasonable results. Similarly, estimate building construction, land, utilities, and other costs from experience.

When making estimates, always try to err on the conservative side. That is, the project's calculated return based on your analysis should be less than its actual return as often as possible. That way, after implementation, the chosen alternative is unlikely to produce disappointing results. Always remember that an economic analysis is only as good as its estimates and assumptions.

One more caution: The difficulty of constructing economic justifications and engineers' reluctance to tackle them have encouraged vendors (and consultants) to create software modeling tools. Your data go in one end of a black box, an economic analysis comes out the other. This convenience has several drawbacks. First, your situation may differ in some fundamental way from the software's inherent assumptions. Also, vendor-created software models necessarily contain a bias toward that vendor's products. This bias may not even be intentional, but vendors naturally believe in their own products and in the validity of their assumptions. Considering the uncertainty that accompanies any economic analysis, a bias is usually unavoidable. Perhaps most important, however, is that some engineering decisions still require human judgment and analysis. Again, economic modeling cannot avoid "slop"—uncertainties in calculations and assumptions. Software can make only binary decisions at each point along the way. Human beings can see the shades of gray in this grayest of engineering endeavors.

10.8 Assessing the Costs

Examining the costs during economic justification serves two purposes. First, it gives you some idea of what the whole endeavor will cost the company, and therefore allows you to set budgets and estimate expenditures before you begin. It also suggests what your selling prices will have to be to ensure a profitable enterprise. Second, and often more immediate, it provides tools with which you can compare alternatives using one of the above techniques.

Unlike many other economic analyses, when comparing alternatives the "do-nothing" scenario incurs serious costs. As mentioned earlier, scrapping products that fail often represents the largest (often by far) drain on a company's economic resources. Reducing or eliminating that scrap becomes an unarguable economic imperative. Therefore, you *must* analyze the various strategic alternatives (including *none*) and determine their economic implications, regardless of what you eventually decide to do.

Although the cost of this evaluation affects the total cost of the project, for comparing alternatives it makes no difference whether you spend $5 or $500,000. All you care about at this point is the *difference*. You can ignore any quantity shared by all of the alternatives. If you know that you must choose a test solution, and the solutions are fairly comparable in their impact on the factory, then you can also ignore such factors as site preparation, network setup, and ongoing network and data analysis. Similarly, if you determine that a cost difference is small enough to get lost in the "noise" of the model's uncertainty, you can ignore that as well.

Economic analysis can prove a formidable, complicated task. Any simplifications that do not affect the final result are worth taking.

This chapter introduces the concept of economic justification and provides some tools. It does not attempt to cover the subject completely. Interested readers can find much more comprehensive discussions in my *A Six-Step Economic-Justification Process for Tester Selection* (1997) and *Economically Justifying Functional Test* (1999).

10.9 Summary

Any test problem permits a variety of technically "correct" solutions. Each of those solutions also has economic consequences. The best test-strategy selection offers acceptable product quality at an affordable price.

Determining an alternative's economic implications requires understanding the individual contributors to overall manufacturing cost as well as the methods for comparing and evaluating them. In addition to helping test-engineering managers select the best strategy, the analysis provides necessary tools that company managers often demand to justify spending the money.

Costs include sunk costs such as strategy-evaluation and tester-evaluation costs, which a manufacturer must spend even if the project is later abandoned. Operating costs reflect day-to-day expenses, amortized capital expenditures, test-program-development costs, and other large items.

In addition, every test strategy generates benefits, primarily in cost avoidances compared with not testing and enhanced customer goodwill. Managers must estimate benefit differences among candidate strategies, then compare costs and benefits using a break-even-analysis technique, such as payback, accounting rate of return, net present value, or internal rate of return. The first two methods are easy to calculate but ignore the time value of money. The last two are more complicated but can be much more accurate.

Any economic analysis is only as good as the accuracy of its estimates and assumptions. Nevertheless, taking time to evaluate test alternatives' costs and benefits can contribute substantially to a manufacturer's long-term financial health.

CHAPTER **11**

Formulating a
Board-Test Strategy

One primary challenge of building a successful board-test strategy is that
"conventional wisdom" changes as fast as the rest of electronics technology. Therefore, test managers must consider not only the best guidelines from the past but
also the criteria that led to those guidelines in the first place. For example, consider
the following traditional approach:

> In most cases, in-circuit test followed by functional test gives the lowest
> costs.
> One hundred percent component screening lowers the cost of any strategy.
> Functional-only is cheaper than in-circuit-only for prescreened components.
> In-circuit-only is cheaper than functional-only when volume is low or when
> functional-fault incidence is low.
> Component testing, followed by in-circuit testing, followed by functional
> testing minimizes test costs.

Although once generally valid, these propositions make many implicit
assumptions about test situations that no longer hold true. For example, many
manufacturing operations have reduced their reliance on MDA and in-circuit
testing in recent years. The transition from through-hole to surface-mounted components has made bed-of-nails testing much more difficult, more expensive, and
less accurate. Inaccessibility of individual components and the resultant need to
evaluate device clusters eliminate one big advantage of this test technique—that
the failing test automatically identifies the fault. In addition, although shorts-
and-opens and MDA-type testers easily detect most shorts (once the scourge of
board manufacturing), surface-mount technology, BGAs, flip-chips, and other
technological advances are much more susceptible to open-circuit problems, which
are more difficult to detect. The migration to increased use of inspection in addition to or instead of test reflects this trend.

In addition, as manufacturing processes get better, the number of faults that
testing must find declines. Despite the higher cost of finding each one at a later test

step, there may simply not be enough total failures to justify a screening test, and in some cases a functional-only test may suffice.

An in-circuit-only strategy or in-circuit coupled with inspection verifies merely that the manufacturing process works correctly. It cannot detect racing problems, additive-tolerance errors, and other board-wide anomalies. For mature designs or designs in which engineers have confidence, however, a bed-of-nails test (where board layout permits one) followed by a system or hot-mockup test may be enough. Most disk-drive manufacturers, for example, use this approach, as do some PC makers. Other products, especially complex digital products, require some kind of formal functional test.

When the first edition of this book came out in the mid-1990s, it appeared that as design-for-testability, statistical process control, and other preventive practices gained in popularity, the need for postprocess testing would decline further. I expected that in high-volume facilities, sample testing to ensure that the process remains in control would replace comprehensive testing of every product. Unfortunately, ever-advancing board technologies have frustrated the effort to reach that laudable goal. (Based on these stellar predictions, I have put away my crystal ball.) New challenges have arisen, from coplanarity (flatness) problems that degrade the quality of BGA boards to the now commonplace customer assumption (still often contrary to fact) that a product should work trouble-free from purchase to death. Successful test strategies must still reflect the maturity of the constituent components and modules, the design, and the process, as well as factors such as production volumes, quality targets, previous test-strategy experience, and (of course) budgets. Certainly for low-margin (sometimes vanishing margin), high-volume products like cell phones, routinely testing every one could prove prohibitively expensive. Nevertheless, most test engineers cannot expect routine comprehensive testing to go away any time soon.

11.1 Modern Tester Classifications

As part of the test industry's evolution, dividing lines between traditional tester types are breaking down. Today's MDAs often offer capabilities that rival those of in-circuit testers. In-circuit testers, with their higher speeds and cluster-testing techniques, begin to resemble functional machines. Infrared inspection blurs the line between inspection and test. Functional testers can provide stimulus/response, emulation, and system-test-like solutions. Some combinational testers began life in either the in-circuit or functional domain. Product enhancements that customers demanded permitted them to find problems that previously came only from the other camp.

Today's board testers break down roughly into the following categories:

Low price, low performance (shorts-and-opens testers, MDAs)
Low price, medium performance (benchtop functional testers and low-end in-circuit and combinational machines)

Medium price, medium performance (traditional in-circuit testers, automated-optical inspection systems)

Medium price, higher performance (some faster in-circuit and traditional functional machines)

High price, high performance (high-end functional and combinational testers and automated x-ray inspection systems)

Custom, rack-and-stack, and VXI-based systems follow the same pattern.

Each of these solutions, singly and in combination, has value in certain situations. There are no panaceas, however. No single strategy will solve everyone's test problems. Remember, *all generalizations are false, including this one.*

11.2 Establishing and Monitoring Test Goals

Strategic planning is difficult enough under the best of circumstances. Beyond technical and economic considerations, test-strategy planning has managerial implications and political consequences. Unless everyone's goals are compatible, people will be working at cross-purposes, increasing any test strategy's costs, reducing its efficiency, or (most often) both.

Creating a coherent strategy requires establishing overall quality goals within departments, divisions, and entire companies. This step must involve designers, manufacturing people, test people, and service people and does not necessarily relate to particular projects or products. It is imperative that managers to the highest levels of an organization support this effort. Without their support, many a well-intentioned program disintegrates into petty squabbling.

At this earliest stage, facilities that maintain hard lines between departmental responsibilities experience more trouble than those with more integrated corporate hierarchies. If each department tries only to improve its own cost picture and product throughput, the organization as a whole suffers as problems pass from hand to hand. That attitude resurrects the stone-age days when engineers threw product designs "over-the-wall." (Before modular offices, designs came "over-the-transom.") Manufacturing and test engineers had to figure out how to build them reliably. This imperialistic attitude created semiautonomous fiefdoms at war with one another; any hope of a common goal got lost in the process.

A comprehensive quality plan should cover acceptable quality levels, available tester types, test priorities, lines of responsibility for the plan's elements, and budget targets. For each product, the design/test team should attempt to predict failure levels, circuit areas most likely to fail, and tactical steps to detect and correct those problems (relatively) easily. The accuracy of these predictions will significantly affect the resulting test strategy's success.

The plan's authors should provide *written* documents outlining its elements in detail. These documents should specify necessary quality training, as well as operational guidelines for designers and manufacturing people.

Written documents offer numerous advantages over less formal alternatives. They minimize misunderstandings and reduce the likelihood that participants will inadvertently violate agreements. They also help prevent project-flow discontinuities (quality or throughput) resulting from personnel turnover.

In-person quality training sessions provide face-to-face forums for clarifying agreed-upon goals and procedures with all responsible groups. Centering such sessions around the written documentation can be more effective than offering written versions alone. The sessions generate camaraderie, a sense that all departments really are working together, and encourage feedback to help policymakers revise the documents if they do not offer the best compromise among competing and often conflicting needs.

A design-policy manual, based on the quality plan and created by designers and manufacturing people *together*, establishes standards that designers are expected to follow. It should specify CAE equipment and the documentation that this equipment provides. Including preferred test approaches helps ensure, for example, that designers accommodate test nodes for beds-of-nails or for guided-probe fault isolation. Access and logic-depth guidelines facilitate test-program development.

A series of formal reviews throughout each product-development cycle helps participants understand and apply these established goals and procedures for that product. Such reviews tend to prevent designers and marketers from specifying product features that are really "frills" and add unacceptably to production and test burdens. Managers must make everyone accountable for any decision that complicates test, manufacturing, or product-support operations.

At the same time, however, formal procedures are at all times cooperative—not dictatorial—documents. No procedure is "cast in stone." *Written procedures must not prevent on-the-fly guideline violation to reflect new information*, either in general or on a case-by-case basis. For example, suppose that established design goals prohibit components on both sides of a board. A designer decides that, in a specific situation, there is no acceptable way to make the product work properly while limiting components to one side. Perhaps a one-sided board would require traces that are too long to achieve timing targets or too close together to avoid crosstalk. The responsible design engineer makes the test engineer aware of the problem as early in the cycle as possible. Together, they must decide among several options:

> There is another design alternative that satisfies the stated design goal and reaches the necessary level of performance.
>
> The design must violate the stated goal, but the designer can take steps to reduce the impact on testing; for example, by adding test pads or placing components so as not to bury existing circuit nodes.
>
> The designer was right all along, and there is no viable alternative.

If analysis shows that the third condition is true, at least the test engineer has that information early enough to consider it when formulating a test plan. Again, the ultimate aim is not to decrease costs and residence time in a particular department but to release a quality product with a development/manufacturing/test strategy that minimizes overall costs and time to market.

Another example involves a surface-mount design that for functional or performance reasons must include a few through-hole components or a few components that cannot tolerate the temperature inside the reflow oven. To avoid adding wave solder, hand assembly, or some other extra step, the manufacturing engineer can specify surface-mount (and heat resistant) sockets for those troublesome parts. Manual or automated insertion after reflow provides a lower-cost, higher-reliability solution.

Creating a formal design-policy manual is not generally a popular suggestion. Designers dislike writing the documents and fear that manufacturing departments will regard them as requirements, sanctions, and prohibitions, rather than guidelines. Managers struggle with the necessary document maintenance that provides all involved departments with the latest versions. Nevertheless, written documents represent the best assurance that everyone consistently follows whatever policy an organization has established. They also offer the best targets for objectors to particular policy elements to shoot at, so they can actually be easier to change than less codified rules.

Implementing design policy within an organization demands that managers include the triumvirate of maximum product quality, minimum overall costs, and minimum time to market as part of everyone's job-performance evaluation and that salaries and promotions depend on each person's cooperation and contribution to the entire project's success. Therefore, designers and test engineers who remain intransigent about accommodating one anothers' legitimate needs will experience reduced access to the corporate fast track. When everyone begins to look at the "big picture," the entire company (and all participants) will benefit.

Once the product enters early production, *monitor procedures and results* to permit easy modification of manufacturing or test-strategy elements as the situation warrants. If integrated design and test strategies have evolved simultaneously, perhaps both must change.

11.3   Data Analysis and Management

The importance of test-data management and analysis for tracking the manufacturing and test process, especially in its earliest stages, cannot be overstated. Failure reports and other feedback furnish the ultimate verification that a strategy works. Many deficiencies that data analysis identifies can be easily corrected.

Consider a post-paste inspection step that finds pads with too much or too little solder, or where solder deposits off the pad, potentially causing shorts. Each of these problems can be traced back to a faulty or clogged solder stencil. Correcting or cleaning the stencil will prevent future problems, raising manufacturing yields—thereby lowering burdens on test-and-repair activities and overall costs, while raising both product quality and the customers' perception of that quality. Clearly, the inspection made an enormous contribution to the "test" strategy.

Note that test-data analysis is distinctly different from test-data accumulation. When disk space was scarce, the need to make room for new files provided a strong incentive to analyze data and purge raw data from hard disks at regular intervals. Unfortunately, the enormous capacity of today's hard disks discourages this useful practice. As a result, disks become cluttered with huge numbers of unnecessary files, making specific data difficult to find and increasing the time required for directory listings and other routine disk-related activities.

In addition, constant disk access scatters both individual files and empty space across the disk surface, as does disk storage of temporary "bookkeeping" files by the operating system. Drives must work harder during read and write operations than if the files resided in contiguous storage blocks. Average information-access time increases, as does the probability of misreads and other data errors. Conventional disk-defragmentation programs can alleviate this particular manifestation by defragmenting the files, but this operation also takes much longer if the disk contains many more files than necessary.

Part of the problem rests with the fact that testing generates lots of raw data. Information (analyzed data) takes up much less space. Figures 11-1 and 11-2 present information from analyzing the same data in two different ways. Figure 11-1 sorts failure information by assembly. Engineers responding to this failure pattern can achieve the most significant quality improvement by addressing why

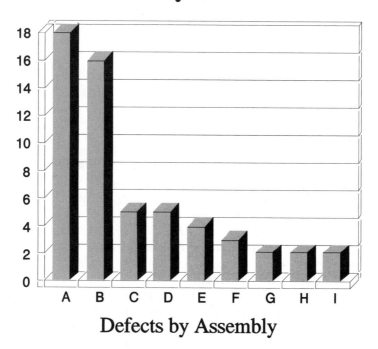

Figure 11-1   Test-failure data sorted by assembly.

Cumulative Failure Causes

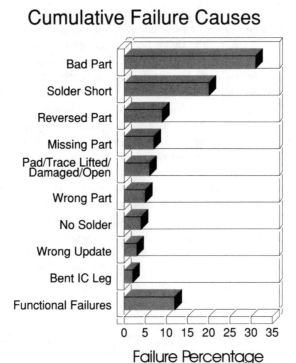

Figure 11-2 The same failure data as in Figure 11-1, sorted by fault type.

assemblies *A* and *B* fail so much more often than the others. Figure 11-2, which sorts by fault type, leads to correcting vendor problems (bad parts) and wave-solder or other automated-assembly errors.

After generating these reports, test engineers rarely return to the raw data files themselves. Therefore, allocate time at least twice a year to evaluate the disk's transient files. Archive and delete unnecessary files. Files that no one has accessed in 6 months or a year should be considered dead. A CD-ROM or DVD writer provides high-capacity, nonvolatile archival storage, giving peace of mind. In the unlikely event that you need access to these files, they are available at a later time without taking up primary hard-drive space.

Engineers should use data-analysis results to suggest manufacturing design changes as well as process modifications. In addition, discovering which test steps uncover which failures can allow modifying tests to improve product quality and test-strategy efficiency and effectiveness.

For example, suppose that system test uncovers a certain collection of digital components that, when working together, often violates the design's timing specifications. Detecting that particular failure may be relatively easy at functional test, which usually reduces overall testing costs. A functional test for that fault may not already exist simply because finding it requires an unusual combination of vectors that the automated simulation and test-generation tools did not provide. Therefore, someone must insert the test manually.

Some cases may call for adding or even eliminating an entire test step. If in-circuit or functional test finds a preponderance of shorts or missing components, prescreening with an MDA or an inspection station may lower test costs by reducing the load on the more expensive machines downstream, freeing up their time for other products or projects. (Of course, correcting the process problem will generally succeed even better.) Similarly, if an in-circuit test rarely uncovers any failures, eliminating that step entirely will reduce costs and work-in-process inventory (there will be one less hold-up step), but will not likely have a deleterious effect on system-test yields or field quality. In that case, an in-circuit or inspection "spot check" ensures that process problems do not resurface to cause large numbers of failures.

Data analysis may indicate that the product's failure pattern does not justify online fault diagnosis. An offline approach may work as well without interfering with product throughput. Also, test-after-repair may reveal that a repair procedure is causing an inordinate number of secondary failures. Popping a failing component off the board, for example, may create opens on adjacent components. Modifying repair tools or procedures can minimize these extra problems.

Whether or not analysis suggests changes in test strategies or tactics, engineers should prepare a written report to company managers and test personnel on the strategy's success. Such a report provides a benchmark to ensure that a process in control remains in control. Subsequent reports will show whether any implemented changes improve results.

11.4 Indicators of an Effective Strategy

An effective board-test strategy requires relatively easy programming, if possible. Of course, in some cases, detecting a particular fault class or testing a certain board design requires a tester that violates this condition. Still, a strategy that specifies a more difficult solution when there are alternatives is less efficient than a strategy that employs those alternatives. (This principle is one of the primary factors driving the industry toward inspection.)

As an example, a few years ago a consultant visited a manufacturing facility containing a number of large combinational testers that were not installed in the production line. In fact, they were not even plugged in! When he asked the test engineer what the big testers were for, the engineer replied that they were there to fulfill a contract obligation that the facility have such machines. Because programming them was difficult and time-consuming, and because the contract did not specify that the boards be *tested* on them, testing occurred exclusively on benchtop equipment. For the same reason, many facilities buy big combinational testers, then take advantage primarily or exclusively of their in-circuit capabilities. Time-to-market constraints often do not tolerate the 6 months or more that creating a comprehensive functional program can take. If a tester's programming process is too complicated to be practical in a given situation, then it is the wrong tester for that situation.

An effective strategy requires high fault coverage, with a minimum of α- and β-type errors. An α-type error is an "escape"—a bad board that the tester thinks is good. A β-type error is a "false failure"—a good board that a test identifies as bad. Such errors represent a tradeoff. That is, tightening test conditions to reduce the likelihood of escapes increases the chances that a marginal board in the passing range will fail, and vice versa. Test engineers must decide case by case which type of failure is more damaging and skew the balance in that direction.

High fault *coverage* is of little value if *identifying* faults requires extraordinary effort. Before automated testing, companies had huge "bone piles" of boards whose simple shorts defied identification. As both boards and their faults have become more sophisticated, fault isolation—even with modern test and diagnostic tools—remains a major concern. For example, a racing condition that inappropriately sets a flip-flop inside a microprocessor could trigger an incorrect instruction or cause the processor to hang or crash. Emulation testing might find such a fault because of its ability to stop on a failure and display register contents and other conditions present on the board at that time.

11.5   Yin and Yang in Ease of Tester Operation

In the beginning, all tester manufacturers offered proprietary machines. Learning to operate one of them provided no advantage when migrating to any of the others except for their general similarity of function. Tester vendors supplied all or virtually all software for tester operation, network control, equipment monitoring, and test-data analysis. An engineer walking through a test trade-show floor in 1980 found that each vendor's software was unique. Even when two vendors shared the same hardware platform, such as a 6800-based architecture or a Data General Nova computer, no commonality existed above a very basic level.

The personal-computer revolution changed that. Today, a comparable trade show contains hundreds of different pieces of equipment based on a relative handful of computer platforms. Most common is the IBM-type PC. A few testers rely on Sun workstations or similar machines running UNIX. Still fewer operate from an Apple Macintosh, while a handful retain their proprietary character. Instruments and other test capabilities generally attach to the engine via one of the standard buses—RS-232 (becoming less common), IEEE 488, VXI, MXI, and so on.

At the software level, the ease-of-use revolution began with the icon-based interface on the Apple Macintosh. (Its predecessors, the Xerox Star and the Apple Lisa, had little impact outside of office environments.) Microsoft Windows extended that convenient graphical user interface (GUI) to the vast PC-compatible world. Now almost anyone with a PC and a mouse can control sophisticated test equipment, instruments, simulators, and test generators with the identical point-and-click technique. For UNIX-based systems, X-Windows provides similar convenience.

At their best, GUIs have enormous advantages. Software all looks the same at the operator level. The tester or test-environment selection process can concentrate on test functions, rather than on command-execution techniques. Training costs decline. Operator, programmer, and test-engineer experience transfers relatively easily between otherwise alien pieces of equipment. The resulting lower barriers to new test strategies or changing strategies add flexibility to test operations. This advantage is particularly handy for contract manufacturers (CMs) who often cater to customers with diverse needs.

Unfortunately, GUIs' greatest strength is also their greatest weakness. When evaluating equipment and vendors to implement a strategy, software-interface similarities make understanding the differences in underlying logical activity difficult or impossible. When most testers contained proprietary software, the human interface's personality generally reflected actual test behavior. Tester vendors do not even develop today's interfaces themselves. Most buy them as packages from Microsoft and other third parties, who generally know little of the unique needs of the test world. If executing a complex input vector on two testers appears identical or nearly identical to a programmer, how can that programmer appreciate that differences in skew, trigger timing, and signal order may lead to their not producing the same test result? More critical is the way in which an icon-based interface creates test programs in response to a test engineer's prodding. At the operational level, the code that two automatic generators produce from a point-and-click sequence may be quite different.

Complicating this problem is the fact that some tester salespeople are enamored with the "ease-of-use" features of their products. Customer demonstrations often consist primarily of showing how effortlessly drop-down menus and point-and-click activities can create test programs, execute test activities, integrate unrelated instruments on a bus, and manage test results. Prospective customers never get a clear picture of what is really happening at the tester's most basic level.

This tirade is in no way meant to denigrate the value of GUIs. The ubiquitousness of Windows-like environments in computer-based applications is a very positive development, especially where the computer is merely an engine driving the "real" work.

Nevertheless, human interfaces aside, test equipment is not interchangeable—that is, no two testers or two instruments behave in exactly the same way. Each machine performs some tasks better than others. Test professionals must take care to fully understand the differences between strategic and tactical alternatives, as well as their similarities, before committing to a specific solution.

11.6   More "Make-or-Buy" Considerations

What happens if you decide to forego the minefield of vendor-supplied equipment in favor of something that your own engineers have designed? What are the advantages and disadvantages of this approach?

A properly designed in-house tester exactly matches a company's test requirements. Engineers design what they need and do not include what they do not need. The tester need not compromise on capability. In contrast, vendors design equipment that best serves all of their customers' needs. Everyone must accept the vendor's accuracy specifications, test and failure-analysis techniques, and other machine characteristics. Once their design is complete, material costs are lower for in-house testers than for vendor equipment because vendors include labor and other burdens in a system's price.

Designing a unique tester may offer manufacturers a competitive edge by permitting better product performance or higher quality than commercial testers can. Tester designers have access to intimate knowledge about the product, so their solution may be better than a more generic alternative.

In-house testers can incorporate a collection of "standard" custom functions, such as instrument modules controlled by a computer engine over an IEEE 488 or VXI bus. Even when testing a particular product demands some unusual procedures, available mix-and-match modules will likely furnish most capabilities, leaving tester designers more time and energy to concentrate on truly unique requirements.

Tester support for maintenance, training, and enhancements is available within the manufacturing facility (or at least within the organization), so response times are (theoretically) faster than vendors can provide. There is no third party, so no one is jockeying for position or shifting blame when problems arise. Local control of test features and easy communication with tester designers facilitate optimizing tester operation for each application.

On the downside, initiating an in-house design project means paying for all nonrecurring engineering (NRE). A vendor spreads NRE costs across a machine's entire installed base, lowering the impact on each customer's bottom line.

In addition, many companies do not account for in-house tester-design time in the same way as direct materials costs or capital-acquisition costs, further reducing the *apparent* costs of the in-house approach. As an example, a couple of years ago, a manufacturer of large capital equipment was evaluating the relative merits of make-or-buy. The commercial tester under consideration had a selling price of about $200,000. *Materials alone* for the in-house alternative would cost $50,000. The lead project engineer tried to convince his management that if he added labor, lost productivity, and other direct costs of tester design, the real cost of the in-house tester would exceed that of the commercial machine. Management overruled him. What transpired not only confirmed the engineer's worst fears, the custom tester could not meet the required completion date, and its performance fell considerably short of the design specifications.

Designing a test system requires specialized skills. Design engineers are experts at designing their company's products. A tester vendor's engineers are expert at designing testers. A computer vendor's expertise lies in computers. As a result, an in-house-tester design may contain errors and inefficiencies that a team with more *tester*-design experience could have avoided.

In-house development often takes 1 to 2 years, so tester planning must lead product planning. Schedules frequently slip. Slippage can delay a new product's introduction, certainly eroding its profitability and possibly compromising its competitive position. A vendor procurement cycle, by contrast, can take as little as 6 months. Promised capabilities are generally available, and customers can be fairly certain of timely delivery.

Also, completely debugging a custom-designed tester presents a considerable challenge. The more copies of any system that exist and the more users that it has, the more opportunities there are to exercise it sufficiently to remove the kinks. A commercial system's other customers will likely uncover problems that you never encounter because the vendor has already corrected them.

In-house support, although close at hand, may not always be available. If tester designers are now working on other projects, getting their attention and quick response may be difficult. If a critical person leaves the company, a piece of the support puzzle may be weak or missing. A vendor, by contrast, generally employs a number of people who can help solve any particular problem. If one of them is unavailable for whatever reason, someone else can fill in. Customers who need it can also generally purchase support beyond the standard package.

Perhaps the most dangerous part of developing custom testers is the need to allocate some of the company's scarce software resources to creating system software, self-tests, and other low-level programs. Despite all of the theories and tools that have emerged over the past few years—computer-aided software engineering (CASE), high-level design languages, shells for bus-based systems, and the like—software development remains something of a black art. Although software today is more likely than in the past to conform to specifications and contain the features and personality that its designers intend, schedules are still uncertain. Test and debugging procedures are far from perfect, and the whole process is, at best, merely painful. Again, the small number of custom-system users makes anything resembling bug-free code less likely. (Always remember the old adage, "Any program more than three pages long contains at least one bug.") Tester vendors employ a staff dedicated to debug tasks, and the larger installed base of users can find problems more easily.

Vendor products generally follow a standard design. That is, a new tester from Vendor A will likely feature a similar architecture and much of the same technology as earlier products from that vendor. The vendor's hardware and software designers can adopt the proven bits and concentrate on the enhancements. Therefore, even a new tester from an established vendor will likely work as advertised.

Commercial equipment generally has a proven and verifiable performance record. On the other hand, until an in-house tester is completed and operating, determining whether it works properly and conforms to specifications is impossible. Nothing prevents you from contacting a vendor's other customers to find out whether they are satisfied with the tester, software, and vendor-customer relationship. Be aware, however, that vendor-supplied customer lists generally contain only "happy campers." Locating the darker side of a company's installed base may take more work. Trade shows and other forms of networking may yield the necessary information.

Because custom test equipment is often dedicated to a certain test problem, operating, programming, supervising, and managing costs are usually lower than for the more flexible commercial machines. Test-floor people participate in the project from the beginning, so they need less training. Unless your staff has previous experience with a particular commercial-tester type, however, introducing a new vendor's equipment or a new model requires a learning process.

Ultimately, the decision to make or buy test equipment depends on how closely commercial solutions match a project's needs economically, technically, and philosophically. As more shell-like tools become available for controlling multivendor modular environments, they will reduce the effort of building custom and semicustom equipment, making this option accessible in more situations.

11.7 General-Purpose vs. Dedicated Testers

Whether you make or buy a tester or assemble one from modules or individual instruments, the result can be a general-purpose machine or a dedicated solution optimized for a single application or class of applications.

General-purpose testers cater to a wide variety of board types. Changing types can be as simple as swapping fixtures and invoking a new test program or as complicated as rearranging instruments or modules and completely reloading software. Such solutions adapt easily to production facilities requiring flexibility in throughput or board mix. Most conventional commercial testers fall into this category. These testers need more highly skilled operators than more task-oriented machines require. To accommodate extra features, they generally take up more floor space, as well.

Dedicated testers are usually rack-and-stack types or in-house–built monolithics. By design, they efficiently test a particular board or class of boards and cannot be easily reconfigured for a different application. For vendor-supplied varieties, customers choose necessary features from a catalog. The vendor then assembles the appropriate system either at the factory or at the customer site. These testers can be faster than their general-purpose counterparts and aim primarily at high-volume, low-mix applications. Hot-mockups represent a special case of the dedicated approach.

One common use of dedicated machines is in the military, where a vendor designs a tester in response to written specifications for a particular task. Adding a test method for a single product can require a number of different physical machines, independently designed and assembled, possibly by different vendors. For example, when the F-111 fighter aircraft was introduced, it contained ten subsystems requiring ten separate test stations at a total cost approaching *$30 million.* (Your tax dollars at work.) Private manufacturers cannot afford to spend so much money to test just one product type. Fortunately, as the pace of technology continues to accelerate, even government and military applications will have to find their solutions with commercial equipment. Also, as modular solutions such as VXI continue to emerge, this dedicated-tester extreme is becoming less common.

Midway between general-purpose and dedicated testers is a growing trend toward *application-specific testers,* which derive advantages from each side. These testers are optimized for a particular class of test problems through specific measurement capabilities or special software. They do not provide the broad spectrum of test capability of the general-purpose machines but aim at niche markets. They can be in-house-built or vendor-supplied.

The earliest examples of this tester category were the memory-board testers of some years ago, designed to deliver very long test patterns at high speed. Their popularity faded as memory in most systems shrank into a small corner of one board.

Today's application-specific testers work with classes of boards, such as disk-drive controllers. These electromechanical boards, although fairly similar to one another, require mixed-signal testing at voltages and currents that are quite different from those required for other board types. Some vendors also offer machines with special embedded tools for testing PC motherboards, PC peripherals, and telecommunications technologies.

So-called service-bay diagnostic systems that many car dealers use to analyze difficult problems also fall into this category. These testers generally contain a standard PC, a touch screen to enhance usability, and measurement devices. The computer engine often contains a CD-ROM or DVD that includes the entire set of documentation manuals as well as video help demonstrations to enable service technicians to search more easily for information to help pinpoint a particular failure.

11.8   Used Equipment

Not every test problem demands the latest-and-greatest, state-of-the-art test equipment. For most companies, only the last few percent of the test job requires the most elaborate solutions. One way to reduce costs is to purchase used equipment to perform some of the less critical testing.

Some people resist the idea of used equipment, claiming that it is obsolete. That statement is accurate but irrelevant. With the pace of development in the electronics industry, there are only two kinds of products—prototypes and obsolete. Product life cycles are too short to allow time for anything in between.

Used equipment has numerous advantages. The downsizing and consolidation that many manufacturers have undergone during the past few years have released a significant amount of equipment from regular service, so there is a substantial supply and a wide selection of models and features.

For sellers, the used market raises some capital and reduces warehouse storage. For buyers, this approach is inexpensive. Product technology is generally mature, so it contains fewer hardware and software bugs than newer versions, where vendors are constantly adding features and capabilities. In some cases, where a buyer already has several of a certain tester model on the manufacturing floor, and where the vendor no longer offers that model, the used market is the only way to add capacity without changing a strategy that works.

Some years ago, a test manager in a major computer company faced a mini-revolt when the vendor who had been supplying all of his in-circuit testers discontinued that equipment family. He replaced them with competitive machines that his staff regarded as much more difficult to program and operate to achieve the same results. Because there was no significant new technology to test, the used-equipment market could have provided a much smoother transition. Adding used testers to handle extra capacity on existing products and phasing in the alternative testers beginning with new products would have better navigated the learning curve, increasing the department's productivity and boosting morale.

In addition to electronics manufacturers selling excess testers directly, several companies have sprung up that buy and sell used machines, providing customer support, training, and other services themselves. One major ATE vendor even takes used equipment in trade and then resells it with factory support, hoping to establish a relationship with cash-strapped customers that will someday translate to new-machine sales.

Although the latest tester models are not usually available on the used market, most machines were originally built by current tester vendors who can offer spare parts and other help if necessary. If the machines come from used-equipment dealers, those dealers furnish software updates and maintenance. Fixture kits, fixture construction services, and program generation are available from the same third parties supporting the customers who originally bought those testers new. In many cases, finding a programming house to support a used tester is easier than for a more modern machine. The older version has been on the market long enough for contractors to gain experience with it. If you buy a used tester and have trouble finding third parties to help support it, the used-equipment vendor knows where they are.

11.9 Leasing

Leasing test equipment sometimes represents an attractive alternative to purchasing because the lessee can "trade up" easily during the lease period and can, in some cases, avoid the tax consequences of ownership, maintenance, insurance, and property taxes. At the end of the lease, the lessee can renew it, return the equipment, or purchase it at its then fair-market value.

Test-equipment leases generally take one of two forms. A *direct finance capital lease* requires that the lessee capitalize the asset, as with a purchase, if it meets at least one of the following criteria:

Ownership transfers to the lessee either before or at the end of the lease term.
The lease includes a bargain purchase option.
The term is at least 75 percent of the tester's estimated economic life.
The present value of the minimum lease payments at the lessee's borrowing rate equals or exceeds 90 percent of the tester's fair-market value.

Otherwise, the lease is an *operating lease*. Operating leases can range from a few hours or days (as with car or instrument rental) to several years. The lessee's balance sheet need not capitalize the asset, improving the company's performance as measured by return on assets. Operating leases for test equipment can be financially advantageous because a tester's useful life is often much shorter than its depreciable life according to tax codes. Therefore, purchasing the machine would mean disposing of it before fully depreciating it. An operating lease might also provide a way to get a machine in-house quickly, circumventing some companies' long and elaborate capital-acquisition approval processes.

Enhancing a purchased asset such as a tester requires determining a new depreciable life for the upgraded system. With a lease, the lessee merely amortizes the upgrade over the remaining lease term.

Deciding whether to lease or purchase a tester requires a net-present-value analysis like those in Chapter 10. Note that lease payments are fully expensed for tax purposes. The amount of lease-payments depends on whether the lessee or lessor retains the tax benefits of asset investment and depreciation.

Categories of lessors include bank-related, tester-vendor captive companies, affiliates of other institutions such as insurance companies, independent leasing companies, and commercial finance companies. Not every instrument is available from every lessor. Some concentrate on a particular industry, equipment type, credit quality, lease term, or other market segment.

In selecting a leasing company, Herman (1985) recommends asking the following questions:

Can the leasing company help the lessee navigate the minefield of the leasing process?
Does the leasing company offer several leasing options?
Does the leasing company concentrate on a particular industry, or is it a financial institution such as a bank or insurance company?
Is the leasing company well capitalized? (The "Will it be there next week?" test.) Does the leasing company understand the electronics manufacturing and test industries?
Does the company stock the equipment, or does it serve simply as a broker?
Does the company help the lessee evaluate and select equipment?
Does the lessor provide maintenance, insurance, or other services?

Not all equipment is available for lease. Test managers should formulate test strategies first, including tester types along with all options and software, then investigate whether leasing is possible. Leasing logistics can add several months to the normal tester-procurement cycle.

Many companies never consider leasing, regarding this choice as overly complicated and unnecessary. For others, the tax and business benefits make this somewhat unusual approach worthwhile.

11.10 "Pay as You Go"

Another alternative to a straight equipment purchase involves buying a basic tester model, but taking delivery of a much more elaborate version. Using only the functions of the basic tester incurs no additional charges. If, however, you need some of the advanced capability that the tester offers, you pay for that extra testing on a per board basis.

This alternative offers several advantages. The purchased portion of the tester costs considerably less than the tester with all the "bells-and-whistles," lowering the customer's capital outlay. Customer and vendor share the consequences of market (and therefore manufacturing) fluctuations. By converting part of what would normally be fixed costs into variable costs, the overall cost of running the operation falls significantly in times of slow production. Combining this approach with an equipment lease reduces variable costs further when production volume drops.

Aside from the lack of familiarity with this financial alternative, its biggest drawback to date is that only one vendor offers it. Other companies should consider adopting this technique as well, to make their equipment more attractive to small, startup, and highly variable or cyclic manufacturers. Small companies tend to grow, and a proverbial foot in the door can smooth test-equipment adoption and justification later on.

11.11 Other Considerations

Before making a final determination of the best test strategy for your operation, be sure to review how upstream and downstream process decisions affect test options. For example, do tight process controls make some postassembly testing unnecessary? Will testing lot samples suffice to ensure that the process remains in control? Will data-analysis procedures notice out-of-control processes before the production line generates a lot of low-quality boards? Do manufacturing-process decisions restrict test-method or test-equipment choices? For example, is material handling automated?

After assembly, is burn-in a routine part of the operation? Does the product's application encourage elaborate environmental stress screening, as with automotive electronics and avionics? Do product prices and profit margins permit the expense of such extra steps as ESS, or will the cost of those steps unacceptably erode profitability?

11.12 The Ultimate "Buy" Decision— Contract Manufacturing

Perhaps the most fundamental change in the industry since the first edition of this book was published is the shift away from actually manufacturing and testing your own products. Original equipment manufacturers (OEMs) today design, develop, and market their products, but production often rests with contractors. This practice has become so common that it has spawned the term "virtual corporation" to encompass all activities relating to a company's products, regardless of who performs them. The brand name of the products you buy no longer identifies the company that makes them.

Contract manufacturers fall into three basic categories. Most common and most familiar are the dedicated contractors, contractors for whom that is their primary business. These companies are generally regarded as the most desirable. They range in size from the proverbial "garage shop" to huge corporations enjoying revenue in the billions.

Some vertically integrated companies serve as contract manufacturers, offering part of their excess capacity to all comers. These contractors contend that their experience with all phases of a product's life allows them to more easily meet customers' needs. The only question about this class is who will receive priority if changing requirements within their own companies begin to strain their manufacturing capacity.

The third type of contractor springs from component distributors, who contend that they can offer cost advantages because they know how to buy components inexpensively. One wonders what expertise they claim for building systems.

Why do so many companies outsource? Reasons vary widely, but certain trends emerge. The earliest call for CMs came in the form of "peak management"—that is, managing the manufacturing cycle fluctuations. The CM represents the preferred alternative to forcing a staff sized for mean times to cope with a peak rush, hiring enough staff to handle peak times and having them sit idle when times are slow, or varying staff by hiring and firing people as the workload changes. This last practice is expensive and consumes personnel resources. It also damages employee morale, which can reduce productivity and increase staff turnover, which in turn carry their own costs.

A CM can offer a fresh look at a product's manufacturing and testing needs. Most CMs deal with many more board types than their customers do. They may have a better test approach with which the customer is unfamiliar. Similarly, engineers at CMs offer a wide variety of capabilities and experiences.

When asked for their primary rationale for outsourcing, many companies cite lower costs and higher quality. On the other hand, when asked why they *don't* outsource, companies claim higher costs and lower quality. Clearly the jury is still out on this one.

Outsourcing requires tighter management practices over the entire manufacturing operation, as well as better process documentation. Both are imperative to ensure that what the customer wants and what the CM builds coincide. Although many engineers regard these requirements with disdain, tight management can minimize the likelihood of design errors and ensure that production processes remain in control, thereby maximizing product quality and functionality and increasing chances for market success.

Outsourcing also holds attractions from a business perspective. If a startup company plans to farm out its production, the cost-of-entry in that business falls dramatically. Venture capitalists and other investors sensitive to the volatility and uncertainty in electronics manufacturing like "spreading the risk," becoming increasingly reluctant to fund the expensive infrastructure that manufacturing demands. In fact, the vice president at one major contract manufacturer referred to his business as "infrastructure for rent." (He also referred to vertically integrated companies as "relics of a bygone era.")

It is important to remember that not all contractors are alike. Some specialize in high-volume, low-margin products. Others excel at the quirky world of specialty products. Many CMs offer engineering services—helping customers to redesign boards and systems to make them easier to manufacture. Such a service is "win-win"—customers get higher-quality, more reliable products, while the CM can make the product at lower cost.

The primary drawback to contracting is loss of control over the manufacturing process. Geographical distances affect delivery schedules for both inventory and final product. Calling in a product designer or manufacturing engineer to solve a production problem is generally slower and more expensive when the process involves a CM. Distances crossing different time zones may aggravate delays. Outsourcing's transportation costs for shifting product are higher than with in-house operations.

Operations whose design and manufacturing facilities are half a world apart underscore the need for local control. Test-strategy engineering and modification must also occur locally because manufacturing departments cannot afford to wait 2 days or longer for answers to engineering problems. The logistics of such companies encourage the breakdown of the wall between engineering and manufacturing that most test experts advocate.

OEMs should permit their CMs a measure of autonomy over the production process. A problem that the contractor can resolve independently of the client reduces the engineering load on the manufacturer, and the time saved reduces the product's time to market.

If you decide that a CM will perform manufacturing or test, selecting the *best* candidate for the job is no less difficult than formulating an effective strategy for in-house implementation. A contractor must offer the right mix of services for the projects that you need done. The company should generally be small enough that your project receives sufficient attention and large enough that you do not provide too high a percentage of its total business.

A preferred test strategy or short list of test strategies should dictate contractor choice. Do not permit a contractor to steer you to a test strategy different from what you want unless you are convinced that it represents a better solution for your specific situation. Each vendor obviously has an installed base of test systems and certain "pet" manufacturing and test procedures. Those capabilities must match your needs. Surrendering control to the contractor is like going to an investment advisor who recommends the same vehicle for everyone. There are no "canned" solutions.

The first 6 to 12 months in a relationship with a contractor are critical. That period establishes procedures and defines work habits and communication paths. Schedulers should allow for longer turnaround times to implement product and process changes than with in-house projects.

The information gathered and the analysis performed during evaluation of potential test strategies permit constructing a short list of options. The next chapter will finally put it all together and offer some guidelines for choosing among various test-strategy alternatives.

11.13 Summary

Formulating a board-test strategy relies less on traditional guidelines and platitudes than on a careful evaluation of a particular manufacturing-and-test situation. With the high quality and high complexity of today's devices and boards, the old order-of-magnitude rule no longer holds. In addition, dividing lines between adjacent tester types are beginning to break down.

Manufacturers must establish coherent quality goals within departments, divisions, and entire organizations. These goals should be codified into written documents that provide a common ground to ensure that all participants understand the goals, and they furnish a known target to shoot at when someone wants to circumvent the guidelines. Periodic reviews throughout the design/manufacturing/test process provide a forum to allow flexibility when the situation demands it. In all cases, written procedures must remain cooperative, not dictatorial, documents. Strategies and procedures can change according to situational needs or as the result of data analysis, which might indicate a more efficient alternative.

Implementing any strategy requires evaluating tactical choices. Test programming should be as fast and easy as possible. When examining competing equipment, engineers should be careful to understand the tester activity and technology that lies beneath the "point-and-click" Windows-like user interface that most equipment shares.

Sometimes, the best strategy is a custom-designed tester, because it provides capabilities that exactly match the product's requirements. At other times, vendor-supplied equipment ensures on-time delivery with a minimum commitment of the system-manufacturer's resources. Used equipment or leasing may minimize costs or maximize return-on-assets assessments of company performance. Farming out the entire assembly and test operation to a contract manufacturer can offer economies of scale for small companies and peak management for large ones.

CHAPTER **12**

Test-Strategy Decisions

Much of the discussion to this point has centered around individual parts of creating a test strategy. Chapter 1 provided an introduction. Chapters 2 and 3 described various test and inspection techniques. Chapter 4 offered guidelines for a cost-effective test operation. Chapters 5 through 9 presented material on specific related issues. Chapter 10 constructed the economic basis for deciding among test strategies and tactical choices within them. Chapter 11 brought up the organizational issues that affect how a particular company makes test-strategy (and, for that matter, all strategy) decisions.

Where does all of this information lead? Is there a single set of rules that can help any test engineer or manager identify the best strategy?

Unfortunately, the answer to that question is *no*. Industry analysts differ widely on what constitutes the best strategy for any particular situation, and each situation is necessarily unique. For example, several years ago, I came across a supposed test "expert" who took the position that it is cheaper to incur the cost of testing to weed out faulty products than to improve manufacturing processes to produce only good ones. That simple statement, which violates everything that design-for-testability, statistical-process-control, and concurrent-engineering proponents have been trying to teach for years, was the single impetus that prompted me to write this book and teach seminars on this subject from New York City to Singapore. If a recognized expert can say that to conference attendees and consulting clients, how many other people involved in the manufacturing-and-test process on a daily basis believe it also? Contrast that with W. Edwards Deming's philosophy that you should strictly control the process and perform no testing at all. Clearly, the reality lies somewhere in between.

12.1 A Sample Test Philosophy

The first step in constructing a test strategy is to identify the tests necessary to ensure that the product will work. For most of today's products, that analysis must begin during design. If designers include a self-test to uncover some of the more pernicious faults, it can reduce the number of conventional tests that the rest of the strategy must include, but *only* if the manufacturer executes it during production and makes the effort to eliminate the redundant tests.

In many respects, self-tests offer fault coverage similar to that of emulation tests. In both cases someone who intimately understands the board's function creates the test, and both tests verify the board's logical (digital) kernel.

A central processor or microprocessor generally calls the self-test, which resides on a ROM or EPROM. After self-test, remaining faults lie primarily with peripheral logic, analog components, and other less-complex circuit areas. In most cases, a bed-of-nails can test for remaining failures without accessing nodes within the kernel itself, where through-holes and other conveniences are less common. Of course, to be useful during production, self-test should locate actual problems, not merely make go/no-go decisions.

Generally speaking, the more faults that a self-test can identify, the less likely that the product will require elaborate test equipment to find what remains. Unfortunately, self-test sometimes conforms only too well to the Pareto rule—that the last 20 percent of any task requires 80 percent of the work. The self-test may do only the easy part of the job, leaving test programmers to grapple with subtler problems. Finding them may still demand sophisticated test equipment. Even in these cases, however, programmers can spend their test-development resources on that fringe of the fault spectrum, confident that by the time the product reaches conventional board test, simple problems are already gone.

Similarly, introducing inspection permits much more easily finding board faults that generally elude testing. Surface mount opens provide a classic example. Again, eliminating these failures during an inspection step simplifies subsequent test considerably.

When you know what tests the product requires, is it safe to execute them? Historically, this question has meant removing shorts before applying power to the board (unless the test operation is testing the department's smoke alarms). A self-test requires power, so process control, inspection, or a bed-of-nails prescreener must ensure that the board is short-free before invoking it. Because self-test reduces the functional-test load, it, too, serves as a kind of prescreen.

An efficient test strategy generally isolates failures only to the next-lower replaceable assembly. If the fault lies on a multichip module, an ASIC, or some other complex form, for example, production test does not linger to discover *why* the part has failed. That activity is left for a more comprehensive offline failure analysis or for the vendor.

Of course, this statement, too, does not apply to all situations. A locomotive manufacturer or the maker of medical magnetic-resonance imaging equipment works in lots of one very expensive product at a time. Boards under test may contain daughterboards or complex modules whose cost and nonexistent spares inventory demand repair rather than replacement. Here, the extra failure analysis does not compromise throughput, and repair avoids the necessity to build another board or module before completing the system.

Some test strategies emphasize catastrophic problems only. The entire intent of ESS is to create catastrophic faults out of sporadic ones. Even so, as a process approaches zero defects, narrowing specifications and taking other steps to improve quality beyond removing those defects that remain receive higher priority.

Ideally, each test within a test tactic should identify a single failure or a unique group of failures. The program should recognize related problems and should know whether one fault means a high probability that other specific faults exist as well. This "single-failure testing" drastically reduces fault-diagnosis efforts. A certain test result indicts a particular circuit element and initiates a specific repair activity.

With inspection, the validity of this approach is indisputable. For bed-of-nails testing, conforming to these guidelines is relatively straightforward, assuming that there is adequate node access. Testing with bench instruments, rack-and-stack equipment, or some kind of dedicated tester also often permits this luxury. For many functional tests, however, especially where diagnosis occurs by guided-fault isolation, achieving that goal is impossible.

A strategy should also avoid redundant tests, either at one test stage or at different stages. There is little point in constructing a functional test to ensure that a 74163 counter works, if an in-circuit test has already verified it.

Tests that merely check the manufacturing process need not occur at speed. A short is a short is a short. On the other hand, detecting that a stray reactive impedance compromises circuit performance generally requires testing very close to the product's normal operating speed.

Interactive tests that require human intervention, from probe placement to pot adjustment, should request only one operator action at a time. Instructions should use standard terms, acronyms, and abbreviations. These precautions reduce the likelihood of human error and therefore a bad test.

Some strategists prefer to test the highest-likelihood failures first. Others ensure that the circuit contains no trivial faults (such as shorts) before looking for more complex problems. Deciding between these two approaches depends largely on whether the manufacturing process is strictly in control, as well as on the board's overall quality level and expected fault spectrum. The lower the overall yields, the more attractive the prescreen alternative becomes.

Tests should invoke long operator activities, such as GFI, last in a test sequence, and then only as a last resort. A fault-dictionary diagnosis, although it may be imprecise, may get close enough to the actual problem to reduce the time required for subsequent GFI steps.

Before formulating a strategy, be sure to know the board's maximum and minimum operating speeds, both by design and—if possible—by experience. Understand the specifications of all testers under consideration and (as discussed in Chapter 8) how they interact. Be aware of how bed-of-nails capacitance, test-circuit or boundary-scan propagation delays, and other test-related elements of circuit architecture affect test speeds.

12.2 Big vs. Small

Formulating a board-test strategy involves numerous tradeoffs. For example, do you need a big tester, or will one or more smaller testers suffice? Small testers offer lower prices. A large tester can cost hundreds of thousands of dollars. Half-a-dozen small testers at $50,000 each can cost considerably less.

Less-expensive machines generally offer targeted capability. That is, each tester in this category addresses only a subset of the universe of test needs. An AOI system, for example, generally checks component presence and absence, as well as the integrity of visible solder joints. Programming and managing such systems can prove less complicated than with their larger siblings.

Perhaps the biggest advantage of smaller testers is production-line flexibility. Consider a product containing six boards, tested in lots of 100. One large functional or combinational tester for all boards must test 501 boards (one lot of each of the first five boards and the first board of the sixth type) before assembling the first complete system. In contrast, if the line features six small (and inexpensive) machines, each dedicated to one board type, it need test only *six* boards before building a system. The result is reduced inventory costs as well as a lower capital-acquisition cost.

In addition, final programming and debugging activities must occur on the testers themselves. A single big tester means either interfering with production or inconveniencing programmers by having them work off-shift. With small machines, a facility might buy not six but *seven* testers, assigning one to test-program development.

What happens in these two instances if a tester goes down? If there is only one tester, the whole line is down. With a cadre of six, a bit of rerouting of the manufacturing flow can cover for the missing machine until it gets fixed. As an alternative, the program-development machine can replace its broken counterpart, virtually eliminating production-line disruption.

Test engineers can redistribute small testers among different projects as test needs change. At the initial stages of a project, for example, several machines can support program development. Where small testers are appropriate, these factors generally combine to reduce time to market.

Such flexibility presupposes that all test equipment of the same model from the same vendor will perform identically. Unfortunately, reality often violates this assumption.

Large testers generally offer higher speeds, higher throughput, and higher fault coverage, as well as a wider range of permissible board sizes and pin counts. These machines provide more *test* flexibility, allowing engineers to change test tactics at short notice and rotate equipment through different projects more easily. Fewer test stations facilitate robotic board handling and other forms of automation.

Deciding between large and small testers also requires knowing how much design and manufacturing information is available on computer-aided engineering (CAE) equipment and in what form. Simulation data, for example, can reduce the test-programming penalty of a large tester. Although both large and small testers can benefit from the information, the more complicated and comprehensive programming process of the large tester gains a greater advantage.

Are designs electrically and mechanically stable? Electrical stability permits developing a functional test program. Mechanical stability means that component placement, size, and pinouts are fairly well fixed, permitting in-circuit test and bed-of-nails fixture generation, as well as inspection program development. Although relatively simple for AOI systems, creating programs for x-ray inspection can take considerably longer. A head start will facilitate the process, and can significantly shorten time to market.

Does the product's design address testability? Are test nodes accessible to a bed-of-nails or guided probe? Can a tester control and observe manageable functional blocks from the board edge or other convenient point? Does the board incorporate scan circuitry, conforming to IEEE 1149.1 or some other scheme such as level-sensitive scan design (LSSD)? Can the tester's program-development tools take advantage of that fact?

How much data must the test generate to ensure high product quality and that the process remains in control? How much analysis is necessary to make sense out of the accumulated data? Generally speaking, the more design-level tools that are available and the more necessary post-test analysis, the more likely that the process will benefit from a large-tester solution.

Inspection equipment is also available with a wide range of prices and capabilities. Some inexpensive machines serve as operator aids and neither decide what to inspect nor evaluate the results. Most manufacturers dedicate such machines to examine certain critical spots on the board. Mid-priced machines offer decision-making software and some automation capability. High-end systems offer better resolution and fault analysis, albeit at a price.

Engineers must evaluate all test options with these points in mind.

12.3   Do You Need a High-End Tester?

Unfortunately, some of the pressures influencing the decision of large vs. small equipment have little to do with the technical merits of either alternative. Many vendors push customers toward large testers and inspection systems because that is what they sell. Selling a large tester requires little more effort than selling a smaller one, and the large machine is more profitable for vendors both at purchase and for continuing support. Because most small-equipment vendors are also smaller companies, they generally lack the marketing and sales resources to make as strong an impact on customers as their more powerful competitors.

Other issues relate to management psychology. Suppose that an engineer recommended $1 million worth of test equipment last year but this year selects a cadre of smaller machines totaling half the price. The department head may ask whether the larger expenditure was necessary last year or whether this level of spending is sufficient to do the whole job. Sometimes the answer is that the less-expensive option was unavailable last year. Perhaps the engineer simply did not know about it. In any case, the justification for the lower expenditure request may be more trouble than simply asking for another acquisition of the same class and price range as the one the department already has.

One reason for resistance to small-tester solutions could be called "the 95 percent dilemma." A test engineer or manager can reasonably assume that large testers and inspection systems can perform all that a particular situation requires and that it can continue to do everything even as the job changes. Individual brands of large systems may differ in their strengths or in their convenience for tackling certain tasks, but they are all capable of performing these tasks. Therefore, there is comfort in buying a big system.

Small systems, on the other hand, cannot do everything. If they could, they would carry the same price tag as the big machines. They rarely do more than, say, 95 percent of any job, and systems from two vendors will cover a different 95 percent. As a result, selecting *which* equipment to buy is much more critical with small systems than large ones. The effort necessary to find the remaining 5 percent of possible faults on a board can make or break companies and careers. Engineers who choose machines that address the wrong 95 percent can quickly find themselves transferred from the production line to the unemployment line.

Managers feel more comfortable with large systems than with their smaller siblings. It is ironic that electronics professionals spend their careers trying to shrink their own products onto the head of a pin, but unless a test system is the size of the state of California, they perceive that it lacks sufficient horsepower to get the job done. Several years ago, one vendor, frustrated by his attempts to convince customers that his small machine could solve their problems, suggested to a colleague that he would have more success if he simply installed the tester's existing electronics in a large mainframe to *look* more like a big tester and tripled the price. Although he never actually implemented the idea, he was only half-joking.

In many organizations, recommending a less-expensive solution means sticking your neck out because it is an alternative that the company has never tried before. In addition, if this year's budget was set up to accommodate a large-tester purchase, justifying a decision not to spend all of the money may be no easier than trying to exceed the budget would be. Many engineers perceive that if they install a small-machine solution now, when a project comes down the pipe that really demands the big system, company managers will resist spending the larger sum.

12.4 Assembling the Strategy

Ultimately, managers hate test and inspection equipment. They feel that it adds no value to the product, and it is very expensive. Therefore, any steps that reduce the need to test or reduce test complexity help to sell the entire strategy. Assess all costs of testing, including the cost of process changes that reduce test efforts and the cost to the company of not testing and bad product reaching the field.

Consider minimum *necessary* test requirements. Some test engineers specify equipment based on needs 5 years down the road. This approach represents overkill. Looking ahead a year or 18 months is probably justified. Farther than that, by the time a project appears that requires the specified capability, three later tester generations will have emerged that do the job better at a lower price. If you are uncertain as to what you actually will need, select modular or expandable testers, such as VXI-based solutions. Such offerings permit not only adding capability but also improving existing capability as vendors introduce new versions of existing modules. This architecture also allows adopting solutions from more than one vendor, which can lead to the best overall test-system performance.

With inspection equipment, decide whether you want to start with less-expensive systems and maintain human control over the inspection process, or "bite the bullet" and plug into the world of high-end automated alternatives.

Examine all strategic alternatives, from farming out test activities to using or expanding equipment that you already own. Look at new vendors and new tester types. Do not hesitate to incorporate or eliminate entire process steps as information from the chosen strategy begins to accumulate. Most important, avoid rejecting a strategic option simply because you have never tried it before.

Before choosing a strategy, estimate the probable first-pass board yield from information about the quality at individual process steps according to the following equation:

$$P(n) = P_1, P_2, P_3, \dots P_n \qquad \text{(Eq. 12–1)}$$

If all of these probabilities are the same and the failure rate for each step is r_f, this equation becomes:

$$P(n) = (100 - r_f)^n \qquad \text{(Eq. 12–2)}$$

Figure 12-1 shows this relationship for different values of r_f, assuming only component failures. Notice that even where component faults are only

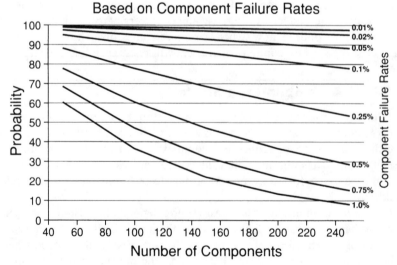

Figure 12-1   The relationship between component failure rate and resulting board yields based on that rate.

0.01 percent, a few percent of boards still fail. With 99.9 percent good parts (a failure rate of 1 out of 1000 or 1000 ppm), yield of a board containing 300 components is 0.999^{300}, or 74 percent. Adding solder-joint failures in this calculation lowers effective yields, as discussed in Chapter 1. Recent research suggests typical failure rates for solder joints alone that hover around 1000 ppm, which will produce somewhat discouraging product quality out of manufacturing.

Resist the temptation to base a test-strategy's economic justification on cost per fault found. As products and processes mature, they show fewer and fewer failures. In the extreme, a perfect design and perfect process will generate no faults at all, so the cost per fault found would theoretically be *infinite*. If you had assurance that the process would remain perfect, then testing would certainly be unnecessary. Unfortunately, absolute perfection is an unattainable goal. Even a 6σ operation sees failures occasionally. People make mistakes. Machines fail or drift out of adjustment. In addition, rare faults tend to be the most difficult to detect and diagnose, so a process that seldom fails may demand the most expensive test equipment, raising the cost per fault found even higher.

One obvious implication of this exponential relationship is that lowering the number of components (and consequently the number of solder joints) on the board will increase board yields. But the push to reduce parts count comes from other quarters as well. ASICs and other highly integrated parts consume far less power than an equivalent function constructed from discretes. Power consumption (and resulting heat generation) takes on considerable importance in the world of PDAs, notebook computers, cell phones, and the like because lower power consumption translates directly into longer battery life.

Despite ever-improving performance of ever-smaller electronics, battery design has defied all efforts to find a technological breakthrough. Oh, there have been incremental improvements. Certainly lithium-ion batteries represent a considerable advance over the older nickel-cadmium (NiCd) types. But a true leap forward has eluded our best efforts. (That statement is one of the few in this book that I *hope* will soon be obsolete.) One recent attempt has created a radical new design that shows considerable promise. To date, it suffers from two small problems. It is a mercury-based design, and mercury is highly toxic. Perhaps more inconvenient, overcharging such a battery tends to make it explode. One conjures up images of an airplane full of businesspeople and their notebook computers plugged into the power sources that many planes now provide. No one wants to see the fireball that results. So, to date, the only way to increase battery life has been to reduce power consumption.

Highly integrated devices fail much less often than their discrete equivalents do, which reduces overall board failures. Shifting much of what once would have been board-level test reduces test complexity (and therefore test-development time) of the boards. And, as mentioned earlier, the consolidation can provide additional access for test points and other aids.

12.5 The Benefits of Sampling

One drawback to the need for information at so many points during production is that gathering the information inherently slows the process down—an intolerable constraint in today's flat-out, high-volume production. In a well-controlled process where board-to-board consistency is very high, checking board samples may prove as successful as testing all of every board. Sampling also allows examining individual boards more thoroughly without burdening throughput. Sampling every tenth board, for example, means that each inspection or test can take up to ten times cycle time. Alternately, to stay within the allotted time, a manufacturer may opt to examine only a subset of the board area, concentrating on so-called "trouble spots" on every board. Both approaches represent a tradeoff between test or inspection comprehensiveness and manufacturing throughput.

Examining only high-risk areas also allows "spending" the cycle time where test and inspection will produce the best results. A typical candidate for this approach would be a board containing a few quad flat packs (QFPs) and BGAs, with primarily large discrete components or conventional through-hole or gull-wing surface-mount ICs elsewhere. Inspection steps could concentrate on the big chips as trouble spots, leaving the rest of the board to bed-of-nails test. If the time required to verify all high-risk areas exceeds cycle time, the manufacturer must either prioritize the areas and examine the worst ones or again resort to sampling.

The question remains—*How* do you sample? How many boards do you sample and under what conditions?

Sampling can occur inline or offline. Offline sampling removes a few boards from the line on a regular basis, perhaps every hour or from every batch. The same test or inspection equipment can serve several manufacturing lines, which minimizes floor space requirements and associated costs. Since any failures on board samples will likely stop the line for further investigation, test and inspection steps must keep false calls to an absolute minimum. Unfortunately, carrying boards to offline test and inspection stations can introduce faults that did not originate in the manufacturing process. In addition, offline sampling creates a considerable lag time between the manufacture of the faulty board and stopping the line for process modification and adjustment, increasing the number of bad boards produced, lowering yields, and correspondingly raising costs.

Inline sampling comes in several flavors. In addition to sampling the nth board, you can adopt a "section-by-section" strategy. In post-paste inspection, for example, you inspect a different subset of solder paste sites on each board to detect random skips in the paste operation. The inspection equipment examines a portion of every board through the process. When the allotted time has elapsed, that board proceeds to the next production step, while inspection continues from that point with the next board. In this way, after every n boards you have checked the entire surface. By extension, this technique can also apply when examining only high-risk areas, but where the number of high-risk areas exceeds the ability to inspect all of them within the cycle time.

Which approach works best? The answer depends a great deal on what you are making, whether you are applying automated or manual assembly techniques, and how easily you can tolerate failures at system test or in the field. In an automated facility, where the process is usually in tight control, systematic faults do not usually appear on single boards. They result from miscalibrated equipment, incorrect component reels loaded onto pick-and-place machines, and similar situations. Generally speaking, in a particular board lot either most boards are good or most (sometimes all) are bad. In that case, thoroughly testing one or a few boards from each board lot should suffice to determine board quality and—if faults exist—their cause.

Assuming a controlled process that produces very few failures, some manufacturers extol the virtues of "adaptive" sampling. This scheme begins by testing every board during preproduction and early production. Process adjustments, engineering change orders (ECOs), and other learning-curve activities gradually bring both the process and the product under control, and the number of failures may dwindle to nearly zero. Thereafter, only samples undergo a thorough test, perhaps coupled with a check of trouble areas on every board. Theoretically, *none* of the boards in the sample should fail. A single failure stops the line and triggers a new round of thoroughly testing every board and analyzing all failures. Process and product modifications and adjustments reduce the probability of future failures. As yields again near perfection, the manufacturer gradually reduces the number of boards tested, perhaps sampling every second board, then every third board, and so on until reaching the equilibrium sampling ratio.

The theoretical basis for adaptive sampling seems sound. Unfortunately, many boards include portions that rarely fail, but few processes produce anything close to perfect boards, so manufacturing operations may never support the optimum sampling rate. As a result, although many manufacturers have investigated the adaptive-sampling technique, few have actually implemented it.

12.6 Tester Trends

Trends in traditional tester types include improving fault coverage, improved handling of the latest board technologies, and reduced test-programming efforts. New in-circuit testers are faster than their predecessors. Tester designers have made a conscious effort to reduce wire lengths drastically between pin drivers and boards under test. Lengths that once measured in feet have shrunk to mere inches as tester manufacturers place drivers in the enclosure directly below the fixture receiver. One vendor promises 3 inch typical and no more than 1 inch on a limited number of critical nodes. This development significantly improves tester signal quality and diagnostic capability.

Bed-of-nails fixture technology, too, is rising to meet today's challenges. Today's version of the dreaded clamshell solution, although still awkward, at least performs better than its predecessors did. Some fixture makers have replaced the rat's nest of wires between the receiver and the nails with a printed-circuit board, improving signal integrity tremendously. The one drawback to this approach is that wired fixtures are much easier to change, so PCB types work best on mechanically mature boards. Recent developments in fixture technology also make both fixtures and boards under test less susceptible to static discharge, which can cause problems from inaccurate tests to real damage on the board itself.

Programming techniques have improved as well. Cluster testing with fault-dictionary-like diagnostic tools can cope with board areas where individual components are inaccessible to a bed of nails. Some analog circuitry allows complete fault detection with as little as 50 percent access, as discussed in Chapter 2. Software tools increase test-programmer productivity. Links to CAE equipment help test programs to better reflect a designer's perception of how the board works.

Software linking inspection and the various stages of test can keep track of what faults each step can detect, so that the strategy can eliminate them from downstream steps, thereby simplifying each test and shortening test development and time to market.

Tester modularity provides much more flexibility than previous-generation monolithics did in terms of both capability and price, so customers more often get the specifications they need but only the features they want. Future descendents of most traditional monolithic machines may be VXI-based, permitting even multiple-vendor test solutions within a monolithic environment.

Of course, no good deed ever goes unpunished. In-circuit testers containing additional features have become more expensive. At the same time, low-cost versions are emerging. They lack advanced capabilities, but for strategies that do not require such sophisticated verification of the assembly process, they adequately prescreen functional, hot-mockup, and other holistic techniques. And companies can consider lowering overall test costs by turning to an inspection technique along with process control in place of the in-circuit step.

Functional testers are also becoming more capable. Speeds have increased to keep pace with board technologies. Pin counts are rising to accommodate boards that interface to high-bit systems and to handle functional beds-of-nails that must address pin-grid-arrays and other complex parts that by themselves often contain 256 pins or more.

Third-party vendors offer sophisticated software that translates between design simulators and test-program generators. The software considers not only constraints of board design but also *tester* capabilities, such as resolution and skew.

Functional fault diagnostics are improving as well. Fault dictionaries are getting better, and many testers combine a fault dictionary with guided-fault isolation for better fault identification and shorter probing sequences. Probe software often recognizes misprobes without requiring the operator to start an entire sequence again. Many testers combine traditional functional techniques with in-circuit or emulation techniques to increase fault coverage without additional equipment or handling steps.

12.7 Sample Strategies

This section presents several classic strategies and some considerations that lead to selecting them. These examples address only the test side of a design/manufacturing/test operation. They by no means cover all available options but merely provide a brief guide to the reasoning that leads to assembling a coherent strategy.

When a manufacturing process is under strict control, from high vendor bareboard and component quality to accurate pick-and-place and solder operations, overall board yields are generally quite high. In military and other high-reliability situations, boards may contain a conformal coating. A bed-of-nails would pierce the coating, requiring an additional step after testing to reapply it. Inspection and process control keep manufacturing failures to a minimum. In these cases, a functional-only strategy, as in Figure 12-2, may represent the best choice. The functional tester could execute stimulus/response or emulation tests, or both.

This strategy assumes that the board has been built correctly (or that inspection has found any problems) and verifies that it actually works. Automatic program-generation tools and efficient CAE-linking software reduce the costs of this approach in terms of both time and money. Boundary-scan technology allows in-circuit-like fault diagnostics from functional test, although its serial character may preclude taking advantage of it in high-volume production.

Most companies that select a functional-only strategy carefully monitor failure rates. An increase indicates a process problem that may require additional diagnostic steps, including a bed-of-nails test or manual failure analysis, and appropriate corrective action.

For operations where overall yields are lower and boards are unlikely to contain design or performance problems, an in-circuit-only strategy, as in Figure 12-3, may work. Here, the tests uncover shorts, missing, backwards, or incorrect

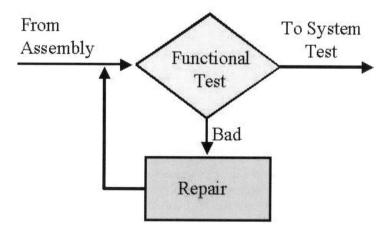

Figure 12-2 A functional-only test strategy assumes that boards have been built correctly and verifies that they actually work.

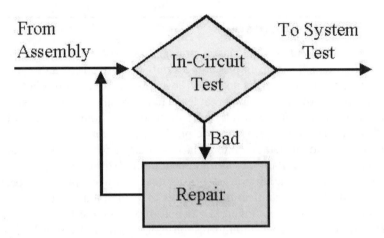

Figure 12-3 For operations where overall yields are lower and boards are unlikely to contain design problems, an in-circuit-only strategy may work.

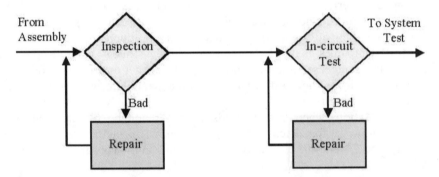

Figure 12-4 For boards that include primarily surface-mounted components and where access to some circuit nodes is impossible, the manufacturing-defects test step may include inspection as well as in-circuit test. (This figure is representational only. The inspection can actually occur at any of the three process points discussed in Chapter 3.)

components, and other process-related problems. Complex devices such as ASICs have already been tested by vendors or third parties or at incoming inspection. The board offers node access to a bed-of-nails, preferably from only one side. For boards that include primarily surface-mounted components and where access to some circuit nodes is impossible, available nodes must permit cluster testing or the manufacturing-defects test step must include inspection as well, as in Figure 12-4.

Products can proceed from the in-circuit step directly to a test of the final system or to a hot-mockup. As mentioned, manufacturers of disk drives and other products with complex analog and mixed-signal components often adopt this approach, as do some PC makers and makers of products that do not

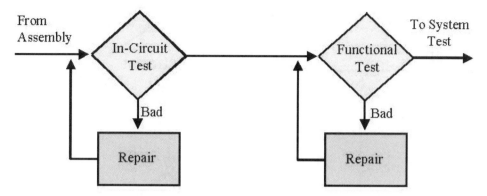

Figure 12-5   Products whose processes are immature, whose designs are suspect, and whose overall yields are fairly low may apply a bed-of-nails technique followed by a functional test.

push the edge of available electronic technology, such as elevators or washing machines.

Products whose processes are immature, whose designs are suspect, and whose overall yields are fairly low may combine these approaches, applying a bed-of-nails technique followed by a functional test, as shown in Figure 12-5. Board repair from the two steps may occur separately or together. Repaired boards return for retest. If test managers expect secondary failures from repair operations, all boards should rejoin the process stream at in-circuit test. Otherwise, the board returns to the test level that previously failed. Bed-of-nails constraints from the in-circuit-only strategy also apply here.

If process problems include exclusively or primarily shorts, missing components, backwards components, and other relatively simple problems, a shorts-and-opens tester or an MDA may provide a cost-effective alternative to a full in-circuit tester. A so-called MDA that also permits guarded analog measurements may suffice when complex process and component failures are confined to analog circuit elements. On the other hand, if test engineers anticipate analog-component tolerance and digital-component functionality problems, a full in-circuit test is necessary, despite the added cost.

Where access is spotty but component problems still plague the product, an inspection station can prescreen the in-circuit test at any of the three process points discussed in Chapter 3.

Low production volumes may permit substituting emulation for stimulus/response-type functional alternatives. In this case, the test department must have access to experts for program generation. Both testing and fault diagnosis are slower than with more traditional functional testing. Where throughput is not a concern, however, this arrangement substantially lowers capital costs.

Failure rates, fault types, and throughput considerations may suggest the most efficient path for bad boards. Online diagnostics, especially during functional

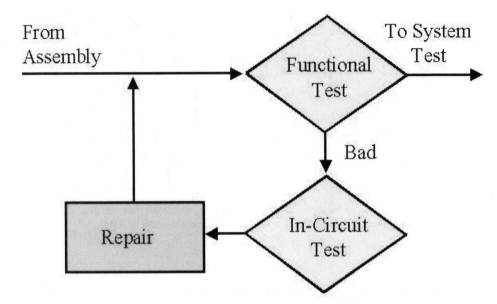

Figure 12-6 Companies with relatively low overall failure rates may perform all initial testing on functional testers. The in-circuit tester gets only boards that fail.

testing, may significantly increase test times and correspondingly reduce maximum available test capacity. Offline approaches generally take much longer per board, but they do not interfere with pushing products out the door. For the same reason, comprehensive failure analysis, usually as part of statistical process control, also generally occurs offline. The fewer failures that a company experiences, and the more likely that finding them is difficult using guided-probe and other tester-aided techniques, the more likely that an offline repair station will be preferable to online alternatives.

Another consideration in maximizing throughput is where to put in-circuit test in a strategy that includes both in-circuit and functional types. Functional test provides the fastest test times for good boards. In-circuit test is slower on individual test runs, but bad boards take little longer than good ones. Companies with relatively low overall failure rates may perform all initial testing on functional testers, feeding passing boards directly to system test and out the door. The in-circuit tester gets only boards that fail, identifying as many faults as possible and sending only the most sophisticated problems for benchtop analysis. Figure 12-6 illustrates this arrangement.

Managers must allocate scarce test-development resources prior to the start of production. Completing an inspection program requires only a good board or a layout simulation. Completing an in-circuit test requires a debugged physical board and a working fixture. Completing a functional test may require only a debugged design, and that design can exist only in simulation. If your strategy includes all three and if functional-test development begins as soon as possible, at least a rudimentary test may be available before a fabricated board arrives to permit developing the in-circuit test. This incomplete strategy implementation allows at least some verification of engineering-prototype and very early production boards before ramp-up to final production volumes. Once a real board arrives, engineers can develop the in-circuit test and fixture, finishing the functional program and fine-tuning the inspection step after gaining some actual production experience. This scenario helps to minimize time to market.

If the manufacturing process includes ESS, it should generally reside after board assembly and before the conventional tests. During early production, some managers test both before and after screening to determine how many failures the extra step is uncovering, and whether it is actually worth the considerable expense.

12.8   A Real-Life Example

The challenge of testing cellular-phone technology demonstrates both the decision-making process of setting up a test strategy and some of the challenges that specific product characteristics can bring with them. It also shows how desperation can push manufacturers toward unconventional solutions.

The growth of cellular phone use over the last decade has been nothing short of staggering. Cell phones have proliferated around the globe, demonstrating half of the Negriponte Principle, formulated by Nicholas Negriponte at MIT. When looking at technological advances, he observed, "The technology that once was wired is becoming wireless. What was wireless is becoming wired." (Cable television exemplifies the other half of the principle.)

In addition to high production volumes, most manufacturers keep profit margins very slim. Profits in the industry depend on customers' buying the service. In fact, some service providers give the phones away in exchange for a year's service commitment.

Compounding the problems is the telephone's very small size. Sets must fit comfortably into the caller's hand. The electronics must broadcast on the phone's assigned frequency, in the neighborhood of 900 MHz, without interfering with adjacent channels in its own cell, then switch to another frequency seamlessly when the caller passes to another cell (such as while driving). Other product features include stringent timing accuracies, multipath equalization, and signal encryption. Also, some countries (such as Germany) demand that cell phone manufacturers accept faulty phones and recycle as much of the phone's material as possible. Obviously, minimizing such an expense by achieving the highest possible product quality has become a high-priority goal.

In addition to testing that the phone works, certification in Europe and elsewhere requires "type-approval" testing, which verifies that the phone conforms to each country's regulatory standards. In fact, difficulty in creating this critical test left manufacturers throughout Western Europe unable to produce enough certified handsets to meet the promised introduction date for GSM (global system for mobile communications) cellular products in 1991.

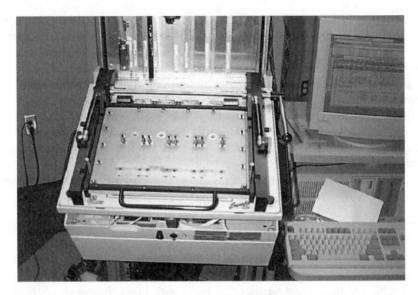

Figure 12-7   A "modified" bed-of-nails fixture for functionally testing four cell-phone boards at a time. (Courtesy GenRad.)

Reliance on almost 100 percent surface-mounted components precludes testing handsets with a conventional bed-of-nails. In addition, attaching radiofrequency (RF) circuitry to a bed-of-nails creates considerable interference and crosstalk because the nails act as antennas.

Therefore, manufacturers have been relying primarily on a functional-only approach. With the huge production requirements, this strategy has survived only because of a Herculean effort to raise product quality and reduce the number of bad boards and systems that reach the test step.

One functional-test solution employs the "modified" bed-of-nails fixture in Figure 12-7. The board's edge connector provides primary access to board logic. The relative handful of nails makes contact where possible without encouraging the "antenna" effect. This fixture allows testing up to four boards at a time, which reduces test overhead. The cover shields each board under test from outside interference and from adjacent boards. The vacuum top plate also uses the cover to provide downward force to ensure probe contact. Figure 12-8 shows the boards in place for test, with the top cover open to show the shielding.

Still, if the yield were to drop, the lack of sufficient fault diagnostics would quickly turn this strategy into a major nightmare. To improve access, some manufacturers are turning to boundary-scan-based designs. Unfortunately, boundary-scan cannot serve the RF portions of the circuit. In addition, the functional nature of a boundary-scan test does not meet the need for a simpler MDA or in-circuit solution.

Cellular-phone manufacturers are investigating other options. Automated optical inspection would verify that the board includes the correct devices, that they

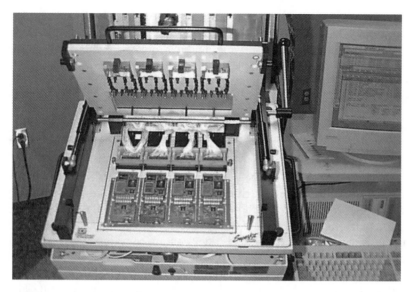

Figure 12-8   The fixture in Figure 12-7 showing the boards ready for test, with the top cover open to show the shielding. (Courtesy GenRad.)

are oriented properly, and that solder joints appear acceptable. X-ray inspection verifies inner layers of multilayer boards and looks more closely at the joints. For pin access to available nodes, some manufacturers have adopted "moving-probe" fixtureless in-circuit techniques, whereby a mechanical two- to four-point probe contacts solder pads for testing. These ultrafine probes permit better access than even a small-center bed-of-nails. Although this approach permits analog in-circuit-type testing, its maximum rate of 24 tests per second does not address high-volume requirements, and the four-pin limit is insufficient for digital tests. (Note that the maximum test rate does not include x-y positioning.) Currently, manufacturers are struggling with existing test technologies, sticking with functional test and paying extra attention to process yields.

 Another issue in cellular-phone quality is that the products must often survive in hostile environments, such as automobile trunks and dashboards, as cars sit in the winter cold and the summer sun, or drive on cobblestone streets. Many manufacturers have introduced comprehensive ESS programs prior to final test and shipment. Hine (1994) identifies temperatures ranging from −20°C to +55°C as well as shock and vibration at up to 2 g and frequencies as high as 250 Hz. He recommends that product reliability require not only screening under temperature and vibration condition extremes but also evaluating both sets of parameters together.

 At Matsushita's plant in Thatcham, England, GSM qualification tests include swept-sine vibration tests from 10 Hz to 100 Hz and back again in each of the three axes at accelerations up to 3 g. Each cycle takes 10 minutes per axis, and the entire test runs for 4 hours. Every phone must also survive a physical drop test. The company also subjects all phones to a 12-hour soak at 80°C, followed by 3 minutes of demanding random vibration while making and terminating calls using RF instrumentation. Production samples undergo an even more stringent version of these steps. By maintaining diligence during both manufacturing and test parts of the process, Matsushita rarely experiences field failures.

12.9   Changing Horses

Test strategies are more like living beings than like objects. They grow and evolve as test engineers forever strive for the perfect tactical mix. Several very different strategies may coexist within a single facility, reflecting the uniqueness of each particular situation. Consider the company making copiers and fax machines. Because fax machines include modems, they contain considerably more analog and mixed-signal circuitry than copiers do. The two strategies must reflect that fact. They will undoubtedly share tactics and tactical elements (such as test programs or parts of programs), but there will be areas of divergence as well. The fax-machine test might include a full in-circuit test or IEEE-488 instrument add-ons to a conventional functional test.

In addition, managers should constantly monitor strategies even after implementation and production ramp-up to ensure that the existing solution continues to meet current needs. Some companies "lock in" a strategy, then move on to another project until they update the product design or manufacturing process. Unfortunately, that approach ignores any new information that the strategy itself provides.

For example, consider an automated-assembly facility that opts for functional-only testing. From likely board yields and fault spectra, engineers decide that shorts and simple component failures would occur only rarely and therefore the situation does not justify any kind of prescreen.

Now, suppose that production materials begin to fail more frequently. Perhaps vendor processes have drifted slightly or the purchasing department is obtaining components or bare boards from a theoretically equivalent second source. That source may even be a named vendor from the original manufacturing specification. Again, the new material quality is not as high as before. The result is lower throughput, longer times and higher costs for fault diagnosis, and the risk of lower shipped-product quality and dissatisfied customers. Solving the vendor's problem would obviously provide the most cost-effective answer. Sometimes, because of new manufacturing methods, vendor business reversals, or other factors beyond normal control, that option is unavailable. In that case, adding some kind of prescreening step—either test or inspection—becomes necessary.

Other factors can encourage changing strategies, as well. A new product or product line may allow reallocating existing resources or acquiring additional test equipment or test tools. Test engineers can then reexamine all existing strategies to determine possible alterations.

Facility expansion encourages change for similar reasons. Expansion generally includes additional equipment. There is no rule, however, that new equipment must be the same make or model as equipment in the installed base.

Replacing worn-out equipment also does not demand that replacements match their predecessors exactly. Of course, changing tester vendors or architectures might incur extra training and test-programming burdens, so such a decision has broad implications beyond mere capital costs.

New engineers, either additional staff or replacements, often bring with them a wealth of experience and expertise that may suggest a strategy or tactic that managers had not previously considered. New budget constraints and policies may affect both existing and future projects. A difficult business climate, for example, may prompt company management to restrict all capital-equipment purchases to less than, say, $125,000. Suppose that the test department already owns a sophisticated mixed-signal combinational tester and a product that is headed for preproduction would benefit from its capabilities. The test manager could obtain a tester that falls within the stated financial constraints but use it to replace the combinational tester on an existing product line, moving the more expensive tester to the line where it is most needed. This solution satisfies the new product's test requirements within the company's financial guidelines.

The key to adjusting test strategies and tactics is *data collection and analysis*. Analyzing test results helps test managers predict maturing processes, which may not need all test steps in the existing strategy. It represents an important part of *total quality management*.

Some data require analysis in real time, especially in automated manufacturing facilities. Any time several consecutive boards fail in the same way, the line should stop to allow a supervisor to investigate. Suppose, for example, that an operator loaded a tube of digital devices onto a pick-and-place machine backwards. Unchecked, the line would produce 50 or more bad boards. If a *real-time alarm* monitors the process and stops the line after, say, four consecutive identical failures, a supervisor can correct the problem, saving both time and money.

Test managers should maintain constant vigilance over existing production processes, as well as plan new ones. Day-to-day activities will confirm strategic elements that work adequately and those that require modification. In final analysis, no strategy is perfect, and no strategy is "cast in stone." Reacting quickly to a strategy's shortcomings that may emerge at any time carries substantial rewards.

12.10 Summary

There is no single test strategy or laundry list of test strategies that will solve every manufacturer's problems. Each situation is unique. If a process could produce only good products, testing would be unnecessary—as long as it *continued* to produce only good products.

Test managers should try to incorporate all available resources into a strategy. Self-tests can address some of the more pernicious faults, leaving test engineers free to concentrate on problems outside of the self-test's venue. The strategy must ensure that tests will not damage the board—for example, by removing shorts before applying power for functional or self-tests.

Within each test tactic, managers must decide whether to select large or small testers and inspection systems, and choose vendors who can provide both equipment and support. Small testers offer economy and flexibility. Larger, more expensive machines provide higher speeds and greater capability. Sometimes this decision depends less on the situation's technical requirements and more on the realities of management, budgeting, and policy.

Testers in each category are becoming more capable, and the lines between them are breaking down. Although machines with every possible feature are becoming more expensive, test managers who do not need that much power can find less-expensive test or inspection alternatives or modular machines that allow adding capability later.

No strategy is cast in stone. New information, product or facility changes, and personnel turnover may suggest alternatives that no one had previously considered. Fault spectra may not agree with even the most carefully prepared predictions. Vendor and other conditions may change.

The key to adjusting test strategies is data collection and analysis. Reacting quickly to a strategy's shortcomings that may emerge at any time carries substantial rewards.

CHAPTER **13**

Conclusions

Achieving an effective test strategy depends as much on declaring and understanding test goals as on implementing them. Every company, every department, every product-planning activity should include targets for testability, test coverage, and acceptable quality levels in the factory, as well as reliability, testability, and repairability after shipment to customers. Strategic planning must follow concurrent-engineering practices—a "we're all in this together" commitment from designers, marketers, test personnel, production people, and managers at all organizational levels. This level of cooperation is particularly important when a contract manufacturer performs the actual production.

Evaluating a strategy requires determining the effectiveness of each step or tactic. One common approach assesses a particular step by examining the board (or system) next-step yield. This method assumes that there are no new failures from product handling and transfer between test stations and, therefore, that escapes from the preceding test step are the only failures. If there is no subsequent step, customers perform this analysis.

Another technique is fault injection. Primarily useful in high-throughput facilities, fault injection "seeds" the production process with boards containing known problems, then determines how many of those problems the test strategy will catch. In theory, the percentage of known faults that testing finds approximates the overall fault coverage of the chosen strategy.

This approach is time-consuming and reduces test-operation capacity. It is only comprehensive if the set of known faults is exhaustive and the seed faults represent the entire set. It is quite commonly used for *software* testing, however. Software test methods are considerably less precise than their hardware counterparts, and final-product software is almost never completely bug-free. Seeding provides software engineers with a reasonable estimate (sometimes the only available estimate) of the code's remaining problems.

For many operations, fault *simulation* represents the best means to determine test-strategy effectiveness. Its success depends on the efficiency and accuracy of both fault simulators and the design models on which they operate.

307

In final analysis, no one can determine your ideal test strategy without thoroughly understanding the particular situation and processes involved. A test "expert" entering your facility with the "solution to your problems" should be promptly fired. Instead, an expert should come in with questions to which you seek answers together.

You know your situation best. A strategy that works for someone else may not be the best choice for you. On the other hand, an approach that other manufacturers find useless may turn out to be your "better mousetrap."

Time-Value-of-Money Tables

Present Value—Single Payment

Years	1%	2%	3%	4%	5%	6%	7%
1	.9901	.9804	.9709	.9615	.9524	.9434	.9346
2	.9803	.9612	.9426	.9246	.9070	.8900	.8734
3	.9706	.9423	.9151	.8890	.8638	.8396	.8163
4	.9610	.9238	.8885	.8548	.8227	.7921	.7629
5	.9515	.9057	.8626	.8219	.7835	.7473	.7130
6	.9420	.8880	.8375	.7903	.7462	.7050	.6663
7	.9327	.8706	.8131	.7599	.7107	.6651	.6227
8	.9235	.8535	.7894	.7307	.6768	.6274	.5820
9	.9143	.8368	.7664	.7026	.6446	.5919	.5439
10	.9053	.8203	.7441	.6756	.6139	.5584	.5083
11	.8963	.8043	.7224	.6496	.5847	.5268	.4751
12	.8874	.7885	.7014	.6246	.5568	.4970	.4440
13	.8787	.7730	.6810	.6006	.5303	.4688	.4150
14	.8700	.7579	.6611	.5775	.5051	.4423	.3878
15	.8613	.7430	.6419	.5553	.4810	.4173	.3624
16	.8528	.7284	.6232	.5339	.4581	.3936	.3387
17	.8444	.7142	.6050	.5134	.4363	.3714	.3166
18	.8360	.7002	.5874	.4936	.4155	.3503	.2959
19	.8277	.6864	.5703	.4746	.3957	.3305	.2765
20	.8195	.6730	.5537	.4564	.3769	.3118	.2584

Years	8%	10%	12%	14%	16%	18%	20%
1	.9259	.9091	.8929	.8772	.8621	.8475	.8333
2	.8573	.8264	.7972	.7695	.7432	.7182	.6944
3	.7938	.7513	.7118	.6750	.6407	.6086	.5787
4	.7350	.6830	.6355	.5921	.5523	.5158	.4823
5	.6806	.6209	.5674	.5194	.4761	.4371	.4019
6	.6302	.5645	.5066	.4556	.4104	.3704	.3349
7	.5835	.5132	.4523	.3996	.3538	.3139	.2791
8	.5403	.4665	.4039	.3506	.3050	.2660	.2326
9	.5002	.4241	.3606	.3075	.2630	.2255	.1938
10	.4632	.3855	.3220	.2697	.2267	.1911	.1615
11	.4289	.3505	.2875	.2366	.1954	.1619	.1346
12	.3971	.3186	.2567	.2076	.1685	.1372	.1122
13	.3677	.2897	.2292	.1821	.1452	.1163	.0935
14	.3405	.2633	.2046	.1597	.1252	.0985	.0779
15	.3152	.2394	.1827	.1401	.1079	.0835	.0649
16	.2919	.2176	.1631	.1229	.0930	.0708	.0541
17	.2703	.1978	.1456	.1078	.0802	.0600	.0451
18	.2502	.1799	.1300	.0946	.0691	.0508	.0376
19	.2317	.1635	.1161	.0829	.0596	.0431	.0313
20	.2145	.1486	.1037	.0728	.0514	.0365	.0261

Present Value—Annuity

Years	1%	2%	3%	4%	5%	6%	7%
1	0.9901	0.9804	0.9709	0.9615	0.9524	0.9434	0.9346
2	1.9704	1.9416	1.9135	1.8861	1.8594	1.8334	1.8080
3	2.9410	2.8839	2.8286	2.7751	2.7232	2.6730	2.6243
4	3.9020	3.8077	3.7171	3.6299	3.5460	3.4651	3.3872
5	4.8534	4.7135	4.5797	4.4518	4.3295	4.2124	4.1002
6	5.7955	5.6014	5.4172	5.2421	5.0757	4.9173	4.7665
7	6.7282	6.4720	6.2303	6.0021	5.7864	5.5824	5.3893
8	7.6517	7.3255	7.0197	6.7327	6.4632	6.2098	5.9713
9	8.5660	8.1622	7.7861	7.4353	7.1078	6.8017	6.5152
10	9.4713	8.9826	8.5302	8.1109	7.7217	7.3601	7.0236
11	10.3676	9.7868	9.2526	8.7605	8.3064	7.8869	7.4987
12	11.2551	10.5753	9.9540	9.3851	8.8633	8.3838	7.9427
13	12.1337	11.3484	10.6350	9.9856	9.3936	8.8527	8.3577
14	13.0037	12.1062	11.2961	10.5631	9.8986	9.2950	8.7455
15	13.8651	12.8493	11.9379	11.1184	10.3797	9.7122	9.1079
16	14.7179	13.5777	12.5611	11.6523	10.8378	10.1059	9.4466
17	15.5623	14.2919	13.1661	12.1657	11.2741	10.4773	9.7632
18	16.3983	14.9920	13.7535	12.6593	11.6896	10.8276	10.0591
19	17.2260	15.6785	14.3238	13.1339	12.0853	11.1581	10.3356
20	18.0456	16.3514	14.8775	13.5903	12.4622	11.4699	10.5940

Years	8%	10%	12%	14%	16%	18%	20%
1	0.9259	0.9091	0.8929	0.8772	0.8621	0.8475	0.8333
2	1.7833	1.7355	1.6901	1.6467	1.6052	1.5656	1.5278
3	2.5771	2.4869	2.4018	2.3216	2.2459	2.1743	2.1065
4	3.3121	3.1699	3.0373	2.9137	2.7982	2.6901	2.5887
5	3.9927	3.7908	3.6048	3.4331	3.2743	3.1272	2.9906
6	4.6229	4.3553	4.1114	3.8887	3.6847	3.4976	3.3255
7	5.2064	4.8684	4.5638	4.2883	4.0386	3.8115	3.6046
8	5.7466	5.3349	4.9676	4.6389	4.3436	4.0776	3.8372
9	6.2469	5.7590	5.3282	4.9464	4.6065	4.3030	4.0310
10	6.7101	6.1446	5.6502	5.2161	4.8332	4.4941	4.1925
11	7.1390	6.4951	5.9377	5.4527	5.0286	4.6560	4.3271
12	7.5361	6.8137	6.1944	5.6603	5.1971	4.7932	4.4392
13	7.9038	7.1034	6.4235	5.8424	5.3423	4.9095	4.5327
14	8.2442	7.3667	6.6282	6.0021	5.4675	5.0081	4.6106
15	8.5595	7.6061	6.8109	6.1422	5.5755	5.0916	4.6755
16	8.8514	7.8237	6.9740	6.2651	5.6685	5.1624	4.7296
17	9.1216	8.0216	7.1196	6.3729	5.7487	5.2223	4.7746
18	9.3719	8.2014	7.2497	6.4674	5.8178	5.2732	4.8122
19	9.6036	8.3649	7.3658	6.5504	5.8775	5.3162	4.8435
20	9.8181	8.5136	7.4694	6.6231	5.9288	5.3527	4.8696

Future Value—Single Payment at Time 0

Years	1%	2%	3%	4%	5%	6%	7%
1	1.0100	1.0200	1.0300	1.0400	1.0500	1.0600	1.0700
2	1.0201	1.0404	1.0609	1.0816	1.1025	1.1236	1.1449
3	1.0303	1.0612	1.0927	1.1249	1.1576	1.1910	1.2250
4	1.0406	1.0824	1.1255	1.1699	1.2155	1.2625	1.3108
5	1.0510	1.1041	1.1593	1.2167	1.2763	1.3382	1.4026
6	1.0615	1.1262	1.1941	1.2653	1.3401	1.4185	1.5007
7	1.0721	1.1487	1.2299	1.3159	1.4071	1.5036	1.6058
8	1.0829	1.1717	1.2668	1.3686	1.4775	1.5938	1.7182
9	1.0937	1.1951	1.3048	1.4233	1.5513	1.6895	1.8385
10	1.1046	1.2190	1.3439	1.4802	1.6289	1.7908	1.9672
11	1.1157	1.2434	1.3842	1.5395	1.7103	1.8983	2.1049
12	1.1268	1.2682	1.4258	1.6010	1.7959	2.0122	2.2522
13	1.1381	1.2936	1.4685	1.6651	1.8856	2.1329	2.4098
14	1.1495	1.3195	1.5126	1.7317	1.9799	2.2609	2.5785
15	1.1610	1.3459	1.5580	1.8009	2.0789	2.3966	2.7590
16	1.1726	1.3728	1.6047	1.8730	2.1829	2.5404	2.9522
17	1.1843	1.4002	1.6528	1.9479	2.2920	2.6928	3.1588
18	1.1961	1.4282	1.7024	2.0258	2.4066	2.8543	3.3799
19	1.2081	1.4568	1.7535	2.1068	2.5270	3.0256	3.6165
20	1.2202	1.4859	1.8061	2.1911	2.6533	3.2071	3.8697

Years	8%	10%	12%	14%	16%	18%	20%
1	1.0800	1.1000	1.1200	1.1400	1.1600	1.1800	1.2000
2	1.1664	1.2100	1.2544	1.2996	1.3456	1.3924	1.4400
3	1.2597	1.3310	1.4049	1.4815	1.5609	1.6430	1.7280
4	1.3605	1.4641	1.5735	1.6890	1.8106	1.9388	2.0736
5	1.4693	1.6105	1.7623	1.9254	2.1003	2.2878	2.4883
6	1.5869	1.7716	1.9738	2.1950	2.4364	2.6996	2.9860
7	1.7138	1.9487	2.2107	2.5023	2.8262	3.1855	3.5832
8	1.8509	2.1436	2.4760	2.8526	3.2784	3.7589	4.2998
9	1.9990	2.3579	2.7731	3.2519	3.8030	4.4355	5.1598
10	2.1589	2.5937	3.1058	3.7072	4.4114	5.2338	6.1917
11	2.3316	2.8531	3.4785	4.2262	5.1173	6.1759	7.4301
12	2.5182	3.1384	3.8960	4.8179	5.9360	7.2876	8.9161
13	2.7196	3.4523	4.3635	5.4924	6.8858	8.5994	10.6993
14	2.9372	3.7975	4.8871	6.2613	7.9875	10.1472	12.8392
15	3.1722	4.1772	5.4736	7.1379	9.2655	11.9737	15.4070
16	3.4259	4.5950	6.1304	8.1372	10.7480	14.1290	18.4884
17	3.7000	5.0545	6.8660	9.2765	12.4677	16.6722	22.1861
18	3.9960	5.5599	7.6900	10.5752	14.4625	19.6733	26.6233
19	4.3157	6.1159	8.6128	12.0557	16.7765	23.2144	31.9480
20	4.6610	6.7275	9.6463	13.7435	19.4608	27.3930	38.3376

Future Value—Annual Payment

Years	1%	2%	3%	4%	5%	6%	7%
1	1.0000	1.0000	1.0000	1.0000	1.0000	1.0000	1.0000
2	2.0100	2.0200	2.0300	2.0400	2.0500	2.0600	2.0700
3	3.0301	3.0604	3.0909	3.1216	3.1525	3.1836	3.2149
4	4.0604	4.1216	4.1836	4.2165	4.3101	4.3746	4.4399
5	5.1010	5.2040	5.3091	5.4163	5.5256	5.6371	5.7507
6	6.1520	6.3081	6.4684	6.6330	6.8019	6.9753	7.1533
7	7.2135	7.4343	7.6625	7.8983	8.1420	8.3938	8.6540
8	8.2857	8.5830	8.8923	9.2142	9.5491	9.8975	10.2598
9	9.3685	9.7546	10.1591	10.5828	11.0266	11.4913	11.9780
10	10.4622	10.9497	11.4639	12.0061	12.5779	13.1808	13.8164
11	11.5668	12.1687	12.8078	13.4864	14.2068	14.9716	15.7836
12	12.6825	13.4121	14.1920	15.0258	15.9171	16.8699	17.8885
13	13.8093	14.6803	15.6178	16.6268	17.7130	18.8821	20.1406
14	14.9474	15.9739	17.0863	18.2919	19.5986	21.0151	22.5505
15	16.0969	17.2934	18.5989	20.0236	21.5786	23.2760	25.1290
16	17.2579	18.6393	20.1569	21.8245	23.6575	25.6725	27.8881
17	18.4304	20.0121	21.7616	23.6975	25.8404	28.2129	30.8402
18	19.6147	21.4123	23.4144	25.6454	28.1324	30.9057	33.9990
19	20.8109	22.8406	25.1169	27.6712	30.5390	33.7600	37.3790
20	22.0190	24.2974	26.8704	29.7781	33.0660	36.7856	40.9955

Years	8%	10%	12%	14%	16%	18%	20%
1	1.0000	1.0000	1.0000	1.0000	1.0000	1.0000	1.0000
2	2.0800	2.1000	2.1200	2.1400	2.1600	2.1800	2.2000
3	3.2464	3.3100	3.3744	3.4396	3.5056	3.5724	3.6400
4	4.5061	4.6410	4.7793	4.9211	5.0665	5.2154	5.3680
5	5.8666	6.1051	6.3528	6.6101	6.8771	7.1542	7.4416
6	7.3359	7.7156	8.1152	8.5355	8.9775	9.4420	9.9299
7	8.9228	9.4872	10.0890	10.7305	11.4139	12.1415	12.9159
8	10.6366	11.4359	12.2997	13.2328	14.2401	15.3270	16.4991
9	12.4876	13.5795	14.7757	16.0853	17.5185	19.0859	20.7989
10	14.4866	15.9374	17.5487	19.3373	21.3215	23.5213	25.9587
11	16.6455	18.5312	20.6546	23.0445	25.7329	28.7551	32.1504
12	18.9771	21.3843	24.1331	27.2707	30.8502	34.9311	39.5805
13	21.4953	24.5227	28.0291	32.0887	36.7862	42.2187	48.4966
14	24.2149	27.7950	32.3926	37.5811	43.6720	50.8180	59.1959
15	27.1521	31.7725	37.2797	43.8424	51.6595	60.9653	72.0351
16	30.3243	35.9497	42.7533	50.9804	60.9250	72.9390	87.4421
17	33.7502	40.5447	48.8837	59.1176	71.6730	87.0680	105.9306
18	37.4502	45.5992	55.7497	68.3941	84.1407	103.7403	128.1167
19	41.4463	51.1591	63.4397	78.9692	98.6032	123.4135	154.7400
20	45.7620	57.2750	72.0524	91.0249	115.3797	146.6280	186.6880

Acronym Glossary

AC	alternating current
AI	artificial intelligence
ALU	arithmetic logic unit
AOI	automated optical inspection
APG	automatic program generation
ARR	average rate of return
ASIC	application-specific integrated circuit
ATE	automatic test equipment
AXI	automated x-ray inspection
A/D	analog-to-digital (colloquially, also refers to an analog-to-digital *converter*)
BGA	ball-grid array
CAD	computer-aided design
CAE	computer-aided engineering
CARD	computer-aided research & development (not an industry-standard term)
CASE	computer-aided software engineering
CBI	computer-based instrumentation
CM	contract manufacturer
CMOS	complementary metal-oxide semiconductor
CPU	central-processing unit (of a computer or microprocessor)
DC	direct current
DFT	design-for-testability
DMA	direct memory access
DR	data register (IEEE 1149.1)
DVM	digital voltmeter
D/A	digital-to-analog (colloquially, also refers to a digital-to-analog *converter*)

ECL	emitter-coupled logic
ECO	engineering change order
EISA	extended industry standard architecture (personal-computer I/O bus)
EMC	electromagnetic compatibility
EMI	electromagnetic interference
EPROM	electrically programmable read-only memory
ESD	electrostatic discharge
ESS	environmental stress screening
FAST	Fairchild advanced Schottky technology (a family of digital devices)
FET	field-effect transistor
FV	future value (economics)
fpm	feet per minute
GB	gigabyte (literally $1024 \times 1024 \times 1024$ bytes)
GSM	global system for mobile communications (European cellular-phone standard)
GUI	graphical user interface
HAST	highly accelerated stress testing
HDTV	high definition television
HVI	human visual inspection
IC	integrated circuit
IR	instruction register (IEEE 1149.1)
IRQ	interrupt request (VXI/IEEE 1155)
IRR	internal rate of return
ISA	industry standard architecture (the most common personal-computer I/O bus)
I/O	input/output
JTAG	Joint Test-Action Group (the committee that first proposed boundary-scan standards that became IEEE 1149.1)
LAN	local-area network
LSI	large-scale integration
LSSD	level-sensitive scan design (an early scan approach)
MB	megabyte (literally 1024×1024 bytes)
MDA	manufacturing-defects analyzer
MRI	magnetic-resonance imaging (medical equipment)
MTBF	mean time between failures
MVI	manual visual inspection

NPV net present value
NRE non-recurring engineering
NVRAM non-volatile random-access memory (sometimes called EAROM—electrically alterable read-only memory)

OEM original equipment manufacturer

PC personal computer
PCB printed-circuit board
PDA personal digital assistant (a hand-held computer)
PERT program evaluation and review technique
PLA programmable logic array
ppm parts per million
PRV parallel response vector (for functional test)
psi pounds per square inch
PTV parallel test vector (for functional test)
PV present value (economics)

RAM random-access memory
RF radio frequency
ROI return on investment
ROM read-only memory

SCPI standard-computer programming interface
SMT surface-mount technology
SOIC small-outline integrated circuit
SPC statistical process control
SRL shift-register latch
SRV sequential response vector (for functional test)
STV sequential test vector (for functional test)

TAP test-access port (IEEE 1149.1)
TCK test clock (IEEE 1149.1)
TDI test-data in (IEEE 1149.1)
TDO test-data out (IEEE 1149.1)
TMS test-mode select (IEEE 1149.1)
TRST* test reset, active low (IEEE 1149.1)
TTL transistor-transistor logic

UART universal asynchronous receiver transmitter
UUT unit under test

VCR video cassette recorder
VLSI very large-scale integration
VXI VMEbus eXtension for Instrumentation (IEEE 1155)

Works Cited and Additional Readings

Note: This list does not pretend to be comprehensive. Some of the items here address topics covered in the book. Others provide additional topics for further reading.

Ahrikencheikh, Cherif and Michael Spears. "Limited-Access Testing of Analog Circuits: Handling Tolerances," *Proceedings of the International Test Conference*, IEEE, 1999, p. 577.

Andrews, John. "P1149A Extensions to IEEE STD 1149.1-1990," *Proceedings of Nepcon West*, Reed Expositions, February, 1993, p. 189.

Arena, John and Stephen Cohen. "Tester Specs vs. Realistic Test Speeds I," *Test & Measurement World*, November, 1989, p. 54.

Arena, John and Stephen Cohen. "Tester Specs vs. Realistic Test Speeds II," *Test & Measurement World*, January, 1990, p. 40.

Arena, John and Stephen Cohen. "Tester Specs vs. Realistic Test Speeds III," *Test & Measurement World*, February, 1990, p. 55.

Arment, Elmer L. and William D. Coombe. "Application of JTAG for Digital and Analog SMT," *Proceedings of the ATE and Instrumentation Conference West*, Miller-Freeman Expositions Group, January, 1989, p. 99.

Barr, Robert W., et al. "End-to-End Testing for Boards and Systems Using Boundary Scan," *Proceedings of the International Test Conference*, IEEE, 2000, p. 585.

Bateson, John, *In-Circuit Testing*, New York: Van Nostrand Reinhold, 1985.

Beitman, Bruce A., "Find ATE Problems Before They Find You," *Test & Measurement World*, June, 1990, p. 63.

Bergeron, Jay, "Moving From Real-Pin ATE to Multiplexed-Pin ATE," *Proceedings of the ATE and Instrumentation Conference West*, Miller-Freeman Expositions Group, January, 1991, p. 203.

Bleeker [1994]. *See* Stalheim, et al.

Board-Level Burn-In Basics, flyer from Micro Instrument Company, Escondido, CA, circa 1989 (no date).

Bogue, Adam. "Open Modular Configurable Low-Cost T&M for the 1990s," *Proceedings of Nepcon West*, Reed Expositions, February, 1992, p. 767.

Bond, John. "Vibration-Tester Survey," *Test & Measurement World*, April, 1991, p. 83.

Boue, Philippe. "Does Functional or In-Circuit ATE Suit Your Maintenance Strategy?" *Test & Measurement Europe*, July, 1995, p. 32.

Buckroyd, Allen. *Computer-Integrated Testing*, New York: John Wiley and Sons, 1989.

Caldwell, Barry. "Implementing 1149.1 Boundary Scan for Board Test," *Proceedings of the Test Engineering Conference*, Miller-Freeman Expositions Group, June, 1991, p. 143.

Chao, Linda, Daren Dance, and Tom DiFloria, "Get a Handle on Your Cost of Test," *Test & Measurement World*, April, 1995, p. 45.

Charette, Colin. "X-Ray Solutions for Limited-Access Boards," *Proceedings of Nepcon West*, Volume I, Reed Exhibitions, March, 1998.

Charette, Colin. "Understanding Automated X-ray Inspection," *Proceedings of Nepcon West*, Reed Exhibitions, February, 1999.

Cheek, D. and R. Dandapani. "Integration of IEEE Std. 1149.1 and Mixed-Signal Test Architectures," *Proceedings of the International Test Conference*, IEEE, October, 1995, p. 569.

Clark, David. "Sampling vs. 100% Inspection," *Circuits Assembly*, March, 1998, p. 56.

Cohen, Stephen A. and Kathy Shottes Regan, "Electro-Static Discharge and Vacuum Fixtures: A Case Study," *Proceedings of Nepcon West*, Reed Expositions, February, 1993, p. 215.

Corby, Robert. "Digital Functional Test: Refuting the Myths," *Proceedings of Nepcon West*, Reed Expositions, February, 1992, p. 774.

Corby, Robert. "Testing Digital Cards Using Bus Emulation," *Proceedings of the Test-Engineering Conference*, Miller-Freeman Expositions Group, June, 1990, p. 27.

Cortner, J. Max. *Digital Test Engineering*, New York: John Wiley and Sons, 1987.

Cortner, J. Max. "Measuring the Cost of In-Circuit Testing Versus the Cost of Not In-Circuit Testing ASICs," TSSI application note, 1990.

Cron, Adam. "IEEE P1149.4: Almost a Standard," *Proceedings of the International Test Conference*, IEEE, 1997, p. 174.

Daniel, Wayne and Brenda J. Round. "Design Verification using IEEE 1149.1 Boundary Scan," *Proceedings of Nepcon West*, Reed Expositions, February, 1993, p. 904.

Davis, Brendan. "Selecting the Optimum Test Strategy by Life-Cycle Cost Analysis," presented at the TEST '92 conference in Brighton, England, published as "Economic Modeling of Board-Test Strategies" by BD Consulting, Berkshire, England, 1993.

Davis, Brendan. *The Economics of Automatic Testing*, Second Edition, London: McGraw-Hill Book Company (UK) Limited, 1994.

Davis, Brendan. "Board Testers Pin-Point Open Circuits," *Test & Measurement Europe*, Spring, 1995, p. 30.

Davis, Don and Brendan Davis. "The Economics of Stress Screening," *Proceedings of Nepcon West*, Cahners Exposition Group, 1989, p. 1813.

de Jong, Frans, et al. "Power Pin Testing: Making the Test Coverage Complete," *Proceedings of the International Test Conference*, IEEE, 2000, p. 575.

de Jong, Frans and Steffen Hellmold. "Static Component Interconnection Test Technology in Practice," *Proceedings of the International Test Conference*, IEEE, 1999, p. 556.

DeSena, Art. "Guidelines for a Cost-Effective ATE Operation," *Test-Industry Reporter*, ADS Associates, Mineola, New York, January, 1991, p. 4.

Draye, Hugo. "Selection Criteria for Functional Board Test," *Proceedings of Nepcon West*, Reed Expositions, February, 1992, p. 751.

Durickas, Daniel A. "228X AFTM Applications," *GenRad Application Note*, GenRad, Concord, MA, 1992.

Eerenstein, Lars. "Test Complex Digital Boards With Boundary Scan," *Test & Measurement World*, December, 1995, p. 39.

Eerenstein, Lars. "Testing Two Generations of HDTV Decoders—The Impact of Boundary-Scan Test," *Proceedings of the International Test Conference*, IEEE, 1994, p. 911.

Eisler, Ian. "Requirements for a Performance-Tester Guided-Probe Diagnostic System," *Proceedings of the ATE and Instrumentation Conference*, Miller-Freeman Expositions Group, January, 1990, p. 202.

The Environmental Stress-Screening Handbook, Holland, MI: Thermotron Industries, 1988.

Farren, D. and A. Ambler. "Cost-Effective System-Level Test Strategies," *Proceedings of the International Test Conference*, IEEE, 1995, p. 807.

Frayman, Felix, Mick Tegethoff, and Brenton White. "Issues in Optimizing the Test Process—A Telecom Case Study," *Proceedings of the International Test Conference*, IEEE, 1996, p. 800.

Fundamentals of Accelerated Stress Testing, Holland, MI: Thermotron Industries, 1998.

Gamble, Chuck, "Board Inspection: Solder Paste vs. Solder Joint," *Evaluation Engineering*, August, 1995, p. 20.

Geary, Greg. "Which Side Is Up?" *Proceedings of Nepcon West*, Reed Exhibitions, February, 1996, p. 1118.

Gillette, G. "A Single-Board Test System: Changing the Test Paradigm," *Proceedings of the International Test Conference*, IEEE, October, 1995, p. 880.

Goldberg, Joel. "The Future Is Self-Test," *Test & Measurement World*, February, 1996, p. 22.

Goldberg, Joel. "Wireless Fixtures Solve Test Problems," *Test & Measurement World*, February, 1997, p. 26.

Goldman, Jacob. "An Ideal Couple: The PC and ATE," *Evaluation Engineering*, June, 1996, p. 58.

Golla, Laura. "PC Hardware Technology and the Virtual Instrumentation Revolution," *Proceedings of Nepcon West*, Volume I, Reed Exhibitions, February, 1995, p. 487.

Grace, Phil. "GSM Manufacturing Test—The Jury's Still Out," *Test & Measurement Europe*, Winter, 1994.

Grace, Phil, "'Pay As You Go' Reduces ATE Cost of Ownership," *Test & Measurement Europe*, Spring, 1994, p. 44.

Haigh, Dominic. "Why AOI? Why Now?" *Proceedings of Nepcon West*, Reed Exhibitions, March, 1998.

Hamel, Richard. "Managing Life-Cycle Costs," *Proceedings of the Test Engineering Conference*, Miller-Freeman Expositions Group, June, 1991, p. 177.

Hansen, Peter. "The Impact of Boundary Scan on Board-Test Strategies," *Proceedings of the ATE and Instrumentation Conference West*, Miller-Freeman Expositions Group, January, 1991, p. 219.

Harding, Ed. "Experiences Using Vectorless Test," *Proceedings of Nepcon West*, Reed Exhibitions, February, 1996, p. 855.

Harris, Cyril M. (ed.). *Shock and Vibration Handbook*, Third edition, New York: McGraw-Hill Book Company, 1988.

Herman, Edward R. "The Lease-Purchase Decision," *Test & Measurement World*, September, 1985, p. 110.

Hine, Peter. "Why Environmentally Test GSM Phones?" *Test & Measurement Europe*, Winter, 1994.

Hofer, Dave. "Scan Is Not Free, But It Is a Bargain," *Evaluation Engineering*, November, 1992, p. 162.

Howe, Edward. "Improper Environmental Screening Can Damage Your Product," *Test Engineering & Management*, October/November, 1998, p. 22.

HP Boundary-Scan Tutorial and BSDL Reference Guide, Hewlett-Packard Company, 1990, part number E1017-90001.

Hulvershorn, Harry. "1149.5: Now It's a Standard, So What?" *Proceedings of the International Test Conference*, IEEE, 1997, p. 166.

Hutchinson, J. Andrew. "Advantages of Digital Convergence for Functional Test," *Proceedings of Autotestcon*, IEEE, 1998.

IDI Source Book: A Manual on Probe Design and Applications, Kansas City, KS: Interconnect Devices, Inc., 1993.

"IEEE Standard 1149.1 Test-Access Port and Boundary-Scan Architecture," IEEE, Piscataway, NJ, 1990.

Introduction to Multi-Strategy Testing, Concord, MA: Genrad, 1989.

Jacob, Gerald. "Low-Cost ATE Deserves a Closer Look," *Evaluation Engineering*, June, 1992, p. 14.

Jarwala, Najmi and Chi Yau. "A New Framework for Analyzing Test Generation and Diagnosis Algorithms for Wiring Interconnects," *Proceedings of the International Test Conference*, IEEE, 1989, p. 63.

Jensen, Curtis. "Applications of Standard Testability Bus Structures for Board and Subsystem Test," *Proceedings of the Test-Engineering Conference*, Miller-Freeman Expositions Group, June, 1990, p. 47.

Johnson, Carolyn. "Before You Apply SPC, Identify Your Problems," *Test & Measurement World*, April, 1996, p. 43.

JTAG Boundary-Scan Architecture Standard Proposal, Version 2.0, JTAG Technical Subcommittee, 1988.

Kaskel, Robert. "The Use of In-Circuit Testing in Contract Manufacturing," *Circuits Assembly*, June, 1998, p. S6.

Keahey, Julia Ann. "Programming of Flash with ICT: Rights and Responsibilities," *Proceedings of the International Test Conference*, IEEE, 2000, p. 711.

Keller, Muriel and Steve Cook. "Test Probes for Surface-Mount Devices," *Test & Measurement World*, December, 1985, p. 88.

Kerns, Tamara. "The Software is More than the Instrument," *Proceedings of Nepcon West*, Reed Exhibitions, March, 1998.

Kohlberger, John and Lynelle D'Aquilla. "Design for Testability and the Effects on Fixture Fabrication," *Proceedings of Nepcon West*, Reed Expositions, February, 1996, p. 1137.

Laine, Brian T. "New Technologies Expand Role of MDAs," *Evaluation Engineering*, October, 1995, p. 88.

Langston, Kent. "Machine-Vision Inspection—Is the Time Right?" *Proceedings of Nepcon West*, Reed Exhibitions, February, 1996, p. 565.

Ledden, John W. "Test Strategies for the Modern PCB Manufacturer," *Proceedings of Nepcon West*, Reed Exhibitions, February, 2000.

Lenker, James E. "Low-Cost ATE Using Memory Device Emulation," *Proceedings of the ATE and Instrumentation Conference West*, Miller-Freeman Expositions Group, January, 1991, p. 241.

Lesmeister, G. "A Tester for Design (TFD)," *Proceedings of the International Test Conference*, IEEE, October, 1995, p. 886.

Ley, Adam W. "The Integration of Boundary-Scan Test Methods to a Mixed-Signal Environment," *Proceedings of the International Test Conference*, IEEE, September, 1999, p. 159.

Lockheed, Daniel. "Check Your Test Fixture's Problems," *Test & Measurement World*, February, 1998, p. 39.

Lu, Howard and Nancy H. McAndrew. "ChipScan: Theory and Applications," *Proceedings of Nepcon West*, Reed Exhibitions, February, 1996, p. 815.

MacLean, Ken. "Design for Test," Supplement to *SMT Magazine*, July, 1998, p. 8.

Mahoney, R. Michael and Greg A. Larsen. "A Bayesian Approach to Improving Test Effectiveness and Evaluating Test Strategy," *Proceedings of the ATE and Instrumentation Conference West*, Miller-Freeman Expositions Group, January, 1990, p. 431.

Marshall, Julian. "A Low-Cost Boundary-Scan Test Interface," *Proceedings of Nepcon West*, Reed Exhibitions, February, 1993, p. 196.

Mawby, Terry. "Probe Construction Affects Test Performance," *Test & Measurement World*, August, 1989, p. 61.

Mayerfeld, Pam. "Output Monitoring Is New Direction for Burn-In," *Evaluation Engineering*, November, 1992, p. 77.

McClintock, David, Lance Cunningham, and Takis Petropoulos. "Motherboard Testing Using the PCI-Bus," *Proceedings of the International Test Conference*, IEEE, 2000, p. 593.

McCullough, Bob. "VXI meets expectations as a viable test alternative," *Computer Design/News Edition*, January, 14, 1991, p. 15.

McDermid, John. "Limited Access," *Proceedings of Nepcon West*, Reed Exhibitions, March, 1998.

McElfresh, Karen. "Vectorless Opens Testing Using an RF Technique," *Proceedings of Nepcon West*, Volume III, Reed Exhibitions, February, 1995, p. 1930.

Miller, Don. "X-Ray Inspection Systems," *Circuits Assembly*, June, 1998, p. 58.

Minneman, Michael. "The Blurring Boundary Between Standard and Virtual Instruments," *Proceedings of Nepcon West*, Volume I, Reed Exhibitions, February, 1995, p. 500.

Mullen, Dr. Charles J. "Engineering Design Service Firms, OEMs, and Contract Manufacturers," *Circuits Assembly*, June, 1998, p. S12.

Nadeau-Dostie, Benoit, et al. "An Embedded Technique for At-Speed Interconnect Testing," *Proceedings of the International Test Conference*, IEEE, 1999, p. 431.

Nelson, Wayne. *Accelerated Testing*, New York: John Wiley & Sons, 1990.

Oakland, John S. and Roy F. Followell. *Statistical Process Control—a Practical Guide*, Second Edition, Oxford, UK: Heinemann Newnes, 1990.

Olson, Doug E. "Jet into Functional Test with Fewer Opens," *Proceedings of Nepcon West*, Reed Exhibitions, February, 1996, p. 836.

Oresjo, Stig. "Combining PCB Test Strategies," *Circuits Assembly*, June, 1998, p. 28.

Oresjo, Stig. "A New Test Strategy for Complex Printed Circuit Board Assemblies," *Proceedings of Nepcon West*, Reed Exhibitions, February, 1999.

Parker, Kenneth. "Standards-Based Design for Testability," *Proceedings of Nepcon West*, Volume III, Reed Exhibitions, February, 1995, p. 1921.

Parker, Kenneth. "System Issues in Boundary-Scan Board Test," *Proceedings of the International Test Conference*, IEEE, 2000, p. 725.

Portnuff, Colin and Brian Wycoff. "Transforming Data into Information to Improve Electronics-Manufacturing Processes," *Proceedings of Nepcon West*, Reed Exhibitions, March, 1998.

Prang, Joe. "Controlling Life-Cycle Costs Through Concurrent Engineering," Addendum to the *ATE & Instrumentation Proceedings*, Miller-Freeman Expositions, 1992, p. 1.

Pynn, Craig. "Shrinking the Cost of Board Test," *Evaluation Engineering*, June, 1996, p. 38.

Pynn, Craig. *Strategies for Electronics Test*, New York: McGraw-Hill, 1986.

Pynn, Craig. "Vectorless Test Boosts Fault Coverage and Cuts Cycle Time," *Evaluation Engineering*, August, 1995, p. 37.

Pynn, Craig. "Single-Stage Test Strategy," *Circuits Assembly*, June, 1998, p. 48.

Pynn, Craig. "Test/Inspection," Supplement to *SMT Magazine*, August, 1998, p. 40.

Racal-Dana Instruments, Inc., *VXIbus: The New Standard for Test and Measurement*, collection of articles, published by Racal-Dana Instruments, Inc., Irvine, California, second printing, 1989.

Rahe, David. "The HASS Development Process," *Proceedings of the International Test Conference*, IEEE, 1999, p. 566.

Reed, D., J. Doege, and A. Rubio. "Improving Board and System Test: A Proposal to Integrate Boundary Scan and I_{ddq}," *Proceedings of the International Test Conference*, IEEE, October, 1995, p. 577.

Ristelhueber, Robert. "Outsourcing Product Development," *Electronic Business Buyer*, August, 1995, p. 44.

Robinson, Gordon. "In-Circuit Programming and Board Test," *Proceedings of Nepcon West*, Volume III, Reed Exhibitions, February, 1995, p. 1939.

Robinson, Gordon. "NAND Trees Help Board Test, But Not as Much as Boundary Scan," *Proceedings of Nepcon West*, Volume III, Reed Exhibitions, February, 1995, p. 1913.

Robinson, Gordon and John Deshayes. "Interconnect Testing of Boards with Partial Boundary Scan," *Proceedings of the International Test Conference*, IEEE, 1990, p. 572.

Rolince, David. "Simplifying TSP Development and Execution in a Web-Based Environment," *Proceedings of Autotestcon*, IEEE, 1998.

Runyon, Stan. "X-ray May Be PC-Board Key," *Electronic Engineering Times*, April, 21, 1997.

Rupert, Jeffrey, et al. "Laser-Based Scanning for Component Detection," *Proceedings of Nepcon West*, Reed Exhibitions, 1999.

Sadtler, Sam. "Test Strategies for Reducing Product Test Development Cost and Time to Market," *Proceedings of Nepcon West*, Volume III, Reed Exhibitions, February, 1995, p. 1621.

Santa Maria, Vicki. "Delta-Scan—An Analog Junction Technique," *Proceedings of Nepcon West*, Reed Exhibitions, February, 1996, p. 825.

Scheiber, Stephen F. "Breaking the Complexity Spiral in Board Test," *Proceedings of the International Test Conference*, IEEE, September, 1999, p. 155.

Scheiber, Stephen F. *Building an Intelligent Manufacturing Line*, Florence, MA: Quale Press, 2001.

Scheiber, Stephen F. "Concurrent Engineering is Common Sense," *Test & Measurement World*, October, 1992, p. 67.

Scheiber, Stephen F. *Economically Justifying Functional Test*, Florence, MA: Quale Press, 1999.

Scheiber, Stephen F. "Evaluating Test-Strategy Alternatives," *Test & Measurement World*, April, 1992, p. 57.

Scheiber, Stephen F. "Flying on One Wing," *Test & Measurement World*, June, 1989, p. 58.

Scheiber, Stephen F. "Getting the Check Signed When Buying Test Equipment," *Evaluation Engineering*, December, 2000, p. 60.

Scheiber, Stephen F. "It Isn't Just Testing Anymore (Redux)," *Proceedings of the International Test Conference*, IEEE, 2000, p. 718.

Scheiber, Stephen F. "JTAG Cuts SMT Testing Down to Size," *Test & Measurement World*, April, 1990, p. 73.

Scheiber, Stephen F. *A Six-Step Economic-Justification Process for Tester Selection*, Florence, MA: Quale Press, 1997.

Scheiber, Stephen F. "Test Tactics for Partial-Scan Boards," *Test & Measurement World*, April, 1991, p. 69.

Scheiber, Stephen F. "Testing Boards with the VXIbus," *Test & Measurement World*, August, 1988, p. 38.

Schlagheck, Jerry. *Methodology and Techniques of Environmental-Stress Screening*, Cincinnati, OH: ESSC, 1988.

Schoettmer, U. and T. Minami. "Challenging the 'High-Performance/High-Cost' Paradigm in Test," IEEE, *Proceedings of the International Test Conference*, IEEE, October, 1995, p. 870.

Schweighofer, Georg. "Rigorous RF Tests Confront GSM Mobiles," *Test & Measurement Europe*, January, 1996, p. 18.

Shearer, Steve J. "Xpress Yourself—A Mobile Experience," *Proceedings of Nepcon West*, Reed Exhibitions, February, 1996, p. 845.

Shina, Sammy G. *Concurrent Engineering and Design for Manufacture of Electronic Products*, Van Nostrand Reinhold, New York, 1991.

Smithson, Stephen A. "Effectiveness and Economics—Yardsticks for ESS Decisions," *Proceedings of the Institute of Environmental Sciences*, 1990.

St. Onge, Gary. "Fixture-Technology Update—Hitting Small Targets," *Proceedings of Nepcon West*, Reed Exhibitions, February, 1993, p. 220.

St. Onge, Gary and Jeff Sendzicki. "Wireless Test Fixture Considerations," *Circuits Assembly*, June, 1998, p. 34.

Stalheim, Lars, et al. "Using IEEE 1149.1 for Board-Level Programming," *Test & Measurement Europe*, Winter, 1994, p. 29.

Starks, Fred. "Test/Inspection Strategies for 1995 and Beyond," *Proceedings of Nepcon West*, Volume II, Reed Exhibitions, February, 1995, p. 523.

Stasonis, Robert A. "Combinational ATE—Evolution of the Species," *Proceedings of the Test-Engineering Conference*, Miller-Freeman Expositions Group, June, 1990, p. 337.

Stasonis, Robert A. "Combinational Test Strategies," *Proceedings of the ATE and Instrumentation Conference West*, Miller-Freeman Expositions Group, January, 1991, p. 231.

Stasonis, Robert A. "Combinational Testing with Low-Cost ATE," *Proceedings of Nepcon West*, Reed Expositions, February, 1992, p. 787.

Stasonis, Robert A. "How to Implement Functional Test in an Automated Environment," *Proceedings of Nepcon West*, Reed Exhibitions, March, 1999.

Stasonis, Robert A. "PXI in Action," *Proceedings of Nepcon West*, Reed Exhibitions, February, 2000.

Steinberg, Dave S. "Fatigue Life in Temperature-Cycling Environments," *Proceedings of Nepcon West*, Cahners Exposition Group, February, 1989, p. 146.

Stewart, Bret A. "Board-Level Automated Fault Injection for Fault Coverage and Diagnostic Efficiency," *Proceedings of the International Test Conference*, IEEE, 1997, p. 649.

Strassberg, Dan. "Vectorless Test: Process Development Made Simple," *EDN*, September, 28, 1995, p. 49.

Tegethoff, Mick, Kenneth Parker, and Ken Lee. "Opens Board Test Coverage: When Is 99% Really 40%," *Proceedings of the International Test Conference*, IEEE, 1996, p. 333.

Tegethoff, Mick. "The Emerging Fault Spectrum for PCBs," *SMT Magazine*, August, 1998, p. 70.

Tektronix/Colorado Data Systems. *1992 Card-Modular Instruments Information and Ordering Guide*, 3301 W. Hampden Avenue, Englewood, CO 80110.

Terry, Mark, et al. "Reducing Functional Test-Cell Costs: Attacking the Last Major Untapped Cost in Electronics Manufacturing," *Proceedings of Nepcon West*, Reed Exhibitions, February, 2000.

Titus, Jon. "X-Ray Systems Reveal Hidden Defects," *Test & Measurement World*, February, 1998, p. 29.

Titus, Jon. "X-Rays Expose Hidden Connections," *Test & Measurement World*, October, 1999, p. 28.

Toh, Peng Seng. "Design for Inspection and Automated Optical Inspection," *Proceedings of Nepcon West*, 1999, Reed Exhibitions.

Tsui, Frank F. *LSI/VLSI Testability Design*, New York: McGraw-Hill, 1987.

Ungar, Louis Y. "Built-In (Self) Test Economics in a Manufacturing Process," *Proceedings of Nepcon West*, Reed Expositions, February, 1996, p. 218.

Ungar, Louis. "Functional vs. In-Circuit Test: a 21st Century Perspective," *Proceedings of Nepcon West*, Reed Exhibitions, February, 2000.

Van Nguyen, Tan. "Guardband Testing Ensures Reliability," *Test & Measurement World*, February, 1998, p. 49.

Vaucher, Christophe and Louis Balme. "Analog/Digital Testing of Loaded Boards Without Dedicated Test Points," *Proceedings of the International Test Conference*, IEEE, 1996, p. 325.

Victor, Stephen. "A Practical Method for Tailoring Environmental Stress Screens," *Proceedings of ATE West*, Miller-Freeman Exhibitions Group, Anaheim, CA, January, 1989, p. 456.

VMEbus Extensions for Instrumentation—VXIbus System Specification, VXI Consortium, San Diego, CA, Revision 1.3, June, 1989.

VXIbus Project: "Constructing a VXIbus-Based Test System"
    Part I, *Test & Measurement World*, December, 1990, p. 63.
    Part II, *Test & Measurement World*, February, 1991, p. 42.
    Part III, *Test & Measurement World*, April, 1991, p. 48.
    Part IV, *Test & Measurement World*, June, 1991, p. 87.
    Part V, *Test & Measurement World*, October, 1991, p. 89.
    Part VI, *Test & Measurement World*, January, 1992, p. 40.
    Part VII, *Test & Measurement World*, June, 1992, p. 67.
    Part VIII, *Test & Measurement World*, September, 1992, p. 68.
    Part IX, *Test & Measurement World*, January, 1993, p. 57.
    Part X, *Test & Measurement World*, February, 1993, p. 81.
    Part XI, *Test & Measurement World*, March, 1993, p. 48.
    Part XII, *Test & Measurement World*, April, 1993, p. 42.

Whipple, Dave. "PCS Systems Challenge Production Test," *Test & Measurement World*, September, 1996, p. 63.

Whitfield, Jesse. "Automatic Optical Inspection," *Proceedings of Nepcon West*, Reed Expositions, February, 1996, p. 553.

Wolfe, Ron. "VXIbus Software Components," *Evaluation Engineering*, October, 1992, p. 14.

Wolfe, Ron. "VXIplug&play Technology Introduction," *Proceedings of Nepcon West*, Reed Expositions, February, 1996, p. 1587.

Wu, Yuejian and Paul Soong. "Interconnect Delay Fault Testing with IEEE 1149.1," *Proceedings of the International Test Conference*, IEEE, September, 1999, p. 449.

Ziaja, Thomas A. "Using LSSD to Test Modules at the Board Level," *Proceedings of the International Test Conference*, IEEE, September, 1999, p. 163.

Index